Standardization Research in Information Technology:
New Perspectives

Kai Jakobs
RWTH Aachen University, Germany

Information Science REFERENCE

INFORMATION SCIENCE REFERENCE
Hershey · New York

Acquisitions Editor: Kristin Klinger
Development Editor: Kristin Roth
Senior Managing Editor: Jennifer Neidig
Managing Editor: Sara Reed
Copy Editor: Shanelle Ramelb
Typesetter: Michael Brehm
Cover Design: Lisa Tosheff
Printed at: Yurchak Printing Inc.

Published in the United States of America by
Information Science Reference (an imprint of IGI Global)
701 E. Chocolate Avenue, Suite 200
Hershey PA 17033
Tel: 717-533-8845
Fax: 717-533-8661
E-mail: cust@igi-global.com
Web site: http://www.igi-global.com

and in the United Kingdom by
Information Science Reference (an imprint of IGI Global)
3 Henrietta Street
Covent Garden
London WC2E 8LU
Tel: 44 20 7240 0856
Fax: 44 20 7379 0609
Web site: http://www.eurospanonline.com

Library of Congress Cataloging-in-Publication Data

Standardization research in information technology : new perspectives / Kai Jakobs, editor.
p. cm.
Includes bibliographical references and index.
Summary: "This book amasses cutting-edge research on the application of standards in the market, covering topics such as corporate standardization, linguistic qualities of international standards, the role of individuals in standardization, and the development, use, application, and influence of information technology in standardization techniques"--Provided by publisher.
ISBN 978-1-59904-561-0 (hardcover) -- ISBN 978-1-59904-563-4 (ebook)
1. Information technology--Standards. 2. Standardization. I. Jakobs, Kai, 1957-
T58.5.S7325 2008
004.02'18--dc22
2007023446

British Cataloguing in Publication Data
A Cataloguing in Publication record for this book is available from the British Library.

Standardization Research in Information Technology: New Perspectives is part of the IGI Global series named *Advances in IT Standards and Standardization Research Series (AISSR)* (ISSN: 1935-3391).

Table of Contents

Section III
Diffusion and Adoption of Standards

Section IV
IS Perspectives

Section V
Cases and Projects

Detailed Table of Contents

This study considers the developments in international standardization over the last 20 years, particularly the status of formal standardization as compared with consortium-based industrial standardization. The report shows that the radical reform of the global formal standardization system that started in 2008, prompted by the loss of interest in formal standardization on the part of large corporations and the sometimes less than satisfactory outcomes from consortium-based industrial standardization in terms of competition and anti-trust considerations, has helped to compensate for the declining significance of national formal standardization. This specifically relates to national governments, and is to be regarded as a clearly positive development, from both the economic and the institutional and political points of view. Global public interests are now catered for by internet-supported information markets. In particular, online documentation has also enhanced the transparency of the formal standardization process and provided freedom of access for small and medium sized companies in particular, irrespective of geographical region. Finally, the study shows that the debate that took place in and around the year 2004 between Europe and the USA regarding the path towards the internationalization of formal standardization processes was superfluous, incomplete and even counterproductive, owing to the hardening of the political divisions between the two sides.

This article summarizes key incentives for vendors, users, government and individuals to participate in the standardization process. It argues that standards can expand markets at all points in the market life cycle with overall impact measured in billions of dollars. The authors hope to encourage standards involvement, and also future research and analysis that might quantify the financial value of standardization for vendors and users.

Linguistic qualities are essential for the fitness for use of every standard. The intentions of the standards developers should become perfectly clear to those who will finally use the documents, but language barriers at several project stages may hinder this. This paper addresses the topic for standards at the global and regional levels using a case study about the linguistic qualities of the standards published by the IEC (International Electrotechnical Commission). Most IEC Standards are bilingual (English and French), and they are frequently translated into national languages. Feedback on standards use, translation practices and user satisfaction has been obtained by means of two questionnaires sent to the IEC National Committees (NCs). These data are assessed with respect to the language skills of the technical experts concerned, to the particular linguistic aspects of the standards, the process of standards development, national translating practices, and to standards user satisfaction. Standards development in two languages adds to their fitness for use, but this advantage should be balanced against the cost of bilingualism. The current practice satisfies more or less all parties involved, nevertheless some improvements can be suggested. The issue bilingualism versus unilingualism also has an important cultural/ political dimension.

This paper presents a simulation framework that analyzes the diffusion of communication standards in different supply networks. We show that agents' decisions depend on potential cost reduction, pressure from members of their communication network and implementation costs of their communication standards. Besides focusing on process-specific market power distributions, the impact of relationship stability and process connectivity are analyzed as determinants of the diffusion of communication standards within different supply network topologies. In this context two real-world scenarios, from the automotive and paper/publishing industries, are used as examples for different network topologies. The results support the thesis that increasing relationship dynamics and process connectivity lead to decreasing competition of communication standards. In certain circumstances local communication clusters appear along the value chain, enabling these clusters to preserve their globally inferior standardization decision.

Market selection of product compatibility standards has long been explained through aggregate positive-feedback theoretical models of economic utility. Explaining aggregate patterns of organizational standards adoption requires two additional steps—not only differences between organizations, but also differences within organizations. Here we present a qualitative study of how organizations do (or do not) adopt a new

and usability. Our findings suggest that the application of the CSS standard can improve data delivery across independent devices with varied bandwidth and resource availability, thereby providing device independence and improved usability respectively. We demonstrate that through their effect on device independence and usability, CSS play an important role in the evolution, expansion, and openness of mobile computing.

This paper looks at implications of the emerging global Intellectual Property (IP) regime for Developing Countries (DCs) and their attempts to improve their technological capabilities. It further highlights the new perspectives for DCs opened up by the emergence of non-proprietary (open source/ free) software such as Linux. A case study of the battle between Microsoft and Linux in China is used to explore the dilemmas faced by China in determining what IP regime (strict or weak) to adopt, and the threats and opportunities that either may pose for indigenous technology development. Based on the case analysis, the paper criticises the simplistic polarised views that have been presented of the implications of the global IP regime and of the potential of non-proprietary software. It explores some of the complex considerations about the interplay between technology strategy and IP protection for China and discusses the policy implications for China and other DCs.

Between early 2004 and 2007, the Cooperation Platform for Research and Standards (COPRAS) deployed a series of activities to improve the interfacing process between ICT research and standardization. This included the conclusion of Standardization Action Plans with a series of projects in the EU-funded 6th Framework Programme, and the development of a set of generic Standardization Guidelines, supporting future ICT research projects in their interfacing with standardization. COPRAS' results show that cooperation and cross-fertilization between research and standardization is perceived as increasingly important by projects, and also demonstrate how direct and indirect support mechanisms are able to increase the amount of tangible results ICT research will be able to submit to ongoing standards work. However, its results also show that structurally improving the research/standards interfacing process will not be possible unless all parties to the process, including the research and standards communities as well as the administrators of the research programmes, take additional—and continuous—action.

Preface

Things have clearly developed positively since the first volume of this series was published. This holds certainly for Europe, where these days even commissioners and national secretaries of state highlight the economic relevance of standards and recognise the importance of research in the field. Of course, it remains to be seen whether or not these grandiose statements translate into increased research funding. However, things certainly look promising.

Practically relevant research is still needed. For example, while I am writing this, a discussion is still continuing on an e-mail list on whether companies fare better with or without basing their products or services on (open) standards. Indeed, research does not yet have a conclusive answer to this question. In fact, a claim from this discussion, that "the academic economy does not try to provide 'practical knowledge'—but rather contribute to the 'theory'," is not really that far fetched. It certainly resonates with my own limited experience: At many universities, tenure, promotion, and funding largely depend on your publication record. And, unfortunately, practical relevance is not exactly terribly high on the priority list of most journals.

The *International Journal of IT Standards and Standardisation Research* has not yet reached 'A-level' status; it does value 'significance to practice.' Thus, this book brings you a selection of papers that among themselves not only represent a fairly good cross section of today's standards research, but many of them should also be of interest to practitioners (or so I hope).

The book is divided into five sections, each of which comprises two to four chapters.

The first section, entitled "Setting Standards," is also the largest one. In Chapter I, "Challenges for Formal Standardization: The Institutional Reforms of 2008-2010 Reconsidered," Ulrich Blum enjoys the benefit of hindsight when considering the hypothetical developments in both formal and consortium standardisation over the last 20 years. He shows that the radical reform of the overall system in 2008 will be beneficial for all stakeholders.

Chapter II, "IT Standardization: The Billion Dollar Strategy," by John Hurd and Jim Isaak, argues that standards have the potential to shape, expand, and create markets. Against this background, the chapter identifies the key incentives for stakeholders to participate, and argues that it would be beneficial for all stakeholders if they participated in the standards-setting process.

Moving away from international to corporate standardisation, Henk H. de Vries' chapter entitled "Best Practice in Company Standardization" (Chapter III) describes a best-practice model for corporate standardisation. Best practices are developed based on a literature review and six case studies of multinational companies. Using insights from IT, quality, and knowledge management, the new findings are benchmarked.

The chapters of Section II, "Specifics of Standards and Standards Setting," provide closer insights into some general characteristics of standards and the underlying standards-setting process. Chapter IV, "Open Standards Requirements," by Ken Krechmer, notes that open standards are one element of an open society. It therefore discusses the various (desirable) characteristics of the much-discussed term *open standard*.

At the end of the day, standards are not developed by companies, but by people. Accordingly, their exact role in the process should be of considerable interest. Jim Isaak has a closer look in "The Role of Individuals and Social Capital in POSIX Standardization" (Chapter V). Using the IEEE POSIX set of standards as an example, and drawing upon more than 15 years of personal experience, Isaak demonstrates that the importance of the individual in standards setting must not be underestimated. In particular, he shows that they do not necessarily represent the goals of their respective employers.

Chapter VI is devoted to a slightly more arcane topic. Hans Teichmann, Henk H. de Vries, and Albert Feilzer look at the "Linguistic Qualities of International Standards." As standards are written documents, these qualities are essential for their proper understanding. Based on a case study about the linguistic qualities of the International Electrotechnical Commission's standards, the authors analyse standards use, translation practices, and user satisfaction. It turns out that the current practice of publishing bilingual versions of the standards is largely satisfactory, and the chapter suggests some improvements.

The next section is devoted to the step that follows standards setting and implementation, and is entitled "Diffusion and Adoption of Standards." Chapter VII, coauthored by Michael Schwind, Tim Stockheim, and Kilian Weiss, is entitled "A Diffusion Model for Communication Standards in Supply Networks." It presents a simulation framework that analyses the diffusion of communication standards in different supply networks. Using real-world scenarios from the automotive and paper and publishing industries, the chapter shows that increases in the dynamics of the relation between entities and of process connectivity lead to a decrease in competition between standards.

The diffusion of standards implies their adoption and deployment. In Chapter VIII, entitled "Scope and Timing of Deployment: Moderators of Organizational Adoption of the Linux Server Platform," Joel West and Jason Dedrick present a qualitative study of how organisations adopt Linux as a new server platform standard. They show that the advantages enjoyed by incumbent standards can be neutralised, for example, through a low price. Reducing switching costs by starting small further reduces the newcomer's disadvantage. In contrast, it is shown that internal standardisation increases the attractiveness of the incumbent standard.

The IS discipline is concerned with the development, use, application, and influence of information technology. There is, therefore, little surprise that many IS researchers are contributing to standards research. The section "IS Perspectives" comprises three chapters from different IS points of view.

Chapter IX, "Standards for Business Component Markets: An Analysis from Three Theoretical Perspectives" by Heiko Hahn and Klaus Turowski, presents an analysis of the conditions that have to be met for the successful realisation of large-scale component markets. It does so from the perspective of system theory, knowledge codification, and standardisation. It is argued that future technological developments and their outcome have a considerable impact here.

Subsequently, the question "Should Buyers Try to Shape IT Markets Through Nonmarket (Collective) Action?" is posed by Kai Reimers and Mingzhi Li in Chapter X. The answer is obtained by developing a new transaction-cost theoretic model of network effects. Application of the model suggests that novel forms of user organisations, providing information exchange and knowledge sharing, would improve users' chances of replacing an established technology with a new, potentially superior one.

Chapter XI, by M. Lynne Markus and Ulric J. Gelinas, Jr., is titled "Comparing the Standards Lens with Other Perspectives on IS Innovations: The Case of CPFR." Collaborative planning, forecasting, and replenishment (CPFR) has not only been labeled a standard, but also a business philosophy, a methodology, and a set of technologies. The authors find that combining the standards perspective with other conceptual lenses provides valuable insights.

The final section, "Cases and Projects" (using the former term loosely) looks at aspects relating to applications of standards in the market. This is a fairly wide field, ranging from standards applied in different engineering areas to those making inroads into national economies, to a project that aimed at a better integration of research and standardisation.

Chapter XII, "Market Response to ISO 9000 Certification of Software Engineering Processes" by G. Keith Fuller and Ilan Vertinsky, looks at the fate of a widely accepted certification standard, ISO 9000 (International Organization for Standardization), in an area where the cost effectiveness of such certification of processes is still debated. Yet, the authors' findings suggest that certification would be beneficial for companies also in software engineering.

Next, "The Value of Web Design Standards for Mobile Computing" is discussed by Matt Germonprez, Michel Avital, and Nikhil Srinivasan in Chapter XIII. Specifically, the chapter investigates if and how Web standards like HTML (hypertext markup language) and CSS can support crucial characteristics of mobile computing: device independence and usability. The authors conclude that CSS especially can play an important role in the evolution and openness of mobile computing.

In Chapter XIV, entitled "Developing Country Perspectives on Software: Intellectual Property and Open Source. A Case Study of Microsoft and Linux in China," Xiaobai Shen analyses the impact of a global IPR (intellectual property right) protection scheme on developing countries. Using the Windows and Linux case in China as an example, the chapter challenges current views of the implication of global IPR regimes, and of the benefits of open software. It also looks at associated policy implications for developing countries.

Finally, Bart Brusse presents findings from the project "COPRAS: Encouraging ICT Research Projects to Produce More Tangible Standardization Results" in Chapter XV. This project aimed to improve the interface between ICT research projects and standardisation. It turns out that cooperation and cross-fertilization between research and standardisation is increasingly perceived as important by projects, and that the level of research and development input into standards setting can be increased through adequate support mechanisms.

And now—happy reading!

Kai Jakobs

Section I
Setting Standards

Chapter I
Challenges for Formal Standardization:
The Institutional Reforms of 2008–2010 Reconsidered[1]

Ulrich Blum[2]
Halle Institute for Economic Research, Germany

ABSTRACT

This study considers the developments in international standardization over the last 20 years, particularly the status of formal standardization as compared with consortium-based industrial standardization. The report shows that the radical reform of the global formal standardization system that started in 2008, prompted by the loss of interest in formal standardization on the part of large corporations and the sometimes less than satisfactory outcomes from consortium-based industrial standardization in terms of competition and antitrust considerations, has helped to compensate for the declining significance of national formal standardization. This specifically relates to national governments and is to be regarded as a clearly positive development from both the economic and the institutional and political points of view. Global public interests are now catered for by Internet-supported information markets. In particular, online documentation has also enhanced the transparency of the formal standardization process and provided freedom of access for small and medium-sized companies in particular, irrespective of geographical region. Finally, the study shows that the debate that took place in and around the year 2004 between Europe and the USA regarding the path toward the internationalization of formal standardization processes was superfluous, incomplete, and even counterproductive, owing to the hardening of the political divisions between the two sides.

THE PARADIGM OF STANDARDIZATION

Terminology

The lack of uniformity in the terminology used in the area of standardization makes it necessary to lay down a set of clearly defined terms, as illustrated in Figure 1. The standardization process as a whole is divided into formal standardization on the one hand, which can be either organized at the national or state level (e.g., DIN [Deutsches Institut für Normung] in Germany), or based on the activities of sector associations with significant state support (e.g., ANSI [American National Standards Institute] in the USA), and industry standardization on the other. Within industry standardization, a distinction is made between consortium-based standards and company standards. In the case of consortium-based standards, a number of companies come together to formulate a common solution, whereas company standards almost always relate to a solution arrived at within the company in question. Both forms of industry standards may either be offered to competitors as open standards or treated as closed industry standards. Standardization in fora may be considered as a mixture of both worlds: It combines a priori openness typical for formal standardization with an entirely industry-driven process.

Retrospective: Economic Expectations

From the perspective of 2004, what were the expectations placed on companies and society in 2020 in terms of standardization? In particular, industry's increasing preference for formulating its standards in fora or consortia of companies rather than in the context of the formal standardization organizations that had existed up to that time had raised the issue of what institutional arrangements—that is to say, organizations and rules—could best satisfy in the long term the expectations placed on a formal standardization system that would address competition and timeliness considerations appropriately. To answer these questions, two scenarios were run at that time and considered in the light of the hypothesis of economic rationality:

- It was assumed that there would be no formal standardization organization—economists call this a "no-budget" situation—thus,

Figure 1. Glossary of standardization

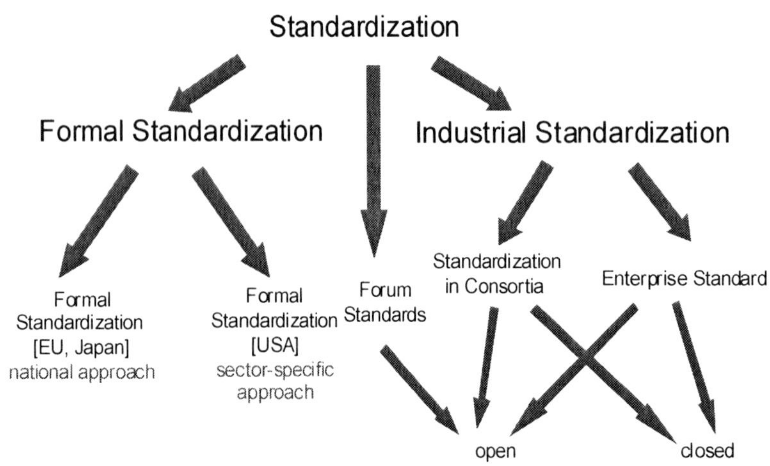

everything could be reinvented from the ground up in line with the anticipated future expectations.
- The parties involved in formal standardization would consist solely of entities subject to a risk vs. return calculation in order to ensure the rationality of actions and outcomes, and thus avoid free riding.

For this purpose, it was necessary to formulate an image of the economy in the year 2020 to determine the general constraints within which formal standardization would be successfully established as a competitive and socially desirable approach in comparison with consortium-based solutions. The scenario formulated as a realistic vision at that time was restricted (using the terminology in Figure 1) to the following standardization arrangements:

1. Standardization is a formal standardization operation meeting the expectations placed on a public process. The goal in this case would be to create an open, generally available system that would provide a platform in terms of an economically viable and proven technology or process.[3]
2. Standardization is a consortium-based process, taking place within a closed circle of participants, whose outcomes would be available either only to the parties involved, as a closed standard, or to the general public, as an open standard. These arrangements required inclusion in view of the clear, efficiency-based challenge they represented to formal standardization in terms of competition considerations.
3. Industry standardization is an a priori open process. In order to avoid problems of antitrust, the benefits of formal standardization (openness) and that of industry standardization (speed) were combined; however, is sacrificed the involvement of all interested parties.

This eliminated the company standard from the picture; as a closed, intracompany process, the company standard does not require discussion here. If a company standard becomes generally available, there is generally an extension to consortium level; that is company standards are subsumed under the discussion on consortium-based standardization.

It was emphasized that, between the three institutional arrangements that therefore had to be considered (formal standardization, consortium-based open standards, fora- and consortium-based closed standards), there had to be competition,[4] the primary criteria for which would be as follows:

- Speed of the process
- Classification of the outcome in terms of precompetition and competition agreements, in particular issues of concentration and the creation of a dominant market position
- Legal status, particularly in terms of warranties and burden of proof
- Nature of the economic goods created as by-products from the process of consortium-based standardization (closed and open) and formal standardization (private, club, public goods)

In fact, formal standardization lost ground to consortium-based standardization, especially to forum standardization around the turn of the millennium precisely on the basis of these competition factors.

Thus, the renewed formal standardization system in terms of its organizational structures and processes had to be set up to be able to compete. Industry at that time saw forum-based and consortium-based standardization as the proven approach in view of the unsatisfactory nature of formal standardization processes in terms of timeliness and content, at both the national and international level. However, it quickly became evident that this involved some major risks, primarily based on antitrust considerations. However, "new approach" policies pushed by the European community favored formal standardization.

Attempts to make consortium-based standards publicly available, often only after a certain time once the industry standard had become established in the marketplace, often failed because the antitrust authorities regarded the process of creating a consortium-based standard in particular as problematic in terms of gaining a dominant market position, owing to the availability of insider knowledge. This trend was reinforced by the construction of the legal exemption in antitrust law that became increasingly prevalent in Europe from the turn of the millennium on.

At the beginning of the new millennium, competition between industry standards produced some remarkable effects, for example, open interfaces, and in many cases the free granting of licenses, but this was not enough on its own to limit market dominance in cases involving additional insider knowledge and a high tempo of innovation with corresponding system changes (e.g., Microsoft). The virtually free issuing of both the major DVD industry standards (R+ and R-) led to a situation whereby most manufacturers incorporated both in their systems, which in practice was tantamount to a market entry blockade for integrated or new standards.

The effects outlined in the previous paragraph clearly demonstrate that a worldwide public-law, formal standardization system may well prove to be preferable to consortium-based standardization since the loss of know-how is quickly counterbalanced by enhanced legal security in terms of compliance with freedom-of-competition regulation requirements and product liability. This insight was an important driver for the later reform and the successful institutional relaunch of formal standardization on a global level.

Development of the Economy Through 2020: Framework of the Analysis

The major developments that led to a significant increase in market integration and transparency were set up back at the start of the millennium. These related to characteristics of the future consumer, which are important for a correct formulation of the expectations that formal standardization has to meet in order to be in the public interest. Markets were increasingly developing toward offerings in the form of systems or system relationships, which was an important change from the traditional categorizations in terms of type (goods, services, rights) and tradability, that is, transportability (local and international goods). Finally, the centralization of strategic company functions as opposed to decentralized operating functions was changing the ways companies were set up, with associated changes in competition structures. The significance of national states was also decreasing since they had not proven to be an adequate mechanism for protecting their populations from fundamental risks in the global economy. In some areas, this had had major impacts on the participants in the formal standardization process. Accordingly, this study will make a further distinction between aspects where little or no change occurred in the years in question, and areas where major system changes took place; the roles of the state and the consumer in the formal standardization process will be particularly important in this context.

SIGNIFICANT STRUCTURAL DIFFERENCES IN THE ECONOMY IN 2020

The end of the 20th century was still characterized by what could be called "Taylorism-based globalization." While there had been some significant changes in the perceived nature of a commercial company, away from a functional understanding and toward approaches based on the analysis of processes, often defined by logistics, during this first phase of globalization the dominant concept continued to be one of territorial expansion rather than the integration of the company and the product.

A major change came with the move toward the "competence society." The concept of the "information society" had been in vogue in the latter years of the 20th century, but the complete lack of any clear definition of this idea soon became apparent, and it was replaced with the "knowledge society," that is, a society that made information useful by evaluation and filtration. However, this potential on its own was a necessary but not sufficient condition for economic success, so it was replaced in turn with the concept of the competence society, that is, a society that makes economically appropriate use of knowledge.

The significant structural aspects are therefore as follows:

- Competence as the essential criterion for economic success: This ultimately relates to the rapid and creative bundling of resources, especially intellectual resources.
- Dissolution of the distinction between supply and demand: As a result of subscription systems, sales by rapid prototyping and distribution via globally accessible auctions, the distinction between these two sides of the market becomes blurred, both structurally and geographically. Targeted marketing in particular becomes difficult since companies do not know who (and where) their customers are.
- The dematerialization of further elements of production: This relates primarily to the use of products as status symbols. Whereas individuals have found it easy to define themselves by their ability to obtain certain goods, this is hardly possible in the case of intangible goods.
- The devaluation of traditional political systems: This also means that the state was increasingly less able to regulate goods specifications, although safety standards played a role in this context through the legal sanction process.

Industry standardization played a central role in this trend as the way in which mix-and-match problems and issues relating to consumer confidence products, and even liability issues (particularly in the case of formal standardization), were resolved. This was also the basis for success in reconciling increasing economies of scale in the mass production of components (e.g., the manufacture of electronic components) with the erosion of economies of scale in final assembly in terms of specific systems (e.g., for the construction of a digital office with the characteristics of a motor vehicle, an intelligent building, etc.).

What Aspects have Largely Remained Stable?

The Markets

The revitalized wave of internationalization processes following the fall of the Iron Curtain continued into the new millennium, with the result that most goods were now internationally tradable. It was of no consequence whether they were actually traded: The underlying production technologies (blueprints) were often exchanged internationally as a way of minimizing transport costs and highlighting the company's market presence. A number of objectives have been achieved through industry standardization:

- Ensuring global delivery capacity in the value-added chain
- Successful production from blueprints in factories
- Provision of safety standards, also from the product liability perspective, thereby reducing information asymmetries

- Ensuring mix-and-match (compatibility) features for the customer

For services, the defining characteristic from the earlier perspective was cited as the impossibility of storing the product, and indeed, the dominant view was that the product could not be separated from the sale: Either the customer had to go to the location where the service was provided, or the service had to be provided at the customer's location. However, already by the end of last century, communication technologies were increasingly making it possible to separate the two, with the availability of standardized communication technologies (particularly HTTP [hypertext transfer protocol], etc.) playing a central role in this process. Growth in this sector has been dominated by services that were communication intensive and allowed the separation of production and use, leading to a new, international division of labor. Accordingly, more and more services that had been tied to a specific location have been globalized by means of communication systems. This has become a central driver of competition between business locations.

Rights can be defined as local when they are valid only within a restricted territory, for example, as national legislative provisions. At this time, international competition and the integration of companies were increasing the pressure for adjustment that was already present, a process that led either to a system of metalegislative provisions administered by international organizations to which national legislative provisions were subordinate, or to the direct creation of international legislative provisions. In some cases, increasing uniformity was driven by the extraterritorial extension of national legislative provisions, particularly by the USA and the EU. Finally, pressure for change was also coming from the freedom to execute binding contracts in third-world countries.

Finally, standardization supports the trend toward the systemic bundling of goods, which was one answer to the problem of increasingly complex requirements and expectations.

The Companies

The world economy is now dominated by globally structured companies, company networks, and cooperation structures as a major support mechanism for GMEs (global medium-sized enterprises) that are entrepreneur owned and internationally active. Their networks have become an important element in counteracting the demand-side power of the retail sector and final producers. With respect to global companies, recent years have been characterized by increasing competi-

Table 1.

	Goods	Services	Rights
Local goods	In fact, there are very few items in this category; a more significant class is that of tradable but locally produced products that are based on tradable rights and services. This category has largely disappeared.	They are often local in the sense of a mandatory requirement for demand at the place of provision or production at the place of use. Increasingly, globalization is promoted by information technologies.	There are legislative provisions with local applications, e.g., administrative provisions, etc.; it is likely that only legislative provisions with little relevance to the economic sphere can be maintained at the local level.
International goods	Standardization provides the basis for global sourcing, capability for production in globally distributed factors, safety standards, and mix-and-match characteristics.	Standards are of central importance for the separation of locations of production and use. These are dominant in the services sector.	Global legislative provisions and extraterritorially extended local systems of legislative provisions will predominate.

tion between company subsidiaries as opposed to competition between different companies. In-house competition between the various locations of one company (or network of companies) has become the key instrument for increasing efficiency, while no such trend has been evident in competition between different companies by harking back to the value of the brand.

The increased competition within companies and increased decentralization of operating entities, in line with the predictions of transaction cost theory, have led to a disproportionate increase in the significance of product standards, and also standards for interfaces. Finally, another essential factor for the increased importance of standardization has been that of the well-known critical thresholds and compatibilities for network systems.

The importance of standardization as such for the utilization of economic externalities as the source of prosperity can be described as follows:

- Scale effects (cost degressive effects) play a central role at the component level; the starting point in this case is product standardization (e.g., mass-production pioneer Henry Ford with conveyor belt manufacture).
- Linkage effects become increasingly important in both production and use; the starting point in this case is interface standardization (e.g., the possibility of importing and exporting data in different systems).
- Network effects are based on standardization in (communication) networks, which exist in both physical and virtual form. For the latter, standardization refers primarily to transfer protocols (e.g., HTTP for the Internet).

The improved ability of GMEs to compete and promote innovation has largely been stimulated by global formal standardization, especially transparency and accessibility to intellectual property rights, in which they cannot be included for reasons of insufficient size.

Areas Where There has been Significant Change

The Consumer

The consumer has finally become empowered, also from a political perspective. In particular, the availability of modern communication systems had dramatically increased market transparency from the turn of the millennium on. Virtual marketplaces have made it possible to bundle product and system qualities from different views, and thus eliminate information asymmetries (constitutional nonknowledge).

Product characteristics became openly disclosed through the possibility of consumers acting before the transaction to find out what others thought of the product, and also later submitting their own impressions; this capability is extended by forums on products and manufacturers, sometimes actually provided by the manufacturers, who thereby gained feedback as the basis for ongoing enhancement of their quality management systems. Finally, the retail sector, after being shaken by scandals such as the bovine spongiforme encephalopathy (BSE) affair, was forced to own the problem of product quality much more than it had done previously.

These consumers of today no longer need the state to empower them or act on their behalf, and, accordingly, the latter has largely withdrawn from the consumer protection role.

States and Communities of States

As had been predicted at the time, competition between locations has reduced the long-term ability of national states to impose taxes to the level of the location rent, thereby noticeably reducing the burden on businesses. This has forced, proactively or reactively, the harmonization of business legislation provisions. In addition to standardization, which was originally in the company's interest, this has raised the issue of a wider form of standardization based on the global public interest, closely linked to the public interest involved in formal standardization.

- **Creation of level playing fields:** The primary purpose of standardization at the company level is to boost the companies' own externalities, that is, to increase their competitiveness. The standard in question may well be in competition with another standard; then competition will decide which standard prevails. A critical role in this process is played by the mix-and-match profiles and network effects (e.g., the failed attempt by IBM to abandon their own PC standard and move to the personal system (PS) standard because of the competition from cheap suppliers). The number of standards that can exist (simultaneously) in the market is given by the competition and the size of the market. If the market can accommodate only one standard, the level playing field has been achieved, but only with the possibility of serious consequences in terms of antitrust requirements (e.g., the problems encountered in the past by Microsoft when it had built up its Windows interface into an essential facility). The major diplomatic tensions between the USA and the United States of Europe in the lead-up to the complete crushing of Microsoft in 2012 are well remembered. The restriction on corporate freedom of movement or the wearing effects of competition between standards may represent a major incentive to enter into a formal standardization process—as long as it was sufficiently rapid and flexible, and also offered the advantage of security against breaching antitrust requirements. It is always necessary to consider the balance of risk between success on the one hand and the associated danger of antitrust restrictions and losing the competitive battle on the other. A third overarching solution is the benefit of a wider market as a result of formal standardization. However, the traditional formal standardization process had operated too slowly to provide an incentive for companies in the modern industrial environment. This was because the disclosure of their know-how and bundling with their competitors seemed to make sense only if this could be expected not only to obviate antitrust problems, but also to lead to a market penetration and expansion rate that far outstripped average growth rates. This meant the need for a form of formal standardization that would effectively preclude the market dominance of globally operating companies, but at the same time compensate for this through the externalities at the level of the wider markets.
- **Using network effects:** Industry had learned some lessons from the fact that the success of the development of the global system for mobile communication (GSM) network in Europe could not be replicated at the global level with universal mobile telecommunications system (UMTS). Whereas in Europe, owing to a formal standardization process, the faster crossing of a critical threshold led to tougher competition because of the open status of the standard, similar success was not achieved in regional markets. In this case, the companies held on too long to their different industry or formal standards, which delayed the process of achieving a critical network size, for which the threshold was crossed only at the beginning of this last decade.
- **Reduction in state intervention:** It is above all a result of formal standardization that allowed reducing the level of state intervention, and in particular to privatize inspections by contracting out arrangements. This was consistent with the deregulation approach that was adopted in Germany and Europe back at the turn of the century under the

regime of the new approach. The essential benefit of this approach as opposed to the state regulation approach was the greater amount of information that was generated in the course of the process: A properly organized formal standardization process does not prevent the operation of market forces. However, formal standardization, as it existed in the years 2004 to 2008, was no longer able to meet this requirement. The withdrawal of companies from the formal standardization process, especially in favor of forum standards, led even state decision makers, though not all of them because there were a few visionary thinkers, to question the legitimacy of a process that left the major stakeholders, that is, industry, out in the cold.

This clearly demonstrated that competition between industry standards (whether closed or open) and formal standards in the new institutional environment led to solutions that reduced the burden on the state, and that were consistent with the demands of the marketplace and increased prosperity.

ELEMENTS OF THE REFORM OF THE FORMAL STANDARDIZATION PROCESS IN 2008

General Requirements

Apart from some culturally determined niches where formal standardization would continue according to the traditional processes based on national or state parameters, and thus different processes in each case, the following were the significant considerations for the establishment of a new formal standardization system in the year 2008 (for the sake of clarity, this new concept will be referred to in this study as "GlobalNorm").

- All forms of standardization define technologies, not markets. Technologies are not subject to frontiers, giving rise to the requirement for an internationally based GlobalNorm organization. A change in people's thinking occurred very quickly, with the recognition that formal standardization would now be required not within regional boundaries, where the aim was to achieve technology-based success, but in terms of success based on technology competence, for which no boundaries existed.
- Industry standardization was to be used, or kept open as a possibility, as a step on the path toward formal standardization. This decision was related to the central importance of companies' motivation to participate in the innovation process, which initially aims to create a private good. In addition, industry standardization bundles a number of different approaches and would therefore often be a central element of the competition between the new combinations.
- It was necessary to ensure the possibility of access to the GlobalNorm process organized spontaneously by the interested parties in the same way as for consortium-based standardization. The criteria would be transparency, nondiscrimination, and the creation of a clearly defined legal space.
- This would accelerate the GlobalNorm process and bring a sustainable increase in transparency, thereby apparently making GlobalNorm competitive vis-à-vis consortium-based solutions for the following reason: Companies taking part in consortium-based industry standardization or GlobalNorm processes receive advance information with the a "club good," which can deliver competitive advantages even while the standardization process is still under way. This significant incentive, of

which the parties were well aware, primarily from their historical experience with national formal standardization processes, could not, however, be accepted in view of the unknown number of potentially interested parties at the global level. This could already be problematic at the level of industry standardization, even if the standard was openly disclosed subsequently. At the time, it was realized that formal standardization becomes attractive in the context of global competition if a rapid transition from industry standards to a level playing field can be achieved through the GlobalNorm process.

- If the GlobalNorm process is therefore to become a function that also bundles the knowledge gained from consortium-based processes, it was clearly appropriate that the new GlobalNorm system should also be given the role of a technology broker. This, in turn, suggested that GlobalNorm processes should be integrated into a new overarching institution dealing with intellectual property rights worldwide.
- Finally, the new GlobalNorm system was also expected to make a contribution in the area of product liability in terms of its increasing use as a reference since the solution of referring to the prior art (which could also be achieved via a formal standard) was often found to be unsatisfactory.

What other characteristics have emerged under the GlobalNorm system? These aspects are summarized in the next few sections of this chapter.

Competition among Intellectual Property Rights and Consequences for Global Norms

Knowledge that is used by only one company, or a single group of companies, provides the opportunity of significant revenue by giving the entity a unique selling point, but also carries the risk of high losses if and when better knowledge is available. Each company is therefore faced with the dilemma of deciding whether to pursue its ideas on its own, or bring them into a common knowledge pool.

Companies generally tend (initially) to hold on to new knowledge to be used for their own benefit, partly in order to test the competition situation under those conditions. A significant incentive for standardization applies where the advantages of such a strategy, primarily faster and more effective market penetration, outweigh the loss of a unique selling point. Standardization and ultimately formal standardization then become significant components of the economic knowledge platform (alongside patents, etc.). This structural characteristic was historically typical of the formal standardization process, but increasingly became devalued as a result of the slow pace of the process—a consequence of the need to take the public interest into account, and also of the strategic attitude adopted by many industrial actors, who tended to use this as a blockade instrument. This meant that the further development and enhancement of the existing national systems was virtually out of the question, and a whole new deal had to be set up.

The choice between industry standardization (closed or open) and (later) a formal standardization system was largely dependent on the incentive conditions. An important role in this context was played by process aspects relating to GlobalNorm standardization, that is, the following aspects.

- Access ("Who gets to sit around the table?")
- How widely the information is spread ("Who else is involved?")
- Consultation rules ("How are decisions adopted?")
- Binding status ("Who decides whether Global Norms can, should, or must be used?")

- Documentation ("How can parties get access to Global Norms?")
- Cost-sharing arrangements ("Who pays, and does that party also derive some benefit?")

Especially in view of ever shorter product life cycles, processes that took 5 years to draw up a formal standard were no longer viable and were therefore replaced by GlobalNorm standardization.

Knowledge Management, European Prototyping, and Global Norms

Requirements from the Perspective of Knowledge Management

The increasingly professional approach taken to knowledge management, and the fact that knowledge has become the most productive and most highly remunerated factor of production, made it necessary to design a new formal standardization system that would be able to compete with spontaneously forming consortium-based standardization systems. The following baseline assumptions were made.

- Technology opportunities are not repetitive; that is, a missed opportunity for standardization is generally lost for good.
- Internationally based companies want a clearly defined access path to formal standards.
- Industry standards and formal standards are a record of the technology knowledge base in the same way as patents and should therefore be publicly accessible to the maximum extent possible.

FUTURE LANDSCAPE OF EUROPEAN STANDARDIZATION

Between 2005 and 2007, the two European formal standardization organizations, the European Committee for Standardization (CEN) and European Committee for Electrotechnical Standardization (CENELEC), drew up a white book on the challenges facing formal standardization in terms of acceptance by customers and by industry, the incentive compatibility of financing, and sustainability in a competitive environment.[5] The most important proposals were the following:

- Formal standardization needs drastically improved visibility and must become a trademark—something historically very typical for many national formal standards and their respective standardization bodies, but washed away by globalization.
- Converging technologies imply institutional reform as sector-specific approaches and procedures that produce slack, against which industry reacted by promoting forum and other industry standards.
- Standardization has to become more oriented toward solutions encompassing industry as well as formal standards where necessary, thus urging formal standards bodies to augment their portfolio with industry standards.
- The scarcity of experts can only be overcome by a more comprehensive approach to standardization; in view of the necessity to speed up standardization processes, formal standards bodies have to give workshop agreements and preliminary available specification more attention as first specifications that may mature to full formal standards, and enforce cooperation in committees against a strategy of blocking decisions by individual companies that want to save time to gain their own monopolistic power in the marketplace.

- The elapsed time of a standardization process is crucial and very much defines interest and support by the most important stakeholder: industry.

CONSEQUENCES FOR GLOBAL-NORM STANDARDIZATION

Accordingly, the GlobalNorm system itself would have to be organized on a competitive basis if it was going to be able to fulfill these roles. This meant there was an antagonism between a uniform system and a competitive structure to avoid the head of the system becoming a bureaucratic structure since the GlobalNorm system had to be able to absorb all available market information. According to economic theory, such systems should be constructed subsidiarily; accordingly, the GlobalNorm system, as is also characteristic of industry standardization, would to a large extent be removed from the sphere of formal standardization institutions and placed directly in the hands of the interested parties.

The Institutional Structure of Global Formal Standardization

On the basis of the above expectations and requirements, the institutional framework of the new global public-law GlobalNorm system was formulated in 2008 under the aegis of the World Trade Organization (WTO) as follows.

1. The GlobalNorm process would take place by personal participation and via any Internet portal. The participants are all parties directly subject to a risk-benefit calculation; that is, they are required to weigh up the extent to which their participation in the GlobalNorm process will be to their economic advantage or disadvantage, to evaluate if they accelerate or slow down the process, and to decide to make a greater or lesser knowledge contribution. Personal participation here means that interested parties would be able to move the GlobalNorm process forward at a location chosen by them. The provision of the Internet portal would mean, first, the possibility of access and input for parties unable to be present (this would refer mainly to SME [small and medium-sized enterprises] and GME companies), and second, that the process would be transparent; that is, no club goods would be generated.
2. This type of GlobalNorm process would be accepted by all participating countries as unproblematic from the perspective of antitrust requirements. The high incentive to achieve rapid standardization in relatively small and dynamic markets was seen as compensating for the fact that no inside knowledge was generated in the formal standardization process as a club good. In fact, this benefit has not been be completely lost. While transparency has been provided, the rapid formulation and implementation of Global Norms still generates benefits today, although these are to be regarded as compatible with the provision of incentives.

This represented a deliberate break with the previous participation model providing the basis for public-law legitimacy since that model was contrary to the required process speed, the profile of today's consumers, and the information access opportunities now available. Consumer representative entities were, of course, entitled to participate at their own cost, but not on a state-subsidized basis. This suddenly created the imperative to set priorities, particularly on the part of the numerous state-subsidized consumer associations. Ultimately, however, experience also showed that entrepreneurs sometimes supported the participation of consumer associations and audit institutions to utilize their specific know-how.

Over time, the structure also accumulated a number of side effects that had basically been intended, but that gained much greater legitimacy from the new institutions than had been envisaged at the time, namely, the function of dealing with further intellectual property rights, particularly patents and open industry standards. This highlighted the concept of Global Norms as a particularly successful innovation.[6]

Funding

It had already become clear at the end of last century that a critical examination of the funding of formal standardization processes was necessary, if only because of the parlous financial position of a number of national formal standardization organizations, and even some international bodies. Whereas the cost of industry standardization was borne by the parties obtaining a direct benefit (expected and quantified), the costs of formal standardization were borne by the companies involved and, through subsidies, the state. In some countries, this led to national formal standardization organizations diversifying in the lucrative auditing sector, but conflicts of interest very soon became clearly evident in this context.

The need to carry the cost of qualified staff, particularly in the case of SMEs and GMEs, created barriers to participation that were sometimes virtually impossible to overcome. In practice, this excluded them from participating in the process and also raised the issue of the legitimacy of the formal standardization process. The large corporations had the freedom to choose from which country and through which formal standardization organization they wished to drive the standardization process, unless they had already decided to use a consortium-based industry standard. In spite of the proven significant benefits accruing to the state from the formal standardization process, in that the increased growth extends the tax base, fiscal constraints led to a decrease in state support.

The main thrust of the 2008 reform was to ensure that the costs of the GlobalNorm process would initially be borne by the companies involved since they had been granted significant tax relief, partly in the course of competition between industrial locations. However, the question arises as to how to put a figure on the costs of Internet participation and the benefits to the state, and what form of incentive is to be provided for the creation of economically successful Global Norms; the analysis of the situation carried out 20 years ago demonstrated that the body of formal standards in existence far exceeded the documents that were actually used for economic purposes. There was a significant incentive for overstandardization in terms of formal standards.

This led to the formulationn of the funding model that still applies today, apart from some minor amendments.

1. The GlobalNorm process is initially funded by the companies personally participating in the process. Since they are generally the parties initiating a GlobalNorm of this kind, and therefore possess information on its future value, it is also this group that sets the option value. The option value is calculated from the discounted future earnings or savings using appropriate evaluation procedures.
2. The costs of this group are billed according to across-the-board charge rates, that is, essentially standard daily monetary amounts. Travel costs are not included in order to limit the possibility of false incentives in the selection of the venue for the meetings.
3. Companies participating on a virtual basis, whose inputs are bundled by one of the GlobalNorm institutions, pay a one-time entry fee and a fee per input, both based on the option price. The option price is set by the personal participants, subject to upper and lower limits specified by the GlobalNorm organization.

4. The GlobalNorm drawn up is made available against financial charges, with the price being based on the option value. From this income received by the GlobalNorm organization, one third is first distributed to the participating companies on a pro rata basis according to their costs incurred, up to the amount of the option value (without interest). Once those costs have been met, all further income flows to the GlobalNorm organization.

This system provides a number of benefits and mechanisms that provide protection against distorted pricing.

- The companies initiating the process are the only parties with a sound basis for estimating the economic worth involved. Accordingly, they set the value.
- If they set the value too high, the option price increases, and there is the subsequent prospect of increased income; however, there is also the risk that the standard will not be used because of the excessive price, with correspondingly lower subsequent payments.
- If the costs imposed on virtual participants, primarily SMEs and consumer associations, are set at the upper end of the participation price range, this may lead to increased payment contributions for the group, or, if they decide not to participate, a loss of their technical input and a less widely disseminated standard.
- Virtual participant institutions are placed on an equal footing with personally participating companies.

At the time, this approach was found to be fully compatible with the existence of incentives for the participants, and indeed this has proven to be the case.

As far as the self-financing and funding of the GlobalNorm organization is concerned, the assumption was that it would have sufficient income-generating possibilities from its additional functions as the body administering global intellectual property and the institution for technology inventory and evaluation.

This was to be based not only on the sale of Global Norms; there was also a possible role as an administrator of open, consortium-based industry standards. Finally, possibly the most promising source of income was from the bundling of activities to provide road maps and innovation consultancy services. Start-up funding was therefore to be provided for a limited time only.

EXTERNAL IMPACTS OF GLOBAL NORMS ON THE KNOWLEDGE SYSTEM

Gradual Evolutionary Changes

Global Norms as a Function Reducing the Role of the State and as Political Correctness

Global Norms become increasingly important in a society in which competition is increasingly extending into public-sector areas. This allows the implementation of best practice, particularly in the services sector. In the period around the turn of the millennium, this mainly involved quality standards. However, as the public sector came under increased pressure to withdraw from activities suitable for privatization, and with the requirement to allow foreign entities to participate in calls for tenders for public contracts in accordance with EU legislative provisions, a growing need was perceived for international process standards in place of the former typically bureaucratic standards.

As the GlobalNorm standardization of products, interfaces, and processes increasingly became a reality, products with such standardization took on a new quality in the eyes of consumers. The result was an interactive process between product characteristics, reference to such criteria in marketplaces and information exchanges, its incorporation in product development and enhancement, and ultimately in the standardization process, particularly in the form of a GlobalNorm. Specifically, in the early market phases, problems of information asymmetry arising from experience- and confidence-based product characteristics often appear to be particularly significant. This highlighted the importance of a fast-track formal standardization process as the only viable way to utilize the potential from reducing information asymmetries with respect to the customer. From the consumers' perspective, this led to a phenomenon that may be characterized as a form of political correctness, whereby the lack of a GlobalNorm was seen as an expression of insufficient customer orientation.

The Creation of Global Norms as a Process Ensuring the Openness of Markets and Freedom of Competition in a Heterogeneous World Economy

The fact that the WTO was one of the founding entities of the International Intellectual Property Rights Agency (IIPRA) marked out the path ahead toward the development of a GlobalNorm system along the lines of the WTO. The traditional benefits of formal standardization—the resolution of freedom of competition issues, dramatic improvement in product dissemination, and clarity with respect to liability issues—have fed back into industry standardization. While industry standardization is still generally a faster process, the significance of creating a broad base has been widely recognized among the parties involved in formulating industry standards so that the consortiums are, on average, getting larger, and the same people often meet up again during the subsequent GlobalNorm process.

Formal standardization at the national level was long seen as an important instrument of strategic national trade policy. However, in the report on formal standards,[7] it very soon became evident that proactive formal standardization that also took account of the type of standardization processes in the target trading countries in a timely manner was particularly beneficial. As stated above, formal standards define technologies, not (national or state) markets. This suggested that even the overcoming of standardization arrangements apparently preventing the possibility of trade would be worthwhile. In particular, it became quite clear that the debate on calls to use national formal standardization policy as a way of gaining trade advantages was no longer viable. Thus, the point of any sort of quasicompetition between formal standardization systems had become eroded.

Major Transformations

Global Norms as the Basis for Drawing Road Maps

The specification of industry standards represents a club good in production operations of the consortium members, and ultimately can even become a private good for individual users if supplemented with additional specifications; this applies to the process even where the specification has been disclosed, where applicable (as provided for in American antitrust law, for example). In contrast, Global Norms as a knowledge platform makes it possible to describe technology development paths and to conduct an effective debate on technology road maps. The GlobalNorm system is controlled to a much greater extent than before via market and competition processes, which means that it also extracts more locally available and relevant information. Accordingly, GlobalNorm organizations have acquired an important role in the feasibility assessment and evaluation of technologies.

These new tasks could not be performed be the old formal standardization organizations. By the beginning of the first decade of this century at the latest, it had become clear that there was a significant legitimacy issue in that studies at the time showed that the economic significance was as high as before, but was at risk of being eroded by the lack of interest shown by the parties involved. One of the unfortunately peripheral phenomena at that time was the fact that participants in the formal standardization process were now playing only a minor role in the various technology and innovations commissions.

Global Norms as an Innovation Driver

Innovation is the primary driver of competition, or, more specifically, is what comes before competition since the transfer, that is, the movement of market share from the company that makes things happen to the company that lets them happen, followed as the second subsequent phase. By providing a platform, industry standards allow the preparation of a level playing field, ultimately defined by the opening of the relevant standards or a GlobalNorm. Particularly as a result of the application of the European antitrust policy of the legal exception in the first decade of this century, companies felt an increasing need to control the increasing risks in terms of competition legislation; this aspect is further discussed below. The higher level of public involvement in the innovation base worldwide has resulted in broader networks, bringing ample compensation for the companies participating in the GlobalNorm process and also opportunities for new start-up entrepreneurs.

In particular, it has been found that such a system effectively fosters the inventor attitude in that the various technology sectors are better marked out than before. The components in this process are as follows:

- Stimulating increasing alertness to new developments
- Serving as enablers, thus making technology sectors easier to access and utilize
- Facilitating a proactive approach by entrepreneurs, who are therefore operating in a less risky environment

Global Formal Standardization as Part of Corporate Risk Management Systems

The adoption of the EU Basel-II equity guidelines by the European Commission in 2004 triggered a vehement debate on the possibility of using Global Norms as a way of reducing risks and making risk identification and assessment transparent. Tougher expectations had already been placed on corporate risk management systems by regulatory measures in Germany (in the Control and Transparency in Business Act, KonTraG), but without any long-term success since the companies lacked operable criteria, and additionally there were no effective sanctions on negligence in this area from the capital markets. However, this pressure suddenly increased to immeasurable levels without any response from the national formal standardization organizations. Ultimately, it was the insurance industry, which had long been including technology factors (and formal standards) in its claim statistics, that showed the way, but it was not prepared to go through to the implementation stage without the establishment of an overarching international formal standardization organization to avoid distortions of competition.

The formal standardization of products, processes, and interfaces along the value-added chain quickly became an important system of criteria for risk inventories and the determination of claim probabilities. Finally, this also made an important contribution to increased rationality in terms of company valuations.

TRAGEDIES DURING THE TRANSITION PERIOD, AND DEBRIS OF THE FORMER FORMAL STANDARDIZATION SYSTEM

The Fundamental Transformation Problem

The transformation operation from old to new faced the following issues.

- Given that the so-called public interest at the time was scarcely compatible with the principles of the liberal market economy, particularly as a result of the artificial construct of stakeholders and that the transparency of products, interfaces, and processes had been dramatically increased in the recent past by modern information technologies, the question arose as to how such a "holy cow" could be slaughtered. It was realized that even new requirements imposed by the international environment would not necessarily force institutions to reform voluntarily and from within. The best and most discussed example from that time was the organizational structures reminiscent of a Soviet-style industrial combine or government-run social insurance system.
- It also quickly proved to be the case that the existing employees found it difficult if not impossible to adapt to their new roles. The decision was then made to implement a revolutionary transformation at a single stroke, namely, to set an international, that is, a global, public-law GlobalNorm organization with regional branches. Employees from the existing institutions were given the opportunity to transfer to the new body after a corresponding assessment. The rest were assigned to positions to meet the residual national formal standardization requirements.

Requirements for National (Cultural) Acceptance

Formal standards have to be accepted by the user culture; along with the subsidiary principle followed regarding responsibility for the creation of formal standards, this requirement highlighted a spatially defined structure that would not be covered by international formal standardization.

This area duly became the preserve of the old formal standardization institutions and standardization specialists, but even here they were repeatedly called upon to justify their existence in view of the transition areas in a dynamic system. Numerous aspects of the Internet-supported participation model developed for the GlobalNorm system were transferred to this residue of national formal standardization structures. However, the triumphant progress of international Global Norms, which ultimately reduced the national formal standardization organizations to the status of regional branches, still left an important area of core business within the purview of national states or national state formal standardization partnerships (e.g., at the EU level). Wherever cultural identities were affected, this restricted the possibilities for internationalization. Accordingly, this impacted more on the formal standards on processes rather than products and interfaces, for example, in the training and education sector and in the health sector.

Typical examples of industrial and formal standardization processes at the local level occurred in the following areas.

- **Health systems:** Owing to different levels of prosperity, with associated differences in the value of life, there were major differences in treatment methods as opposed to medical equipment, for which global formal standards apply. But even here, there is a clear process of diffusion from a regional cultural standard or norm to a GlobalNorm. When, in the middle of the first decade of this century, it was found that health costs in the United Kingdom were around 30% lower than in Germany, but that the average life expectancy was only 5 months (i.e., 0.6%) shorter, the question was asked as to whether the wrong standards were being followed in this country. Ultimately, this led to a best-practice system in medical quality management, for which an international standardization process is currently under way.

- **Training and education systems:** Education creates values and training enables capitalization. The issue of what kind of structures is required in this sector is largely based on culture and identity. This applies particularly to education standards and less so to training standards, which will be forced to standardize to a much greater extent, owing to international competition between locations.

CONCLUSION

As these comments show, the institutional revolution implemented in the formal standardization system from 2008 onward was destined to be a major success based ultimately on the open participation model made possible by improved communication systems. The globalization of the formal standardization process created a competitive good in the form of the GlobalNorm, which is highly attractive to companies of every size, owing to the high transparency of the process and the security provided against interventions by the antitrust authorities jeopardizing the company's strategic objectives. The European initiative to push its formal standardization system to the global competitive edge proved to be a major reference case. There have also been significant positive spillovers into other systems, particularly the patent system and the notification and monitoring of technology developments, which is why the formal standardization function was placed in the hands of an overarching organization with regional branches, but with a global structure that deals with intellectual property rights. Since the commercial exploitation of these other rights can be bracketed with that of Global Norms, it was possible to set up a funding system that freed GlobalNorm processes from state budgetary constraints. Technologically and economically significant corridors and sustainable development avenues have been opened up, increasing prosperity in the global village.

ENDNOTES

1 Adapted from a report to the director of the German Intellectual Property Rights Agency (GIPRA), Berlin, Germany: [BLUM, U. (2002). *Challenges for the Future: Public Standardization in the Spotlight*, (GIPRA/ IIAPRA working papers). Berlin: GIPRA, Bangalore: International Intellectual Property Rights Agency (IIAPRA).]

2 Between 1999 and 2001, the author was involved in drawing up a report on the benefits of public standardization for the DIN (Deutsches Institut für Normung), and in connection with the Hanover Trade Fair in 2004 and the ensuing debate, not confined to public standards circles, prepared a submission on changes in the public standardization system against the backdrop of globalisation. Between 2005 and 2007 he was convenor of the working group of the European Landscape of Standardization set up jointly by CEN and CENELEC to reorganize European standardization; see BLUM, U.; 2007; Whitepaper on the Future Landscape of European Standardization.

3 Under the traditional view of the public standardization process, public interest imperatives are generally deemed to be met by giving the circle of "interested parties"

access to the public standardization process. A high priority is also put on the principle of consensus.

4 This competition has been taking place for some years. The success of consortium-based solutions led to less significance being attributed to public standardization. This did not call into question the positive impacts from public standardization up to that time—the macroeconomic benefits alone to Germany have been estimated at around €15 billion – the situation was rather that all parties involved, particularly those in industry, were quite clear that changing competition conditions (global market and competence leadership) required standardization instruments that were designed accordingly. For a discussion on the necessity of reform, see Blum, U.; 2006; Bedarf Europa einer gemeinsamen Normungsstrategie? CEN, CENELEC und ETSI, DIN-Mitteilungen Nr. 12, 16-21.

5 The public standardization institutions were therefore designed from the outset for the management of intellectual property rights, since their role was to go beyond standardization as such, or in order to give them development opportunities.

6 Cf. BLUM, U. ET AL.; 2001; Gesamtwirtschaftlicher Nutzen der Normung—Unternehmerischer Nutzen Teil 1: Wirkungen von Normen; Teil II: Unternehmensbefragung und Auswertung, Beuth-Verlag, Berlin.

7 Cf. BLUM, U. ET AL.; 2001, Nutzen der Normung, DIN-Mitteilungen Nr. 5, 350-361.

Chapter II
IT Standardization: The Billion Dollar Strategy

John Hurd
Southern New Hampshire University, USA

Jim Isaak
Southern New Hampshire University, USA

ABSTRACT

This article summarizes key incentives for vendors, users, government and individuals to participate in the standardization process. It argues that standards can expand markets at all points in the market life cycle, with overall impact measured in billions of dollars. The authors hope to encourage standards involvement, and also future research and analysis that might quantify the financial value of standardization for vendors and users.

"Standards are like keys always hung at the same nail—they free up your mind for more useful thoughts." -Anonymous

INTRODUCTION

Information technology standards are commonly perceived as having evolved from what was once a cooperative effort by engineers and standards groups (to achieve a consensus aimed at the common good) to a cutthroat struggle by a single company or group of allies to gain market dominance by manipulating standards.

This view is an invitation to anti-trust challenges; to rejection of standards in public procurement or as a foundation for public policy; and for preemption of standards initiatives by other industries or regulatory processes. Standards can grow markets, bring sustained value to users, encourage productive government relations and improve engineering excellence; but not if the IT industry fulfills the above perception.

WHY STANDARDIZE?

The IEEE 802 local area network standards support a \$15 billion per year market. The POSIX/UNIX standards provide an additional \$18 billion per year in the UNIX market[1] (Unter, 1996). Internet and World Wide Web standards exceed these in terms of both financial and social impact, and are totally dependent on a foundation of standards (Cargill, 2001). The social impact of the Internet, and the value to both government and diverse industries, has led to active legislative and regulatory processes affecting the information technology (IT) market in the United States (U.S.) and other jurisdictions. The information technology industry must develop a coherent and responsive approach to IT standardization requirements and associated policy implications. If we fail in this, we will not enjoy the full market growth that is possible, and we risk intervention by government. Here we will look at the incentives for standardization by providers (to make money), users (to save money), government (for the public good) and individuals (professional recognition).

PROVIDER INCENTIVE

Standards are one way to have a multibillion-dollar impact on IT markets over an extended period of time. For example, Figure 1 is the classic market curve: market value over time. The area under the curves is the cumulative market value. Early in the curve (a), standards can be used to increase the pace of adoption, thereby increasing the market size more quickly. The consumer electronic industry has learned this, and in order to enable rapid growth, introductions of new technology (for example, CD or DVD) are preceded by agreement on key standards. In the IT industry, the early acceptance of POSIX/UNIX standards and adoption of these in procurement policies accelerated commercial acceptance (FIPS 151, 1986).

Figure 1. Potential for standards impact on market curve

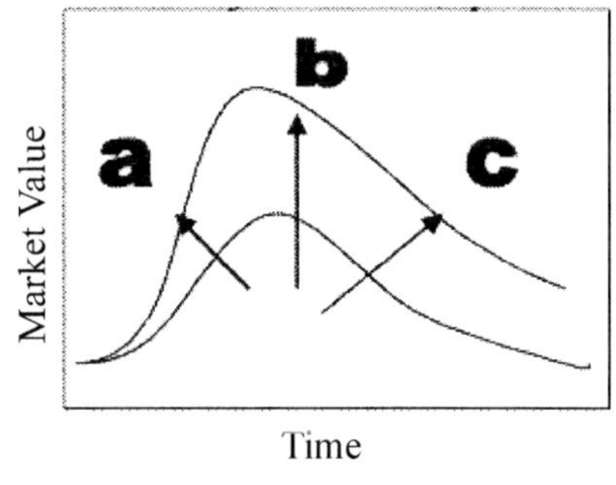

As the market is established, standards can increase the overall market size (b). This is how the local area network standards have sustained their impact: by expanding functionality with technology while maintaining compatibility. Ongoing evolution of these standards and their adoption at the international level have provided for ongoing market growth and increased acceptance. As the market goes into its mature phases, standards can be used to extend the life (c) and provide incremental return on investment. We see this with the ongoing, albeit slow, evolution of standards for Cobol, Fortran and CD-ROMs. In mature areas, standards can be augmented to address specific niche segments and opportunities. The influence of the POSIX standards on the creation of Linux (Torvalds, 1991) provides an interesting example of the extension of a market as a result of standards availability.

The classic market curve operates over multiple years, and providers should measure the impact of the standardization investment in multiple billions of dollars of market growth over a 3+-year period. If the market is not measured in the billions, or targeted as an investment over many years, it is probably not a candidate for standards.

Billion Dollar Impact of Standards

- **IEEE Std. 802 local area network:** $15 billion per year
- **IEEE POSIX and X/Open UNIX:** $18 billion per year
- **IETF TCP/IP:** The Internet
- **W3C HTML:** The World Wide Web

How do standards have this impact? Standards establish a clear communication of capability between the customer and the supplier. We see this every day as consumers: from light bulbs and shoe sizes to electrical codes and highway signs. In some cases, conformance testing and/or branding (such as the Underwriters Laboratories label) can be used to complement the actual specifications and references (for example, DIN 400 for film speed). In some areas, customers are a driving force behind standards, and in other areas, they participate to ensure their needs are met.

Why should a company invest in standards? Standards do not happen without some industry investment. In areas where product development proceeds concurrently with standards development, participants have an inside view of the changes and issues that will affect conformance. Also, participants have a voice in the process to ensure that the standard accommodates their plans and are not unnecessarily limiting.

"Being there" provides a view into the market: the market leaders, the customers and the forces that are establishing the substance and perceptions that will affect market acceptance. Proactive players can help to set the tone, as well as the content, of both the standards and the resulting market environment. Standards are neither necessary nor sufficient in this context, but one of the tools for building a market presence and influence.

Finally, as standards move into the view of the customers, the participants in the standards process have a path to the customer, providing insight and guidance based on the standards experience. Customer confidence and purchasing is governed by a combination of the value of the standard and the credibility that suppliers have in terms of delivering that value.

Vendors can gain insight by looking at the standards from the consumer perspective. In what areas will buyers tolerate a lower price or "innovation" over standards conformance and an open market in purchasing? Answers will differ, both by the area of consideration and by the individual making the decision. Consider examples such as: a new integrated circuit that uses nonstandard pin-outs or power; a PC board with a different bus, a peripheral that has a unique interconnect, media or format, a new removable media form factor or a different keyboard layout.

USER INCENTIVE

- **Y2K Example:** Users experienced the benefits and costs of applying standards in addressing the year 2000 problems. The prevalent use of standard Cobol made it easier to deal with corrections, while the lack of strategic application of software engineering standards made the changes necessary.

Incentives for standards' users complement those of the provider. Clear communications about capabilities, confidence in functional characteristics and participation in the larger-size market benefit the consumer. The buyer also will want to encourage competitive supply and differentiation based on true value added, not just on incompatibility of supplier products. User investment in retraining, re-engineering and acquiring new capabilities can be minimized though effective adoption of standards.

Standards will not happen if users are not going to buy! Vendors are looking for users who can articulate their need for, and commitment to, purchasing standards-based products. There is a catch-22 here if vendors wait for user demand,

and users wait for vendor supply. However, establishing a purchasing intent can be done, and participation in the standard definition provides a clear message about needs and commitment to vendors. Some vendors will prefer locking in users with higher profit margins in a smaller market rather than competing in an open market. Without user demand for standards, users cannot expect vendor compliance.

Standards' participation allows users to define the objectives for the standards in terms of the problems they need to solve. They can keep the standards work on course as it progresses, and verify that they have what they need in the balloting process.

Ultimately, users vote with their pocketbooks. When purchasers require that products "must conform to xxx," vendors get the message (particularly if the user truly values that specification). If users are willing to accept less for a lower price or neat features; vendors will hear that message, as well. User purchasing trends can either reinforce the value of standards investment or erode that value. The broader the user mcommunity demands for conformance, the greater the value to the user community in terms of immediate applicability, and also in ongoing compatibility.

Users who consider the life-cycle cost of their investments must be willing to invest at the front end to ensure that the costs are lower for the inevitable maintenance and replacement cycles. Standards are a strategic investment for users. Application of standards should be based on expectation of multiple millions of dollars of lowered costs and/or increased productivity over a multiyear cycle. This must be the basis for user investment in formal standards (participation, procurement and application). Intra-industry standards tend to get this attention, with buyers like GM actively involved in SAE standards along with suppliers, or computer vendors active in physical layer communications standards along with communications suppliers. This is less common with areas such as IT standards, which may be of high value to non-IT organizations, but they fail to invest in the process.

Once the motivation is clear, participation provides additional insight for the user. Users can ascertain the competence and commitment of the vendors though interaction in the meetings. Also, user contact with other experts can provide for increased effectiveness and productivity for the user in application of the standard.

GOVERNMENT INCENTIVE

Government players have an additional set of incentives to make sure the needed IT standards happen (Garcia, 1993). The world of convergence is the market of the future. National competitiveness, employment, new industry, educational reform, improved healthcare and delivery of government services are dependent on this standardized infrastructure. Moreover, in a worldwide context, this means hundreds of government bodies with strong interests; and in many cases, with power and authority to solve problems though regulatory means if the industry does not succeed in establishing the required standards. Most of the barriers to this brave new world are policy and political barriers, and many of these have significant technology components. These include protecting intellectual property, privacy, universal access (rural, handicapped and disadvantaged), protecting children, controlling crime, providing trustable (and taxable) commerce, and so forth. Government addresses these barriers in terms of *the public good.* Government statements on deregulation and market choice are often coupled with a "public good" escape clause, keeping the door open for intervention. In the U.S., OMB 119a, public law 104-113 and the Telecommunications Act of 1996 all reference government application of volunteer consensus standards, where such standards exist as the preferred way to meet government objectives.

The international market is faced with convergence as the boundaries disappear between communications, entertainment, IT and related industries. A simple rule of thumb in this converged market is: *If products or services from some entity can cut into your market (or eliminate it), they are a competitor.* Do not let your traditional competitive assumptions blind you to this reality. Also, every competitor is a potential ally, with strengths and weaknesses that may complement your businesses. Significantly, we are playing by different rules. The IT industry has traditionally avoided and resisted regulatory or legislative involvement, whereas telecommunications and broadcast companies have learned to use these as part of their core competencies. You will not find many advocates of government intervention, but you will find that government is increasingly involved in the converged world of the 21st century. (Note in Rep. Sensenbrenner's quote, it is "how," not "whether" the IT industry is to be regulated.)

"Speed in the legislative and regulatory process is measured in months if not years. A Congressional session is two years. Many bills take more than one session to be enacted into law. In the world of information technology, a month is an eternity. Legislation is often obsolete by the time it is enacted.

"There is no simple means of reconciling these differences. A key, however, is flexibility. Congress and the Administration must avoid burdening the development of information technologies with overly prescriptive regulations.

"The dynamic nature of the Internet is an asset to our nation's economy. It is an asset we cannot afford to stifle."

Rep. F. James Sensenbrenner, Jr.
Chairman House Committee on Science

IT industry partners and customers in diverse industries need a coherent, competitive, robust computing and networking environment to build their businesses. Failure to provide interoperability, compatibility, portability and ease of use may well interfere with their business growth. IT standards are not just a factor in growth of the IT industry, but also a factor in the growth of other industries. These other industries may not be as tolerant of the petty deviations that have been the tradition of differentiation in the IT industry. Government intervention is one tool these affected players may apply to "fix" problems in the IT industry. We have seen this already with legislation on copyright protection (such as with DVDs) and with proposals for new intellectual property protections, and U.S. government (internal) standards for software and Web accessibility (Access Board Standards, 2000).

The IT industry needs to take a mature, credible and proactive role in responding to the needs of complementary industries and the interests of governments. U.S. law and policy give strong preference to accredited standards as a basis for regulatory action and procurement policy. Similar consideration exists in Europe and deference to international standards is a cornerstone of the World Trade Organization in its objective of eliminating trade barriers. It is time for the IT industry to reinvest in the standardization that is required to minimize government intervention and to meet the needs of affected communities of interest.

INDIVIDUAL INCENTIVE

Individual technical experts, often senior engineers directly involved in product development and planning, ultimately are the foundation of the standards process. While some may participate because of personal interest, most are actively involved and supported by their employers for the reasons previously outlined in this article (provider, user and government). Individuals can find standards activities one of the most rewarding of their careers.

Standards meetings occur at off-site locations and bring together technical experts from diverse industry organizations. Dialogue occurs in meeting rooms, in the halls and over dinners. Influence is developed over time, based on mutual respect, technical competence and professional integrity. The give-and-take involved in standardization is one of the validation points of industry peer recognition and respect. This can be a motivating factor that is rarely available for organizations though internal operations.

To the most outstanding team leader in the field of computer engineering standards in recognition of outstanding skills, a dedication to diplomacy, team facilitating and joint achievement in the areas where individual aspirations, corporate competition and organizational rivalry could otherwise be counter to the common good. (Description of the IEEE Computer Society "Hans Karlsson Award")

Continued individual contribution creates opportunities for leadership and the development or expression of valuable communications and project management skills. Effective standards leaders also become visible industry spokespersons. These individuals provide contact points for the press and consultants, and advise governmental procurement and regulatory processes. Contacts with users extend beyond the standards meetings into user-initiated contacts and opportunities for customer site discussions.

Champions/spokespersons are essential as part of building visibility and acceptance for the resulting standards, and for generating the full market value for the investment. Their impact is greatest when the participants also develop articles for publications, books, tutorials, conference sessions and other materials that facilitates broad industry acceptance and application of the standards.

Standards involvement can be one of the most challenging and rewarding tasks for technical experts. It provides them the opportunity to develop new skills and to prove their expertise and collaboration capabilities. The relationships developed through this process can provide the basis for broad industry recognition, increased technical awareness and a vision, as well as a voice into the future of technology.

THE BOTTOM LINE

IT standards are one key to increased growth in the IT industry and also in other industries. There is an ongoing need for standards to facilitate this growth, to provide investment protection for users and to meet government objectives for the emerging information infrastructure. Failure to deliver essential standards may result in government intervention either to meet public sector objectives or to respond to pressure from affected industries. When the standards are done right, we have an economic and social benefit for providers, consumers, government and the individuals involved. The challenge for the IT industry and profession is to control our destiny, lest others do this for us.

REFERENCES

Access Board Standards. (2000). *Electronic and information technology accessibilitystandards*. Retrieved from www.access-board.gov/sec508/508standards.htm

Cargil,C. (2001). Why are we doing this? *IEEE Computer, 34*(10), 116-117.

FIPS 151. (1986). Proposed Federal Information Processing Standard for UNIX Operating System Derived Environments. *Federal Register, 51*(168).

Garcia, L. (1993). A New Role for Government in Standard Setting? *StandardView, 1*(2), 2-10.

OMB Circular A-119. (1998). *Federal Participation in the Development and Use of Voluntary Consensus Standards and in Conformity Assessment Activities.*

Public Law 104-113. (1995). *National Technology Transferer ad Advancement Act.*

Torvalds, L. (1991). Linux History. *Linux International.* Retrived from www.li.org/linuxhistory.php

Unter, B. (1996). The Importance of Standards to HP's Competitive Business Strategy. *ASTM Standardization.*

ENDNOTE

1 Unter ascribes 30% of HP's UNIX revenue to the standards efforts; the $18 billion dollar amount is a projection of this to the total 1996 UNIX systems market.

This work was previously published in The International Journal of IT Standards and Standardization Research, 3(1), edited by K. Jakobs, pp. 68-74, copyright 2005 by IGI Publishing, formerly known as Idea Group Publishing (an imprint of IGI Global).

Chapter III
Best Practice in Company Standardization

Henk J. de Vries
Erasmus University, The Netherlands

ABSTRACT

This chapter describes a best-practice model for standardization within companies based on a process approach to the development of company standards. Per process, a best practice is developed based on an investigation within six multinational companies and a review of literature, if any. The findings are benchmarked against experiences in three comparable fields: IT management, quality management, and knowledge management. Though the number of company standards exceeds the number of external standards by far, they have been neglected in standardization research. The authors hope that standards practitioners will benefit from their study and that it will stimulate researchers to pay more attention to this topic.

INTRODUCTION

By the end of 2003, the People's Republic of China had 20,226 national standards (including adopted international standards), more than 32,000 professional standards, more than 11,000 local standards, and more than 860,000 company standards (Wen, 2004). Most other countries do not have a central registration of company standards, but it can be expected that also in other parts of the world the number of company standards outweighs the number of other standards to a large extent. This huge difference is not reflected in the amount of attention to company standards in scientific literature. Main exceptions are Hesser (2006), Perera (2006), and some German books in the series *DIN Normungskunde* (for instance, Adolphi, 1997; Hesser & Inklaar, 1997; Schacht, 1991; Susanto, 1988). Professional publications on company standardization include Association Française de Normalisation (AFNOR, 1967), Bouma and Winter (1982), British Standards

Society (1995), Nakamura (1993), Ollner (1974), Simons and De Vries (2002), Toth (1990), Verity Consulting (1995), Wenström, Ollner, and Wenström (2000), and Winter (1990). Publications on IT standardization that touch the topic of company standardization include Cargill (1997) and Rada and Craparo (2001). In this chapter, we will contribute to a better understanding of company standardization by investigating how company standards are developed in company practice and in developing a best practice for this.

A company standard is the result of the standardization by a company or other organization for its own needs (Düsterbeck, Hesser, Inklaar, & Vischer, 1995). Company standardization includes developing standards for use within the company and developing standards to be used in the company's relations with its direct business partners (De Vries, 1999). Developing does not mean that each company standard has to be designed from scratch. A company standard may have the form of:

- A reference to one or more external standards officially adopted by the company,
- A company modification of an external standard,
- A subset of an external standard (for instance, a description of the company's choice of competing possibilities offered in an external standard, or a subset of the topics covered in the external standard),
- A standard reproduced from (parts of) other external documents, for instance, suppliers' documents, or
- A self-written standard. (De Vries, 1999)

Companies may prefer external standards, for example, from IEEE, but these do not meet all their needs and, therefore, they complement these with all forms of company standards just mentioned. In most companies, the number of company standards exceeds the number of external standards.

The chapter describes findings from a research project set up to develop a best practice for company standardization. This project stemmed from the wish of five big Dutch companies to improve their own company standardization performance by learning from each other. At a later stage, a sixth one joined the group. The best practice should be established by comparing the standardization activities of the six companies and by subsequently choosing the best way of performing these.

RESEARCH DESIGN

In order to be able to compare the companies, a common model was needed. The model should describe the processes needed for developing and implementing a company standard. For determining what is best, the expected contribution of standards to business results was chosen as a starting point. For the best practice of a process, its expected suitability for contributing to a successful company standard was the criterion. In order to assess this, findings in company practice were completed by insights from (standardization) literature as far as relevant; though the majority of standards are company standards, scientific standardization literature pays hardly any attention to them.

Steps to Company Standardization Success

As we want to determine the best practice, we need to define success: When is it possible to say that company standardization has been successful? In the case of dominant players such as Microsoft, company standards set the requirements for products and systems available on the market. In most other cases, company standards primarily serve the internal functioning of organizations and contribute to effectiveness and efficiency. Benefits of such company standards include the interoperability of systems, quality, reliability and safety of products, services and systems, efficiency in production, less maintenance and stock, and savings in procurement. These differences in possible benefits hindered a common perception of best practice. Therefore, we looked for a more general indicator. In line with ISO 9001 (International Organization for Standardization, 2000b), we can define success as user satisfaction. We will use the term *direct user* for people that read the standard. Other users use products, services, systems, and so forth in which the standard has been implemented. Again, however, because of the diversity of standards and user categories, it is difficult to measure this satisfaction and subsequently relate it to best practices in the way of preparing the standards.

A third method appeared to be more feasible. The success of company standardization results from the processes that constitute it (see Figure 1). Therefore, we relate success to these processes. The definition of success per process resulted from the interviews carried out within the six companies and from discussions with people who work with, or are professionally engaged with, standardization. These findings were completed with insights from literature. We have distinguished three steps for measuring the success of a standard.

Step 1: The standard should be there. The demand for a (company) standard starts either within the organization or arises from external obligations, for example, legislation. Then, this demand should be assessed and it should be decided whether or not to develop the standard. Such a decision should be based on standardization policy, and the possible development of the proposed standard should be balanced against other requests for standards development (prioritizing).

The development of the standards is the next process; this process consists of the composition of a draft version of the standard, commentary rounds, the writing of the final version of the standard, and the approval of this standard. The output is an approved company standard (or normative document). To develop the standard, there is a need for competent personnel (human

Figure 1. Company standardization model

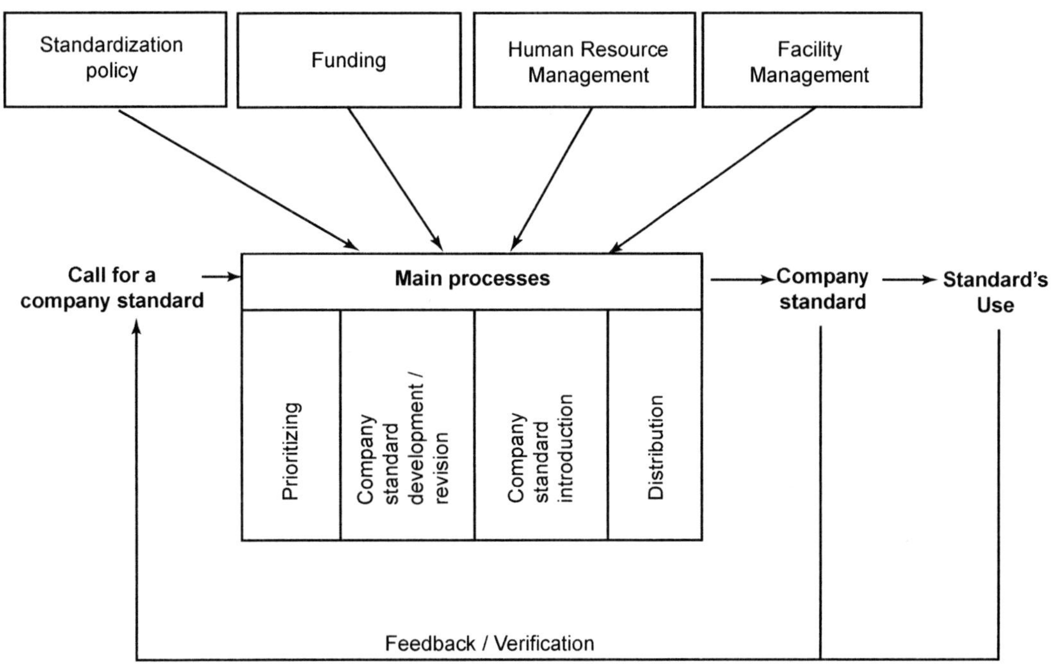

resource management), the standard has to be paid for (funding process), and IT tools should be in place to support standards development (facility management).

Step 2: The standard is known and available. When a standard has been developed (and accorded for), the next step toward success is that the standard becomes available to the intended users and that they are aware of its existence. This is a second prerequisite for success.

During this introduction process the potential direct users should get informed about the standard's existence and its potential fitness for use (Juran, 1988). The benefits of the standard and the reasons for certain choices in the standard can be explained. The more and the better the standard is known, the higher the chance that users will actually implement it and do so in the way intended by the standard's developers. In our model, the output of a process forms the input of the following process. However, practice can be less rigid. For instance, it can be important to start with the introduction (promotion) of the standard already during its development. Also, after the introduction period, the promotion of the standard can continue.

For the standard's availability to the direct users (physically), a distribution process should be in place. This process should assure that the standards reach the direct user in a fast and easy way. This can be done by, for instance, subscription, ordering on demand, or in the form of publishing on demand using an intranet. An extra success factor is that the direct user always works with the right version of the standard. After a period of time, the standard may be revised, so the distribution process has to be defined in such a way as to make sure that the right version of the standard is always being used.

Step 3: The standard is used. Company standardization can only be a success when the standard is used in practice (in the right way). A standard that is of a high quality but that is not used in practice has no value; there has to be a market for the standard. So, the potential direct users must be willing to use the standard and be capable of understanding and using it. We can even define success one step further when we look at our standardization definition. The product of the company standardization, the company standard, has to solve the matching problem. The standard has to be the answer to the demand out of the organization, which was the starting point of the process.

Evaluation of the standard's use may form the basis for withdrawing, maintaining, or changing the standard. The developed standard should be an answer to the question for which it was produced: Are the (potential) users of the standard satisfied? Therefore, user feedback to those who have decided to make the standard as well as to the people who have developed it is essential. The picture shows one feedback loop, but this is a simplification; more might be drawn because the (quality) management of each process needs a form of feedback as in the Deming circle: plan, do, check, act. A process approach that uses feedback and aims at enhancing customer satisfaction by meeting their needs is characteristic for the ISO 9000:2000 quality management approach (ISO, 2000a, 2000b).

Measuring Company Standardization Success

In the research project, it appeared to be difficult to measure success, mainly because of difficulty in measuring the real usage of a standard and its impact on business performance, and the diversity in different standards. Therefore, we have chosen an indirect way to assess best practice, taking the example of Chiesa, Coughlan, and Voss, (1996) who have developed a technical innovation audit. They divide the innovation process into subprocesses (main processes and facilitating processes), and for every subprocess, they define characteristics that are associated with success or failure of the subprocess and the overall innovation process. All these characteristics are put in a statement on a scorecard. The answer on the statement can be put in a scale, for example, from 1 to 5, from *not applicable at all* to *completely applicable*. Innovation and company standardization have a lot in common. Innovation is concerned with the development of a new product or process, or the improvement of an existing product or process. The innovation process can only be really successful when there is a market for this product or process. In company standardization, a standard is developed or revised. This also only can be called a success when there is a market for the standard. As the success of the company standardization process depends on the use of the standards, we have tried to find factors that positively influence the use of the standards. The factors can be related to the elements of our model. By defining these factors, the companies can be compared to each other. The companies can be scored per factor with a score on a predefined scale. This appeared to be an applicable method for the best-practice study.

The scorecard method is also very useful to compare the different practices and to develop a best practice. For every process of the model in Figure 1, we have therefore developed scorecards with propositions that define a (supposed) best-practice situation: best in the sense of the expected contribution of the process to the overall success of company standardization. The statements have been developed on the basis of discussions with specialists from both the practical area (the users and developers of company standards) and the professional area (university staff and experts from the national standards body). Company standardization literature has played a minor role as, in general, it does not provide in-depth best-practice data. Each proposition could be answered on a scale from 1 (*not applicable at all*) to 5 (*very much applicable*). The score of 5 is considered to be awarded to the best practice.

For every company, the scorecards have been filled in both by the companies themselves and by the researchers (objective party). The scores have been put together with the scores of the other companies and a mean score has been determined. These figures have been presented in tables, with the most interesting ones also in graphs. This has been done per process. For every process, the order of companies was made differently so that the companies could not recognize which score belonged to which other company. By comparing their own score with the best practice and with the other companies, it was possible for the cooperating parties to distinguish gaps between their practice and best practice, think about reasons for this, and decide on focus and improvement points for their future policy on company standardization.

Moreover, besides an overall research report for all companies, a small report per company has been made with a description of their actual company standardization and the focus points for them to work toward best practice. This has been supported by a presentation within the company for a group of people involved in the different processes of standardization. In this session, there was also the possibility to discuss the best-practice situation in relation to the possibilities of the company. Of course, best practice can have a different follow-up in each company. Beside the report, these individual sessions have been seen as a helpful method to analyze the status of the

Table 1. Scorecard method example: Three criteria for the company standard development process

Nr.	Description	Score
1.	The party that demands for a new or revised company standard is involved in the development process of the standard.	
2.	During the development process of the standard the potential users of the standard are involved.	
3.	It is possible for contractors or suppliers to give input for a company standard.	

current situation on company standardization and to identify focus points for its future optimization. In this discussion, the presented model for company standardization appeared to be very helpful to make the discussion more clear and structured.

To elucidate the scorecards method, we will here give an example: a graph of three different criteria out of the scorecard for the company standard development process.

These propositions describe the importance of involving the (potential) users of a standard in the development process. This is to make sure that the standard is user friendly, that it can be used in practice (and not only in a theoretical situation), and that it really helps the user. Besides that, the standard must be a solution for the initial demand out of the organization.

The scores of the six companies on these three propositions are visualized in a graph (Figure 2). The picture shows seven different pairs of bars. The first six represent the scores of the six different companies, and the seventh the mean scores on these criteria. So now the companies can compare their scores with those of the other companies, the mean scores, and the best practices. When most of the companies score low at a certain proposition, it is a point of attention for the companies on which they can do better, or it might be a weak point in the best practice.

Figure 2. Scorecard graph example: Three criteria for the company standard development process

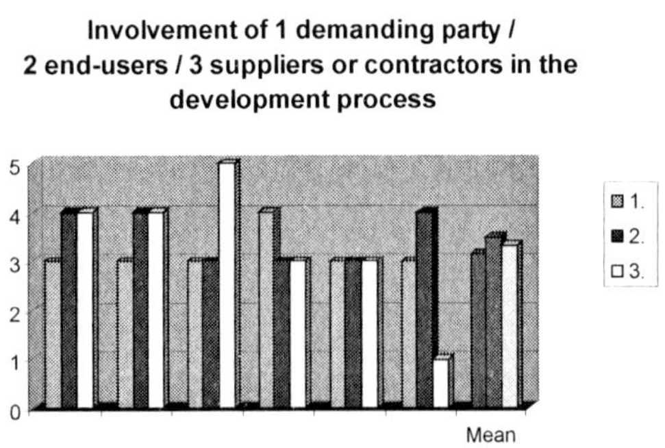

RESULTS: BEST PRACTICE MODEL FOR COMPANY STANDARDIZATION

In fact, the process model presented in Figure 1 can be seen as a first result. Its correctness was confirmed by applying it in the companies.

Next, the results are the best practices for each process. Every company has its own specific needs, culture, and ways of doing things, so the best practices have to be adapted to the specific situation of the company and, therefore, a company's best practice can differ from another's. The overall results are presented in Appendix 1 to this chapter in the form of a listing of best-practice statements. As an example, we will now give an underpinning for a part of the results, namely, some best practices for the first process in Figure 1: policy. Providing underpinning for the other processes as well would extend the available space for this contribution too much.

Standardization work can always be postponed for a day as direct productive labor always comes first. However, postponing may lead to doing nothing at all. Without a clear standardization policy and an organization structure to fulfill this policy, standardization is not effective. The policy making was the only process that has been divided in three parts, namely, the strategic, tactical, and operational levels.

Steering Group

Having a company standardization steering group is an important part of the best practice. Adolphi (1997) is the main scientific study on the organization of company standardization, but neither he nor any other scientific work mentions the idea of a steering group. We found it in Dutch professional studies only (De Gelder, 1989; Simons & De Vries, 2002). The British Standards Society (1995) mentions a standards committee but does not talk about a real steering responsibility. American National Standards Institute's (ANSI) best-practice research (Verity Consulting, 1995) apparently did not find steering groups in the multinational companies they investigated. A few of our six investigated companies have such a group. The steering group develops the policy and the board of directors approves this policy. The standardization policy has to be a deduction of the general company policy. To support this, a general (technical) director in the group can be very helpful. The steering group consists preferably of (technical) managers from the several departments or business units (BUs). The standardization activities have to be linked to the interests of these business units or departments. The line management has to commit itself to the year plan drawn up by the steering group.

Figure 3 presents a possible organizational structure. For making a good standard, the technical expertise of one or more technical experts (who, in general, work within the business units) should be combined with the standardization expertise of a standards engineer (who may work within a standardization department). A networking structure, in one way or another, is expected to be the best way to do this.

Organizational Structure

The preferred organizational structure for company standardization includes a company standardization department (e.g., Adolphi, 1997; AFNOR, 1967; Hesser & Inklaar, 1997). In smaller organizations, this can be one person. The standardization department acts as the steering group's secretary and brings in standardization skills. The standardization department coordinates and supports the standardization activities, and the business unit managers can be addressed for providing their technical experts (with their skills). It can be observed that the number of companies that operate a special (staff) department on standards has diminished, despite all its advantages. This is reflected in the diminishing number of members of national standards users' organizations in most countries and the decrease

Figure 3. Possible organizational structure for company standardization

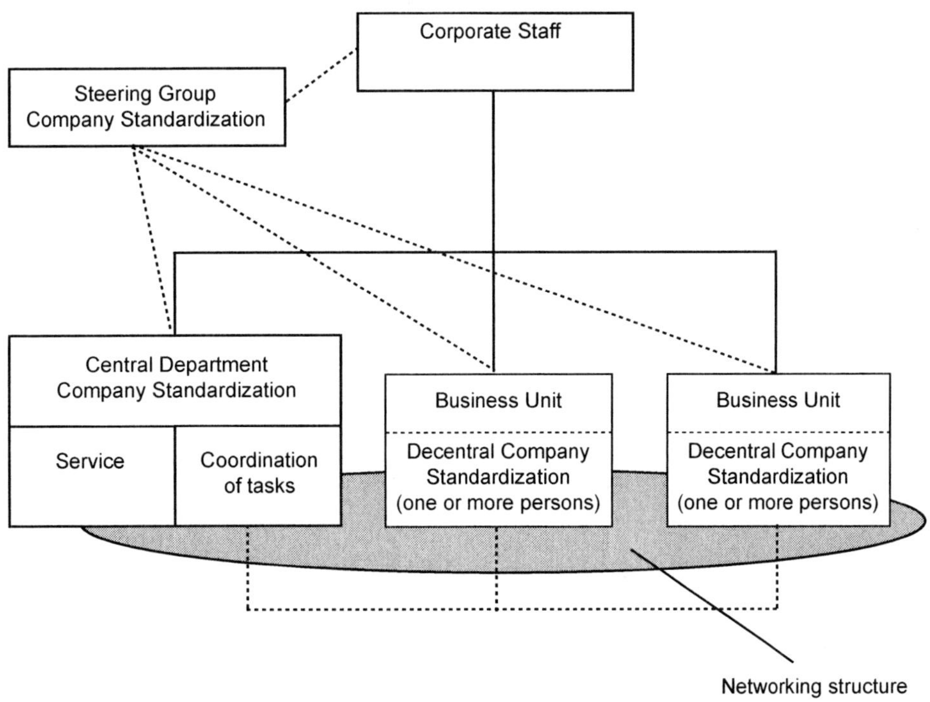

in the total number of these organizations. An indicator for the latter is that in 10 years, the number of countries with a national standards organization that is a member of the International Federation of Standards Users have decreased from 21 (IFAN, 1997) to 10 (http://www.ifan.org). The British Standards Society (1995) only mentions a standards specialist, whereas in an earlier edition (British Standards Institution, 1970), it mentions a standards department. Companies that have maintained a standardization department have reduced their size. Philips, for instance, has a corporate standardization department consisting of 15 employees whereas 20 years ago the department employed 150 people. However, big and medium-sized companies will need one or more full-time or part-time standards specialists somewhere in the organization, and our study confirms that this is best practice, despite the de-staffing fashion.

User Involvement

A third element of the organization of company standardization concerns the involvement of users. From both literature (particularly, Nakamura, 1993) and practice, it is known that involving the users in the (company) standards development process has a positive influence on their actual usage. User involvement makes standards more acceptable and applicable and for this reason, the acceptance rate increases. Because in practice it is impossible to involve all users, a group of representatives is the best option. So, not only does one specialist develop the standard, but an entire committee. After completing a draft version of the standard, other users and specialists can be involved through commentary rounds. The investigated companies recognize the importance of user participation, but often still give it too little attention. Of course, the amount of user involvement and the advantages related to that should be balanced with the cost: Time is money. Diverging user opinions may hinder standards development, but it is better to discover these and reach consensus (including the option to stop further development) than to develop standards the intended users do not want.

Status of Company Standardization

For some of our best-practice elements, we found no evidence in literature; they were just observed in one or more of the six companies. An example is the last statement in the policy process at the strategic level: The strategic policy on company standardization has enough status and is being pursued by the total company. Here we find quite a variation between the scores of the six companies (see Figure 4).

This difference can mean three things: This part of the best practice does not fit within the culture of this company, this is not best practice, or this is a focus point for the company to improve its process. These statements have been the focus points in the discussions with the companies. In this case, all companies saw the statement as best practice and, therefore, as an important focus point in the case on which they scored low. The company experts pointed out that company standardization should have status to make sure that standards are being developed and used. To give company standardization enough status, it is helpful when at a corporate level in the organization

Figure 4. Scorecard graph example: Status of company standardization policy

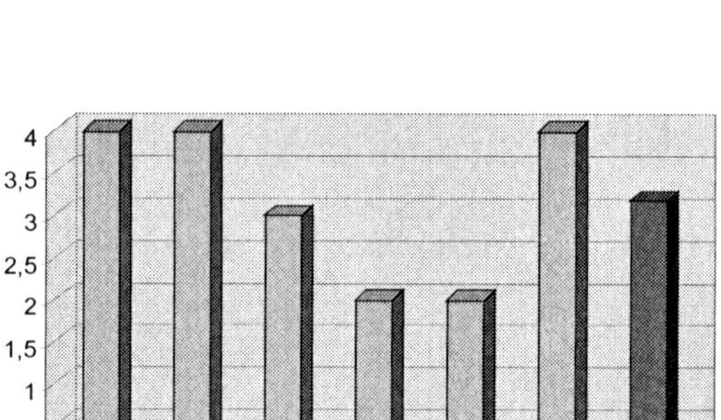

Figure 5. Scorecard graph example: The influence of business units on the strategic policy concerning company standardization

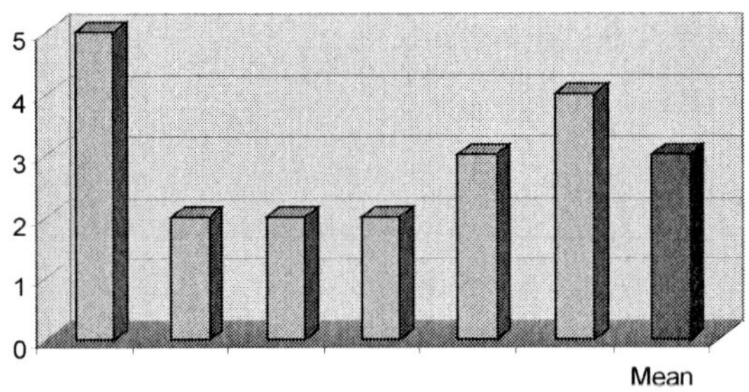

the importance of standardization and the benefits that it can bring are recognized and formalized. When, subsequently, this policy is communicated in the right way, it is more likely that company standardization becomes a success.

Participation of Business Units

Companies that scored low on the status also scored low in the participation of the business units in the definition of a strategic policy on company standardization (see Figure 5). So, though recognition of the importance of standardization at a high level within the organization is an essential factor, this is not enough. In order to make it work also on the other levels, these levels must be involved in the formulation of the strategic policy. In practice, the importance of this fact was recognized but not many of the companies did in fact involve the BUs in strategic policy development.

ANALYSIS

Our best practice in company standardization may be benchmarked with best practices in comparable fields, especially with quality management, IT management, and knowledge management.

IT Management Benchmark

Developing and implementing a standard may be compared with developing and implementing an IT system. Kocks (1997) analyzes the rather common failure of project management in automation projects and discusses two approaches used for such projects: the waterfall approach and the evolutionary approach.

In the waterfall approach, product specifications form the basis for the design of the project. Not only the final product but also some in-between milestones are defined. However, experiences in electronic data processing (EDP) auditing show that often the predefined specs are not met. The main reasons for this are the following:

- A lack of knowledge in extracting user needs from the clients
- A lack of knowledge in deriving a process design from the prespecified product
- Problems in linking people to the process (even with a perfect process design, if people do not do what they are expected to do, the project will fail)

The evolutionary approach reflects the other side of the spectrum of possible approaches. In this approach, the project team starts without knowing the final result. Hope for a positive result is mainly based on trust in the personal skills of the people who carry out the project. Project planning is vague and the budget cannot easily be estimated. The characteristics of a successful evolutionary approach are flexibility, smooth communication, fast decision making, and professional people.

Company standardization naturally resembles the waterfall approach. It may share drawbacks that are inherent to this approach: problems in specifying the standard needed, difficulties in designing the process to arrive at that standard, and difficulties in handling unforeseen human behavior. In the best practice, these drawbacks are tackled as follows:

- During the prioritizing process, the company's standardization department makes a proposal for the project including planning and budget in cooperation with (representatives of) the intended participants. This resembles the functional specification of an automation system and the plan for making it, though in a limited sense.
- Best practice for the standard developing process includes measures to assure a product meets user needs. Milestones are draft standards that already have the form of the final standard. The process usually has a standard design; it is not adapted to the specific requirements of the standard. The only specific thing concerns the selection of the people: representatives of all interested parties join.
- The third drawback is intrinsic to making use of co-designing customers (De Vries, 1999) but can be restricted by having a skilled standardization department. During the whole process, intended users are involved.

So all three reasons for failure of the waterfall approach apply to company standardization, but the best practice includes measures to avoid these, mainly in the form of user involvement.

Standardization practice has at least some characteristics of the evolutionary approach. During the process, unexpected circumstances have to be attended to so that the final result, if any, does not always meet the criteria set at the outset. However, experience shows that unstructured standards development leads to a low priority of company standardization. Company standards themselves provide structure and the process of developing them should not contradict this. However, problems related to the waterfall approach can be solved partly by learning from the evolutionary approach that the project manager, that is, the standards officer, should manage flexibility, informal communication, and informal pre-decision-making.

Luitjens (1997) mentions five key success factors for complex governmental automation projects.

1. **A pronounced assignment:** Without a clear assignment, projects fail. This is reflected in the best practice for the prioritizing process.
2. **Financing schedule from the outset:** From the very beginning, a clear financing schedule for the whole project should be established and agreed upon. This includes giving an estimate of the costs and a list of parties that have expressed their willingness to pay. Salami tactics mostly fail. The prioritizing and funding best practice provide solutions for this.
3. **Understanding complexity:** Complexity consists of product complexity, which is the technical complexity of the system that has to be built, and process complexity, which includes the different stakeholders with diverging interests, mutual relations, and financial possibilities. Not all individual standards are complex, but the whole collection of interrelated standards is. The best practice separates process design and product (standard) design, and provides solutions for stakeholder involvement and resource management.
4. **Creating a basis for the project:** The different actors should support the project. To attain this, it is important to make explicit the arguments for them to start and maintain their participation. Joint conceptualization appears to be a prerequisite for this: The more people are involved in finding solutions and balancing alternatives, the more they will be willing to support the final choice. The policy, prioritizing and standards develop-

ment processes, and user involvement in all processes provide a sound basis for the project.

5. **Personal skills:** Some people manage where others, in comparable circumstances, fail. Common sense and a sense of humor appear to be prerequisites. This is the only missing element in the company standardization best-practice model.

Quality Management Benchmark

Company standardization and quality management can be compared at two levels. First, quality management includes developing, maintaining, and improving a quality management system. This may be compared with developing, maintaining, and improving a standards collection. More than that, such a collection may form part of a quality management system. Secondly, quality management includes the development of procedures and instructions, and these can be regarded as company standards. Therefore, quality management theory can form a benchmark for company standardization at both levels.

At the first level, quality management literature pays a lot of attention to quality as a line function vs. quality as the activity of a separate department or quality manager. Dale and Oakland (1991) and many others emphasize that quality should be managed in the line functions. However:

... there is a role for a quality management representative, at board or senior management level, but this is concerned with facilitation, support, planning, promotion and co-ordination. Quality planning, review, and re-planning are the foundations of quality improvement. Without this basis the development of product and service designs, specifications, operational specifications and instructions, quality manuals, reviews and evaluation procedures will be poorly structured and badly focused. (Dale & Oakland, p. 228)

So, in fact, in order to arrive at quality management for the line, they advise a form of central support. The same can be seen in the best-practice model: Line management establishes policy (policy process) and sets priorities (prioritizing process). Users are involved in all other processes, and these are supported by a standardization department or standards officer. Dale and Oakland (1991) stress the importance of user involvement. The quality circles they advise can be compared with working groups that develop a standard.

At the second level, developing procedures and instructions, our own experience demonstrates the comparability of both. We developed a book and a course on developing procedures for an ISO 9000-based quality management system based on our experience in company standardization (De Vries, 2002). Feedback from dozens of people who have used the book or joined the course confirms that this approach is fruitful for developing procedures. Therefore, conversely, the book was used for developing the company standardization best practice.

Knowledge Management Benchmark

Company standardization can be regarded as a form of knowledge management, where tacit knowledge is transformed into explicit knowledge. Slob (1999) has compared the conclusions of the best-practice study (real practice) with literature on knowledge management (the English-language sources were Brown & Duguid, 1991; Essers & Schreinemakers, 1996; Kriwet, 1997; Nonaka, 1991, 1994; Nonaka & Takeuchi, 1995; Polanyi, 1966/1983; Schreinemakers, 1996; Thayer, 1968) and concludes that knowledge management literature does not add real new insight to the best practice found during the practical research. This underpins that it is really a best practice.

Company standardization can be seen as a way to manage technical knowledge. Figure 6 shows the way company standardization is

Figure 6. Model for company standardization (the old way)

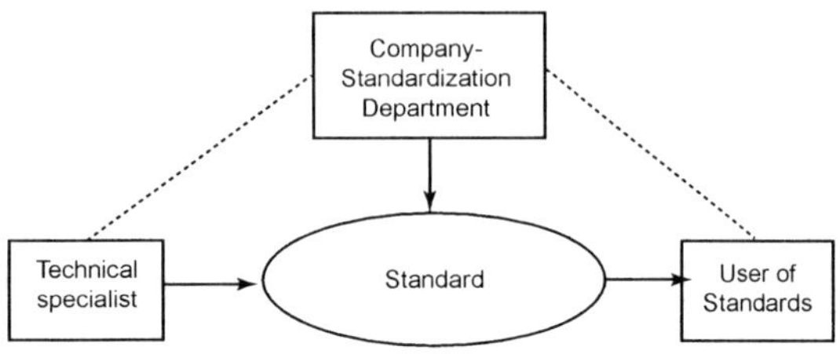

Figure 7. Model for company standardization (best practice)

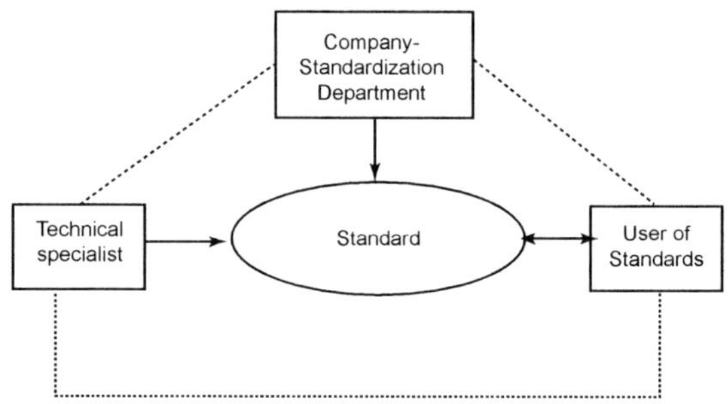

often organized in organizations. The technical specialist(s) together with the standardization department are responsible for the realization of the standard. Knowledge is recorded (in the standard) and transferred to the other workers and users. Knowledge management literature (e.g., Verkasalo & Lappalainen, 1998) distinguishes between the knowledge domain of the providers of the knowledge (in our case, the technical specialists) and the receivers of the recorded knowledge (in our case, the workers or direct users of standards). Both have their own knowledge domain. This has to be an important consideration when the knowledge is recorded.

Standard development appears to be successful when the (technical) knowledge that is captured in standards is available for the direct users and the standards are actually used within the company (Slob, 1999). To achieve this, the standard should be user friendly. To arrive at this situation, the actual user should play an important role in the standard development process. In practice, this is done too little. Arguments for user involvement can be found in several sources of literature, like Adolphi (1997), Brown and Duguid (1991), Gouldner (1954), De Gelder (1989), Nakamura (1993), and Winter (1990). These findings in the area of standardization are supported by experiences and literature in knowledge management (for instance, Nonaka, 1991, 1994). This leads to a slightly different model for company standardization (see Figure 7).

In this model, the practical knowledge and experience of the user is connected with the technical knowledge of the specialist and with the standardization knowledge of the company standardization department. By involving the users in the development of standards, experience and practical knowledge flow into the standards. Even more important is that their knowledge about using standards can help to make the standards user friendlier, which will positively influence their actual use. The presented new model for company standardization (Figure 7) compromises the following conclusions of Slob (1999).

- Company standardization is a way to manage technical (company) knowledge
- Standardization is a structured way to transfer tacit into explicit knowledge
- When tacit knowledge is made explicit, it should be considered by whom this codified knowledge will be used in the future
- It should also be considered that there can be an important difference in the knowledge domain of the specialist(s) or writer(s) of the standard and the intended direct users of that standard
- These users of the standards should, therefore, be drawn into their development

CONCLUSION

In this chapter, we have developed a process-based model for company standardization and a best practice for the processes that constitute company standardization. Best practice has been developed by examining practices in six big companies in The Netherlands, and by using insights from standardization literature. The latter appeared to be incomplete, especially where academic literature was concerned.

The main benchmark for a best practice is the success of the company standard. A company standard is successful when it is used and appears to solve the problem for which it was developed. Success results from a combination of the successful performance of the company standardization processes that can be distinguished and that are shown in the company standardization model. Therefore, for each of these processes, several good practice indicators have been developed. The starting point for this was the way the six companies carry out their activities. The best practice of these was assessed by professionals as well as the research team. Where available, standardization literature was used as an additional benchmark. Afterward, the best-practice findings were confirmed by comparing them with insights from literature in the areas IT management, quality management, and knowledge management.

Findings have been presented in the form of scorecards. Using these scorecards, the individual companies could benchmark themselves against the others, the mean score, and the best practice. The six companies differ in scores. No one is overall the best or the worst: Each one has good as well as bad performances in different areas. The feedback during these presentations in each of the six companies confirmed the findings.

The participating companies and other companies that develop many company standards can use the results to improve the performance of their company standardization processes and, in this way, improve the effectiveness and efficiency of their company standardization activities.

DISCUSSION

The research project did not concern IT standards, but engineering standards used by multinational chemical and petrochemical companies. The following remarks can be made concerning the applicability of the results for other companies and for the typical domain of IT standards.

- The investigated companies use the standards mainly for the quality and safety of their installations, for the compatibility of parts thereof, and for avoiding unnecessary variety of these parts. This can be compared with IT standards used by big companies, though in that case there is more emphasis on compatibility. For example, companies may face the problem of which external IT standards to choose for their systems, or which options within or versions of these standards. So the function of IT standards is to a large extent identical to the one of engineering standards in process industries, though the research in fact does not concern the standards' contents but the process of developing them. Anyhow, it may be expected that results equally apply to IT standards.
- An exception to this conclusion may concern the best way of organizing standardization. In many cases, companies will have a central IT department. This department, instead of a central standardization department, might support the standardization activities. For this choice, the advantage of having specialist knowledge of IT should be balanced against the possible disadvantages of having less standardization expertise and of additional coordination to manage interrelations between IT and non-IT standards.
- In most branches of business, standards related to the products or services are the most important ones. In the case of IT companies, these are product- or service-related IT standards. The main purpose of such

standards is to support the market success of the products or services. Compared to our study, the role and benefits of such standards differ, but the way they are prepared may be identical. Therefore, there is no reason why company standardization and the best practices for it are fundamentally different. However, the participants will differ as standardization in such companies will be more directly linked with product or service development and marketing. This may result in a different organizational setting and governance structure for company standardization (Adolphi, 1997).

- The advantages of some of the company standards increase with growth in company size. For instance, quantity rebates related to preference ranges of software packages are more feasible in big companies. Small companies will not always have the possibilities to have a standardization department, a steering group, an intranet, and so forth. Small enterprises need less formalization so their need for standards is less. However, they will need standards, and the processes described apply for them as well. Best practice for them may be expected to be more low profile.
- Another difference may concern intellectual property rights (IPRs); they hardly apply to engineering standards but may be important for product standards. Therefore, in IT standardization, it may be an additional best practice to have standardization and IPR closely related (Clarke, 2004).
- Possibly, some of the best practices presented here are typically Dutch while in other cultures, other practices would be better. Some differences between Dow Chemical (with a culture that mixes American and Dutch influences) and the other companies suggest this. For reasons of confidentiality, we cannot mention examples. In general, corporate culture of all such multinational companies is becoming increasingly international.

These issues are possible topics for future research. Other future research items may include:

- Practice as well as best practice of IT (company) standardization within companies that produce IT products or services and within (big) organizations that use IT systems to support their functioning
- The relation between best practice and cost reduction, taking into account the cost of (human and other) resource allocation, transaction costs (for standards development and implementation), and total cost of ownership (of the IT systems)
- Balancing between developing a company standard or using an external standard
- Possibilities for modification of the practice model for use by standards developing organizations (in the available best-practice study on national standardization organizations [Bonner & Potter, 2000], such a model is missing)
- The role of company standards in the case of outsourcing activities (this applies to process industries that outsource the engineering or maintenance of plants, as well as to companies that outsource IT system development, support, or maintenance)
- Completion of the comparisons with IT management, quality management, and knowledge management in the form of a more thorough review of the literature
- Comparison of user involvement in company standardization with user involvement in formal standardization and consortia
- (Best-practice) field research in other branches of industry, for instance, mechanical engineering or service industries

The research approach used in this study has been described and analyzed in more detail in another publication (De Vries & Slob, 2008). From a company's perspective, the project can be seen as a rather unique form of knowledge sharing (we have analyzed this in another book chapter; De Vries, 2007).

REFERENCES

Adolphi, H. (1997). *Strategische Konzepte Zur Organisation Der Betrieblichen Standardisierung.* Berlin, Germany: Beuth Verlag.

Association Française de Normalisation (AFNOR). (1967). *La normalisation dans l'entreprise.* Paris: Author.

Bloemen, F. E. M., Van den Molegraaf, J. C. M., & Van Mal, H. H. (1992). Normalisatie vermindert levensduuruitgaven van chemische installaties aanzienlijk. *B&Id, 4*(8), 17-20.

Bonner, P., & Potter, D. (2000). *Achieving best practices in national standardisation: A benchmarking study of the national standardisation systems of Finland, Sweden, Denmark and Italy.* Helsinki, Finland: Ministry of Trade Industry.

Bouma, J. J., & Winter, W. (1982). *Standardization fundamentals.* Delft, The Netherlands: Nederlands Normalisatie-instituut.

British Standards Institution. (1970). *PD 3542: The operations of a company standards department.* London: Author.

British Standards Society. (1995). *PD 3542: 1995 Standards and quality management. An integrated approach.* London: British Standards Institution.

Brown, J., &. Duguid, P. (1991). Organizational learning communities of practice: Toward a unified view of working, learning and innovation. *Organization Science, 2*(1), 40-57.

Cargill, C. (1997). *Open systems standardization: A business approach.* Upper Saddle River, NJ: Prentice Hall.

Chiesa, V., Coughlan, P., & Voss, C.A. (1996). Development of a technical innovation audit. *Journal of Production Innovation Management, 13*, 105-136.

Clarke, M. (2004). *Standards and intellectual property rights: A practical guide for innovative business.* London: NSSF.

Dale, B., & Oakland, J. (1991). *Quality improvements through standards.* Leckhampton, UK: Stanley Thornes Ltd.

De Gelder, A. (1989). *Opstellen van normen.* Delft, The Netherlands: Nederlands Normalisatie-instituut.

De Vries, H. J. (1999). *Standardization: A business approach to the role of national standardization organisations.* Boston: Kluwer Academic Publishers.

De Vries, H. J. (2002). *Procedures voor ISO 9000:2000.* Delft, The Netherlands: NEN.

De Vries, H. J. (2007). Developing a standardization best practice by cooperation between multinationals. In K. J. O'Sullivan (Ed.), *Strategic knowledge management in multinational organizations.* Hershey, PA: Information Science Reference.

De Vries, H. J., & Slob, F. J. C. (2008). Building a model of best practice of company standardization. In J. Dul & T. Hak (Eds.), *Case study methodology in business research.* Oxford, UK: Butterworth Heinemann.

Düsterbeck, B., Hesser, W., Inklaar, A., & Vischer, J. (1995). Company standardization. In W. Hesser & A. Inklaar (Eds.), *An introduction to standards and standardization* (pp. 99-138). Berlin, Germany: Beuth Verlag.

Essers, J., & Schreinemakers, J. (1996). The conceptions of knowledge and information in knowledge management. In J. E. Schreinemakers (Ed.), *Knowledge management: Organization, competence and methodology* (pp. 93-104). Würzburg, Germany: Ergon.

Gouldner, A. W. (1954). *Patterns of industrial bureaucracy.* New York: The Free Press.

Hesser, W. (2006). *Standardization within a company: A strategic perspective.* In W. Hesser (Ed.), *Standardisation in companies and markets* (pp. 177-215). Hamburg, Germany: Helmut-Schmidt-University Hamburg.

Hesser, W., & Inklaar, A. (Eds.). (1997). *An introduction to standards and standardization.* Berlin, Germany: Beuth Verlag.

IFAN. (1997). *IFAN memento 1997.* Geneva, Switzerland: International Organization for Standardization.

International Organization for Standardization (ISO). (2000a). *ISO 9000:2000. Quality management systems: Fundamentals and vocabulary.* Geneva, Switzerland: Author.

International Organization for Standardization (ISO). (2000b). *ISO 9001:2000. Quality management systems: Requirements.* Geneva, Switzerland: Author.

Juran, J. M. (1988). *Juran on planning for quality.* London: The Free Press.

Kocks, C. (1997). Het échec van projectmanagement in de systeemontwikkeling. *Informatie, 39*(4), 6-11.

Kraaijeveld, P. (2002). *Governance in the Dutch banking industry: A longitudinal study of standard setting in retail payments.* Rotterdam, The Netherlands: Rotterdam School of Management, Faculteit Bedrijfskunde.

Kriwet, C. (1997). *Inter- and intraorganizational knowledge transfer.* Bamberg, Germany: Difo Druck.

Luitjens, S. (1997). Interorganisatorische informatioseringsprojecten bij de overheid. *Informatie, 39*(4), 18-22.

Nakamura, S. (1993). *The new standardization: Keystone of continuous improvement in manufacturing.* Portland, OR: Productivity Press.

Nonaka, I. (1991). The knowledge creating company. *Harvard Business Review, 6*, 96-104.

Nonaka, I. (1994). A dynamic theory of organizational knowledge creation. *Organization Science, 5*(1), 14-37.

Nonaka, I., & Takeuchi, H. (1995). *The knowledge creating company.* New York: Oxford University Press.

Ollner, J. (1974). *The company and standardization* (2nd ed.). Stockholm: Swedish Standards Institution.

Oly, M. P., & Slob, F. J. C. (1999). *Benchmarking bedrijfsnormalisatie: Een best practice voor de procesindustrie.* Rotterdam, The Netherlands: Erasmus Universiteit Rotterdam, Faculteit Bedrijfskunde.

Polanyi, M. (1983). *The tacit dimension.* Gloucester, MA: Peter Smith. (Original work published 1966 by Anchor Books, Garden City, New York)

Perera, C. (2006). *Standardization in product development and design.* In W. Hesser (Ed.), *Standardisation in companies and markets* (pp. 141-176). Hamburg, Germany: Helmut-Schmidt-University Hamburg.

Rada, R., & Craparo, J. S. (2001). Standardizing management of software engineering projects. *Knowledge Technology and Policy, 14*(2), 67-77.

Schacht, M. (1991). *Methodische Neugestaltung Von Normen Als Grundlage Für Eine Integration In Den Rechnerunterstützten Konstruktionsprozeß.* Berlin, Germany: Beuth Verlag GmbH.

Schreinemakers, J. F. (Ed.). (1996). *Knowledge management: Organization competence and methodology.* Proceedings of the Fourth International ISMICK Symposium.

Simons, C. A. J., & De Vries, H. J. (2002). *Standaard of maatwerk: Bedrijfskeuzes tussen uniformiteit en verscheidenheid.* Schoonhoven, The Netherlands: Academic Service.

Slob, F. J. C. (1999). *Bedrijfsnormalisatie: De schakel tussen tacit en explicit knowledge.* Rotterdam, The Netherlands: Erasmus Universiteit Rotterdam, Faculteit Bedrijfskunde, Vakgroep Management van Technologie en Innovatie.

Susanto, A. (1988). *Methodik Zur Entwicklung Von Normen.* Berlin, Germany: Beuth Verlag GmbH.

Thayer, L. (1968). *Communication and communication systems.* Homewood, IL: Richard D. Irwin Inc.

Toth, R. B. (Ed.). (1990). *Standards management: A handbook for profits.* New York: American National Standards Institute.

Verity Consulting. (1995). *Strategic standardization: Lessons from the world's foremost companies.* New York: ANSI.

Verkasalo, M., & Lappalainen, P. M. (1998). A method of measuring the efficiency of the knowledge utilization process. *IEEE Transactions on Engineering Management, 45*(4), 414-423.

Wen, Z. (2004). *Reform and change: An introduction to China standardization.* Paper presented at the 11th International Conference of Standards Users IFAN 2004, Amsterdam.

Wenström, H., Ollner, J., & Wenström, J. (2000). *Focus on industry standards: An integrated approach.* Stockholm, Sweden: SIS Förlag.

Winter, W. (1990). *Bedrijfsnormalisatie.* Delft, The Netherlands: Nederlands Normalisatie-instituut.

APPENDIX 1: BEST-PRACTICE MODEL

Standardization Policy: Strategic Level

- There is a clear strategic policy on company standardization.
- At the corporate level, there is a clear framework for operating company standardization.
- At the corporate level, tasks, competencies, and responsibilities for company standardization have been defined.
- Standardization expertise has sufficient influence on the company's strategic policy.
- At the corporate level, management is aware of the importance and benefits of having (company) standards and standardization.
- The maintenance of the existing system of company standards is a part of the strategic policy on company standardization.
- The business units have sufficient influence on the strategic policy on company standardization (to make sure that their needs in this area are met).
- The business units commit to the strategic policy on company standardization.
- The strategic policy on company standardization is derived from the general strategic policy of the company (it supports the general policy and it does not conflict with it).
- In this strategic policy on company standardization, the goals are clearly defined.
- The management is willing to steer company standardization at a high (top) level in the organization in order to minimize the danger of suboptimization.
- The management is aware that by using (company) standards company-wide, the company can achieve cost benefits for the purchasing of materials.
- The management is aware that by using (company) standards company-wide, the company can reduce the cost of engineering and maintenance.
- The management is aware that (company) standards use is needed to assure a specified quality level of the company.
- The management recognizes company standardization as an essential activity and steers this activity at a corporate level.
- Corporate management has authorized the strategic-level company standardization.
- The strategic policy on company standardization has enough status and is being pursued by the total company.

Standardization Policy: Tactical Level

- Managers of experts involved in company standardization are involved in setting the company standardization policy
- People responsible for setting the company standardization policy at the tactical level have enough knowledge to be able to do so
- The company standardization department (CSD) is involved in establishing (separate scores)
    - The way of funding company standardization activities
    - The status of company standards (voluntary or compulsory)
    - Priority areas (for instance, engineering standards or standards to assure and improve environmental performance)
    - Criteria for prioritizing

- Business units (BUs) are involved in establishing (separate scores)
  - The way of funding company standardization activities
  - The status of company standards (voluntary or compulsory)
  - Priority areas (for instance, engineering standards or standards to assure and improve environmental performance)
  - Criteria for prioritizing
- There is a clear company standardization policy deployment to all people concerned
- The CSD informs all people concerned with company standardization issues

Standardization Policy: Operational Level

- The CSD coordinates the operational standardization activities and monitors progress
- The progress of activities is evaluated on a regular basis
- The CSD reports delays, and line management decides on corrective actions, if any
- People involved in standardization are encouraged to propose suggestions for improvement of the standardization policy
- The standardization policy at the operational level is evaluated on a regular basis

Prioritizing Process

- Each employee is authorized to submit proposals for developing, changing, or withdrawing company standards
- Users are encouraged to submit (change) proposals
- The CSD coordinates gathering proposals
- Company standardization expertise is used in prioritizing
- Criteria for prioritizing are in line with general company policy
- Criteria for prioritizing are evaluated on fitness for use on a regular basis
- Optimal overall company results prevail over BU benefits (no suboptimization)
- Once priorities have been set, the CSD makes a proposal for planning and budget in cooperation with (representatives of) the intended participants
- Planning and budget are realistic and performable
- The CSD informs the interested parties on the annual planning and makes it available to them
- The CSD monitors whether set priorities are being realized
- In the case of an absence of agreed-upon expert involvement in developing a company standard, the CSD is responsible to ask the heads of these experts to charge them to pick up their duties

Company Standard Development Process

- Those who have asked for standards become involved in their development
- Intended standards users become involved in their development
- Suppliers and/or contractors can provide input in the company standard development process
- Standards writers communicate with stakeholders during the development process
- Each company standard is assessed on its expected fitness to contribute to business results
- The company has a metastandard that provides criteria for its company standards

- This metastandard is known by all involved in company standards development and they apply it
- On a regular basis, the requirements in this metastandard are assessed on topicality and fitness for use
- Company management has authorized this metastandard
- The company standard is not just based on the personal opinion of one expert, but it is broadly based
- Participants in standard development consider their task as important and urgent
- The status of writing standards equals the status of carrying out projects
- There are enough competent employees for writing new standards and maintaining the quality and consistency of the existing standards collection
- A "why" document is attached to each company standard. It provides the underpinning of the most important choices and decisions that have been made during standards development
- A draft of each new company standard is sent out for comments to a relevant group of people within the company
- There is a procedure for processing comments
- Everybody is allowed to comment on draft standards
- The CSD coordinates comments processing
- Comments, if any, are sent to the development team; they decide on adoption or rejection
- In the case of rejection, they give the reasons why
- CSD checks the standard against the requirements in the metastandard
- Company management authorizes the standard
- IT tools are used in developing and writing the standard

Company Standards Introduction Process

- The CSD is the central help desk for questions concerning company standardization
- The CSD is able to answer questions concerning company standards or to refer people to experts who can answer these questions
- The CSD announces new, modified, and withdrawn standards when necessary via BU officers responsible for standardization or via certain (other) users within the organization
- The CSD is able to communicate with the rest of the organization about new, modified, and withdrawn standards
- There is a procedure for how to announce company standards
- The CSD is able to tailor information to the needs of specific user groups
- The CSD monitors user satisfaction concerning the provision of standards-related information
- Each BU has appointed one or more officers responsible for the diffusion of standards-related information
- The CSD uses IT tools for standards introduction
- The information concerning changes in the standards collection includes reasons for the (new or modified) standard and for major choices within the standard and expected advantages. In the case of withdrawal, the reasons for this are mentioned

Distribution Process

- The CSD makes standards available
- Company standards can be obtained on request or by subscription to the whole collection or to a part of it
- The CSD can advise on searching and ordering standards
- The CSD operates a database with bibliographical data on all company standards
- Outdated versions of existing standards and withdrawn standards remain available for consultation
- The CSD operates a system that shows who has which standard, including the version of this standard
- In the case of a new version, all users of the replaced version get the announcement
- Users can get standards in a form (electronic and/or paper) and format that suits them
- The version of a standard is clear to the users
- The CSD publishes company standards in electronic format and makes them available in a virtual way
- Virtual company standards are, as far as the text is concerned, available in text mode (not in pixel mode only), which enables text search on headwords
- On a regular basis, the CSD monitors the effectiveness and topicality of its subscription system; subscriptions include the department name and name of the employee concerned

Facility Management

- The CSD monitors the market for IT tools that may support its processes
- The CSD has a budget for investments in facilities
- IT facilities to support company standardization should fit with IT used elsewhere within the organization
- IT facilities to support company standardization should fit with IT used by the national standards body for producing, making available, distributing, and searching its standards
- The company operates an intranet and uses it for standards-related information
- Within the company, e-mail can be used for communication about company standards

Funding

- Funding for the maintenance of the system of company standards is assured
- Funding for company standards that are essential for the company is assured
- Fixed costs for the CSD are charged on the company as a whole
- Corporate management decides on the way these fixed costs are charged on the BUs
- Variable costs for company standardization are paid by those that have caused these costs (the part of the company for which the standard has been developed)

Section II
Specifics of Standards and Standards Setting

Chapter IV
Open Standards Requirements

Ken Krechmer
International Center for Standards Research, USA

ABSTRACT

An open society, if it utilizes communications systems, requires open standards. The personal computer revolution and the Internet have resulted in a vast new wave of Internet users. These new users have a material interest in the technical standards that proscribe their communications. These new users make new demands on the standardization processes, often with the rallying cry, "open standards." As is often the case, a rallying cry means many different things to different people. This article explores the different requirements suggested by the term open standards. *Perhaps when everyone agrees on what requirements open standards serve, it will be possible to achieve them and maintain the open society that many crave.*

INTRODUCTION

Open systems, open architecture, open standards, and open source, all sound appealing, but what do they mean?

The X/Open Company, Ltd. provided an early public usage of the term *open*. X/Open Company, Ltd. was a consortium founded in 1984 to create a market for open systems. Initially, X/Open focused on creating an open standard operating system based on UNIX to allow the 10 founding computer manufacturers to compete better with the proprietary mainframe operating systems of IBM (Gabel, 1987). Later, its direction evolved (and IBM joined) to combine existing and emerging standards to define a comprehensive yet practical common applications environment (CAE) (Critchley & Batty, 1993). X/Open managed the UNIX trademark from 1993 to 1996, when X/Open merged with the Open Software Foundation (OSF) to form The Open Group[2].

Perhaps the genesis of the confusion between open standards and open source developed with the similarly named Open Software Foundation and the Free Software Foundation, two independent consortia both based in Cambridge,

Massachusetts, USA. While the Open Software Foundation addressed creating an open standard UNIX operating systems, the Free Software Foundation (FSF) supported software that can be used, copied, studied, modified, and redistributed by creating the GNU (GNUs Not Unix) licensing for software[3]. GNU-licensed software is open source software.

The term *open architecture* also evolved from a related consortium. The Open Group is responsible for The Open Group Architecture Framework (TOGAF), a comprehensive foundation architecture and methods for conducting enterprise information architecture planning and implementation. TOGAF is free to organizations for their own internal noncommercial purposes[4]. Considering this history, it not surprising that there is some confusion between open systems, open architecture, open standards, and open source.

Some definitions are needed: standards represent common agreements that enable information transfer, directly in the case of IT standards and indirectly in the case of all other standards. Open source describes an open process of software development. Often, open source systems make use of open standards for operating systems, interfaces, or software development tools, but the purpose of open source is to support continuous software improvement (Raymond, 2002), while the purpose of open standards is to support common agreements that enable an interchange available to all. Open architecture refers to a system whose internal and/or external interfaces are defined by open standards and/or available under an open source license. Open systems embody each of these concepts to support an open systems environment. Originally, the IEEE standard POSIX 1003.0 (now ISO/IEC TR 14252) defined an open system environment as "the comprehensive set of interfaces, services and supporting formats, plus user aspects, for interoperability or for portability of applications, data or people, as specified by information technology standards and profiles" (Critchley & Batty, 1993,).

A few other definitions are needed. The term *standards setting organization* (SSO) refers to any and all organizations that set or attempt to set what are perceived as standards. The term *recognized SSO* refers to any SSO recognized directly or indirectly by a government. *Consortium* is the term used for any SSO that is not recognized directly or indirectly by a government.

There are many requirements for an IT standard. The basic requirements—that it be consistent, logical, clear, and so forth—are more easily agreed upon. In an attempt to provide a definitive view of the more complex aspects of open standards, this article considers open standards from three vantage points on the Open System Environment.

Standardization consists of more than the process of standards creation; standardization includes implementations of the standard (by implementers) and use of the implementations of the standard (by users). As example, it is common for a user organization to say, "We have standardized on Microsoft Word," meaning that they have agreed to use Microsoft Word software implementations throughout their organization. Microsoft often refers to its implementation of Word as an open standard, meaning that it makes its implementations of Word widely available to users (Gates, 1998). Certainly, Microsoft does not plan to make the software of Microsoft Word open in a manner similar to the Open Software Foundation or the Free Software Foundation. But standards must be implemented and utilized in order to exist; these standards must be without implementations or users are not usually considered a standard. So, the perspective of the implementers and users of an open standard is as necessary as the perspective of the creators of an open standard.

THE EMERGENCE OF IMPLEMENTERS

As recognized SSOs developed in the late 19th century, they focused (often with government approval) on supporting the open creation of standards and not on the open implementation or open use of standards. This was quite reasonable, as often, the only standards stakeholders then were the creators of the standards. For example, railroads, utilities, and car manufacturers were major creators, implementers, and users of standards in this period. In the 19th and early 20th centuries, the significant standardization policy issue was the conversion from multiple company specifications to single SSO standards (Brady, 1929).

After the middle of the 20th century, large integrated organizations (companies that bring together research and development, production, and distribution of their products or services, such as IBM, AT&T, Digital Equipment Corp., British Telecom, France Telecom, and NTT) had engineers who functioned often on a full-time basis as the integrated organization's standards creators. These standards creators supported specific recognized SSOs necessary for the broad aims of the integrated organization (Cargill, 1989, p. 114).

In the latter part of the 20th century, the increase in technology created a need for many more standards. Because of the growth of personal computing and the Internet, the number of implementers and users of standards increased dramatically. The stage was set for major changes in standardization activity and processes. By the middle of the 1980s, a new industrial movement emerged, where larger integrated organizations began to devolve into segmented organizations where the overall organization exerts the minimum unifying management. Each segment of the overall organization focuses only on its own market(s) and, therefore, only supports the SSOs that appear necessary for their specific product development requirements (Updegrove, 1995). This new industrial movement marked the rise of the implementers' activities (independent product development group) in standardization and with it the rise in consortia standardization. At the same time, the overarching integrated organization's standardization organization was disbanded in most cases (e.g., AT&T, IBM, US PT&T's BellCore).

Since the 1980s, technical communications standardization processes have been in transition from being driven by standards creators (standardization participants who are motivated to develop new standards) to standards implementers (standardization participants who are motivated to produce new products that embody a standard). In addition, the users of implementations standards (who usually do not participate in the IT standardization process) have a growing interest in seeing the concept of openness address their requirements. This view is confirmed in the 1994 report sponsored by the US National Science Foundation, which described an open data network as being open "to users, to service providers, to network providers and to change" (NRENAISSANCE Committee, 1994). Considering service providers and network providers, both examples of implementers (perhaps also creators and users), this report identifies the three major perspectives on open standards: creators, implementers, and users.

Individual product development groups in segmented organizations have no history or allegiance to a specific SSO and choose to support any SSO that best fits their specific product development and marketing needs. Often, such a fit is made by sponsoring a new SSO to address the standardization requirements of a specific developer's product implementation (Updegrove, 2004). However, product implementers have very different interests than the standards creators they have replaced. What a product implementer considers an open standard may be quite different from what a standards creator considers an open standard. And they are also different from what a user might consider an open standard.

THREE VIEWS OF STANDARDIZATION

Each of the requirements of open standards relates to one or more of the stakeholders—creators, implementers, and users. To identify all the requirements of an open standard, it is necessary to understand what standards creators, implementers, and users would consider the broadest reasonable requirements of an open standard. Each group of stakeholders is driven by unique economic desires:

- The creation of standards is driven by potential market development and control issues.
- The implementation of standards is driven by production and distribution cost efficiencies.
- The use of implementations of standards is driven by the potential efficiency improvement due to the standard on the user.

While there is some overlap among these economic drivers (e.g., market development and distribution cost efficiency), each stakeholder has a distinct economic motivation. Thus, it is necessary to consider each stakeholder class separately. In the following, the requirements of each stakeholder class are evaluated, and each open standards requirement derived is given a name, usually shown in parenthesis (e.g., open IPR).

THE CREATOR'S (SSO's) VIEW OF OPEN STANDARDS

It is easiest to identify the view of recognized SSOs about open standards. Many recognized SSOs' Web sites state what openness of standardization is to them.

For over a century, the IEEE-SA has offered an established standards development program that features balance, openness, due process, and consensus.[5] (The Institute of Electrical and Electronic Engineers [IEEE])

The European model for telecom standardization allows for the creation of open standards.[6] (The European Telecommunications Standardization Institute [ETSI])

The process to create these voluntary standards is guided by the Institute's cardinal principles of consensus, due process and openness.[7] (The American National Standards Institute [ANSI] National Standards Strategy for the United States [2002])

It is interesting to contrast these views with a view from the European Commission,[8] which, as a government, represents the user and implementers view:

The following are the minimal characteristics that a specification and its attendant documents must have in order to be considered an open standard:

- *The standard is adopted and will be maintained by a not-for-profit organisation, and its ongoing development occurs on the basis of an open decision-making procedure available to all interested parties (consensus or majority decision, etc.).*
- *The standard has been published and the standard specification document is available either freely or at a nominal charge. It must be permissible to all to copy, distribute and use it for no fee or at a nominal fee.*
- *The intellectual property—i.e. patents possibly present—of (parts of) the standard is made irrevocably available on a royalty-free basis.*

- *There are no constraints on the re-use of the standard.*

Most SSOs follow rules to ensure what they consider an open standards creation process by requiring open meetings, consensus, and due process. These represent the standards creators' view of the requirements for open standards. Most SSOs do not suggest that the standard is made irrevocably available on a royalty-free basis (the highest level of open IPR).

THE IMPLEMENTER'S VIEW OF OPEN STANDARDS

Consider commercial developers of software or manufacturers of hardware as examples of implementers. They want the ability to compete on an equal basis with their competitors. This concept often has been termed a *level playing field*. Implementers of a standard would call a standard *open* when it is without cost to them (open IPR, open documents), when it serves the market they wish (open world), when it does not obsolete their prior implementations (open interface), when it does not preclude further innovation (open change), and when it does not favor a competitor. These five requirements ensure a level playing field.

The standards creators' requirements of open meeting, due process, and consensus assist the standards implementers to meet their last two requirements but do not address the first three. Many recognized SSOs are national or regional, while many implementers' markets are international. Most recognized SSOs allow intellectual property to be included in their standards. This is not in the interests of the implementers, who do not have similar intellectual property to trade.

In many cases, consortia have done a better job of addressing the needs of standards implementers. Many consortia offer international standardization and allow IPR negotiation (which recognized SSOs do not). Since consortia address more of the implementers' requirements, this may be one reason for the increase in standardization by consortia.

THE USER'S VIEW OF OPEN STANDARDS

Consider an office or factory that uses implementations of the standard. The simple goal of the user is to achieve the maximum possible return on its investment in the implementation of the standard. While many aspects of the user's return on investment are not related to the standardization process (e.g., the implementation quality), four aspects are. A user of an implementation of a standard would call a standard open when the following apply:

- The implementation operates and meets local legal requirements in all locations needed (open world, open use, open documents).
- New implementations desired by the user are compatible with previously purchased implementations (open interface, open use).
- Multiple interworking implementations of the standard from different sources are available (open interface, open use).
- The implementation is supported over the user-desired service life (ongoing support).
- It is worth noting that users do not often participate in IT standardization. Perhaps that is because many of the requirements specific to users are not even considered in most SSOs.

UNDERSTANDING ALL THE REQUIREMENTS OF OPEN STANDARDS

The previous analysis identified 10 separate requirements that are listed in Table 1. Of course, the names of these requirements are arbitrary, and it is possible to imagine combining different groups of requirements into one requirement category, so there is no importance to the number 10. What is important is that these requirements, by whatever name, are the requirements that standards creators, implementers, and users have rational reasons to desire.

Table 1 shows that the requirements of the major stakeholder classes are sometimes similar and sometimes divergent. Users have little interest in how a standardization process was conducted. The concept that open meetings, consensus, and due process support the development of multiple sources of implementations of a completed standard is recognized but rarely supported by users. Users are focused on being able to purchase compatible equipment from multiple sources. History does suggest that open meetings, consensus, and due process facilitate the creation of multiple sources. In the case of requirement 5 (Open IPR), even though all the stakeholders have an interest, their interests are quite different. Creators appear satisfied to support reasonable and nondiscriminatory (RAND) SSO IPR policies. Commercial implementers require a means to identify and control their IPR cost, which RAND does not offer. This pushes implementers to form consortia. Users often are unwilling to pay high prices caused by multiple IPR claims. So users may be put off by the impact of RAND policies.

Table 1. Creators, implementers, and users see openness differently

Requirements\Stakeholders		Creator	Implementer	User
1	Open Meeting	x		
2	Consensus	x		
3	Due Process	x		
4	Open World	x	x	x
5	Open IPR	x	x	x
6	Open Change	x	x	x
7	Open Documents		x	x
8	Open Interface		x	x
9	Open Use		x	x
10	On-going Support			x

As Table 1 identifies, the first three requirements are oriented to the stakeholders focused on standards creation. The first four requirements are also at the heart of the World Trade Organization (WTO) Agreement on Technical Barriers to Trade, Code of Good Practice.[9] The fourth requirement, Open World, is supported by ANSI but not required. The ANSI open standards concept requires the first three requirements for all ANSI accredited standards organizations (American National Standards Institute, 1998). The fifth requirement, Open IPR, has been formally added to the US standards development process by ANSI and many SSOs.

Currently, the widest interest regarding open standards focuses on Open World and Open IPR. Open World addresses standards such as barriers to trade or enablers of trade. Open IPR impacts the profitability of all communications equipment companies today. The additional five requirements (6 through 10) represent open standards requirements that are emerging but are not yet supported by most SSOs. Table 1 identifies that the five additional requirements are more oriented to the implementation and use of standards.

In the following descriptions of the 10 seaparte requirements, an attempt is made to quantify each specific requirement. Certainly, the details and quantifications proposed are open to further consideration, but some quantification of each requirement is useful to identify how well different SSOs support the concepts of open standards. Using these quantifications, Table 3 summarizes the support of these 10 requirements for several SSOs.

1. Open Meeting

In an Open Meeting, all may participate in the standards development process. Currently, openness of meetings is deemed to be met (i.e., under many SSO requirements), if all current stakeholders may participate in the standards creation process. But, as technology has become more complex, user participation in standards creation has declined significantly (Foray, 1995). When the largest number of stakeholders (users) no longer participates, such a definition of open meetings is no longer functional.

"All stakeholders can participate" is a mantra of many recognized SSOs. But this mantra does not address all the barriers to open meetings. Recent social science research has identified 27 barriers to open meetings and has grouped these into five categories: the stakeholders themselves, the rules of recognized standardization, the way the process is carried out, the role of the technical officers of the committee, and the culture of the committees (de Vries, Feilzer, & Verheul, 2004).

One major barrier to standardization participation is economic. Some recognized SSOs (e.g., International Telecommunications Union [ITU]) and many consortia (e.g., World Wide Web Consortium [W3C]) require membership before attendance. Paying to become a member is a significant economic barrier when a potential standardization participant is not sure it is interested in attending a single meeting. Participation expenses, unless quite low, are part of the real barriers to participation for students, many users, and even startup companies in the field. Currently, only a few SSOs, such as the Internet Engineering Task Force (IETF), the standardization organization for the Internet, and the IEEE, offer low-cost, per-meeting participation.

Economic access is a useful way to quantify Open Meeting. There are two broad classifications of the levels of economic access in Open Meeting:

1. Any stakeholder can pay to become a member (current status of many SSOs).
2. Acceptable cost to join on a per-meeting basis.

2. Consensus

In Consensus, all interests are discussed and agreement is found; there is no domination. Different SSOs define consensus differently. In general, consensus requires that no single stakeholder group constitutes a majority of the membership of an SSO. Consensus may be identified by vote of the standardization committee or may mean without active and informed opposition. Surprisingly, the IETF, which many find to be an example of a more open SSO, does not meet this criterion, as the IETF Area Directors have a dictatorial level of control over the standardization decisions in their area (IETF, 1998).

Consensus is quantified by the requirement (1) or lack of requirement (0) in each SSO.

3. Due Process

In Due Process, balloting and an appeals process may be used to find resolution. Different SSOs describe due process differently. In general, it requires that prompt consideration be given to the written views and objections of any participants. A readily available appeals mechanism for the impartial handling of procedural complaints regarding any action or inaction is part of the Due Process requirement. As explained previously, the three requirements—Open Meetings, Consensus, and Due Process—are considered fundamental by recognized SSOs to the openness of their standardization process.

Due Process is quantified by the requirement (1) or lack of requirement (0) in each SSO.

4. Open World

In Open World, the same standard is used for the same capability worldwide. This requirement is supported by the WTO to prevent technical barriers to trade. The International Federation of Standards Users (IFAN) also supports uniform international standards (IFAN, 2000). However, politically, this can be a very contentious area. There are national standards for food processing that are based on religious beliefs (e.g., halal and kosher). There are standards for the environment, health, medical care, and social welfare that cause an imbalance in cost between countries that implement them (often richer) and countries that don't (often poorer). To avoid these contentious issues, most recognized SSOs currently support but do not require coordination of their standards work with worldwide standards. This allows but does not favor divergent regional or national standards.

In richer countries, the rise of consortia, the decline of publicly funded research, and aggressive commercialism make it more difficult to achieve a single standard for a single function worldwide. The five incompatible wireless technologies of the 3G cellular standards (W-CDMA, cdma2000, UWC-136, TD-CDMA, FD-TDMA) are an example of these effects. Initially, these five 3G versions will operate in different geographic areas, but eventually users will demand worldwide compatible cell phone operation. It appears likely that standardization organizations will continue to create incompatible standards for similar capabilities. This may be viewed as an indication of the failings of recognized standardization (Cargill & Bolin, 2004) or as an indication of the need to increase the support of Open Interfaces (see the following).

Open World is quantified by identifying the geographic operating area of each SSO. International scope is rated 1, and national or regional scope is rated 0.

The first three requirements of open standards have been addressed and, in large measure, resolved in most SSOs. The requirement for Open World is supported by the three recognized worldwide SSOs—ISO, IEC, and ITU—but many nations cling to the view that giving up their national standardization prerogatives would be giving up an aspect of their nation's sovereignty. Consortia standardization, which is unimpeded by such political issues, usually creates worldwide standards.

5. Open IPR

Open IPR is how holders of IPR contained in the standard make available their IPR. Most recognized SSOs and many consortia consider Open IPR to mean that holders of Intellectual Property Rights (IPR) must make available their IPR for implementation on Reasonable and Non-Discriminatory (RAND) terms. Five levels of quantification of Open IPR currently are quantified (0 through 4):

0. Commercial licensing may be the most prevalent way to use IPR legally. It is also the least open. In this case, the holder of IPR and the potential implementer of the IPR agree privately on commercial terms and conditions for the implementer to use the holder's IPR.

J. Band (1995) describes four additional levels of increasing openness relating to IPR:

1. Microsoft believes that interface specifications should be proprietary but will permit openness by licensing the specifications to firms developing attaching (but not competing) products.
2. The Computer Systems Policy Project (CSPP) also believes that interface specifi-

cations can be proprietary but will permit openness by licensing the specifications on RAND terms for the development of products on either side of the interface.

3. The American Committee for Interoperable Systems (ACIS) believes that software interface specifications are not protectable under copyright and that, therefore, reverse engineering (including disassembly) to discern those specifications does not infringe on the author's copyright.
4. Sun Microsystems believes that critical National Information Infrastructure (NII) software and hardware interface specifications should receive neither copyright nor patent protection. This quantification is discussed further under Open Change.

The range of possible refinements in this contentious area is probably limitless. Two further variations that might help to resolve contentious IPR issues are the following:

Approach #2 (the manner of operation of most recognized SSOs currently) might be more acceptable to implementers if an IPR arbitration function existed when IPR was identified during the creation/modification of a standard (Shapiro, 2001).

Approach #4 might be more acceptable to implementers if claims on basic interfaces were precluded but IPR on proprietary extensions were allowed. This could be technically feasible using the concepts of Open Interfaces.

Summary of the First Five Requirements

The fifth requirement—Open IPR—is the most divisive of the first five requirements, as RAND (the current practice of many SSOs) is not sufficient to allow implementers to determine the impact of standards-based IPR on their costs. The semiconductor industry, where manufacturing cost drops rapidly with volume, exacerbates IPR problems for implementers. In the case of semiconductors, IPR costs may be based on fixed-unit charges. Such charges can be the largest single cost component of a semiconductor. Semiconductor implementers must control their IPR costs. It seems only fair that any implementer has a right to determine the exact cost of IPR to them before they accept its inclusion in a new standard. This issue has led many implementers to consortia, since consortia often require joint licensing of related IPR. This practice defines the cost of the IPR to the implementer. While commercial licensing may seem the least open process, it may not be more costly than the RAND approach to specific implementers.

For emerging countries, RAND policies also appear to be causing an undesirable situation. The Chinese are rapidly developing into major suppliers of communications systems and equipment but do not have a portfolio of intellectual property rights that can be used to trade with commercial organizations in more developed countries that have IPR in communications standards. This may cause the Chinese to consider developing non-standard technologies for existing communication systems, so they do not have to pay for previous IPR standardization decisions in which they did not participate (Updegrove, 2005).

The Web site Cover Pages maintains a section on open standards and collects many different descriptions of open standards (including the 1998 version of this article).[10] The view of open standards from the SSOs quoted on this site follows the first five requirements. The view of other organizations is more divergent and includes requirements discussed in the following.

6. Open Change

In Open Change, all changes to existing standards are presented and agreed in a forum supporting the five previous requirements. Controlling changes is a powerful tool to control interfaces when system updates are distributed over the Internet and stored in computer memory. Even with the most liberal of IPR policies identified (4 in Open IPR), Microsoft would still be able to control its Windows Application Programming Interfaces (APIs) by distributing updates (changes) to users that updated both sides of the API interface. But without a similar distribution at the same time, competing vendors' products on one side of the same API could be rendered incompatible by such a Microsoft update.

The only way that interfaces can remain open is when all changes are presented, evaluated, and approved in a committee that supports the first five requirements. Considering today's environment of computers connected over the Internet, identifying and requiring Open Change is vital to the concept of open standards. Surprisingly, this is not widely understood. The original US judicial order to break up the Microsoft PC-OS and application software monopoly did not address this key issue (United States District Court). On March 24, 2004, the European Commission (EC) announced its decision to require Microsoft to provide its browser (Explorer) independently of the Windows operating system and to make the related Windows APIs available to others.[11] This decision did not address the necessity for open change. The EC announced on June 6, 2005, the receipt of new proposals from Microsoft addressing Microsoft's support of interoperability.[12] Unfortunately, these proposals still do not directly address open change.

Open Change is quantified as either supported (1) or not supported (0). Consortia that do not support the first three requirements of open standards are identified as not supporting open change.

7. Open Documents

In Open Documents, committee documents and completed standards are readily available. Open Documents is the requirement for a stakeholder to be able to see any documents from an SSO. The openness of a standardization meeting to outsiders is closely related to the availability of the documents from the meeting. All standardization documentation falls into two classes: work-in-progress documents (e.g., individual technical proposals, meeting reports) and completed standard documents (e.g., standards, test procedures). Different stakeholders need to access these different classes of documents. Standards creators do not require Open Documents, as they are involved in the creation of all the documents. Standards implementers need access to standards work-in-progress documents in order to understand specific technical decisions as well as have access to completed standards. Implementation testers (users and their surrogates) need access to completed standards.

The Internet Society (ISOC) supports a nongovernment-recognized standards making organization—the IETF—which has pioneered new standards development and distribution procedures based on the Internet. While the IETF does not meet the criteria for Consensus and Due Process, the IETF is perhaps the most transparent standardization organization. Using the Internet, the IETF makes available on the Web both its standards (termed RFCs) and the drafts of such standards at no charge. Using the facilities of the Internet, IETF committee discussion and individual technical proposals related to the development of standards can be monitored by anyone and responses can be offered. This transparent development of IETF standards has been successful enough that some other SSOs are doing something similar. In July 1998, ETSI announced that its technical committee TIPHON (Telecommunications and Internet Protocol Harmonization Over Networks) would make available at no charge all committee documents and standards drafts.

Ultimately, as technology use expands, everyone becomes a stakeholder in technical standards. Using the Internet, access to committee documents and discussion may be opened to all. In this way, informed choices may be made about bringing new work to such a standards committee, and potential new standardization participants could evaluate their desires to attend meetings.

Three levels of transparency in Open Documents can be quantified:

1. Work-in-progress documents are only available to committee members (standards creators). Standards are for sale (current state of most formal SSOs).
2. Work-in-progress documents are only available to committee members (standards creators). Standards are available for little or no cost (current state of many consortia).
3. Work-in-progress documents and standards are available for reasonable or no cost (current state of IETF).

8. Open Interface

Open Interface supports proprietary advantage (implementation); each interface is not hidden or controlled (implementer interest); each interface of the implementation supports migration (user interest). Open Interface is an emerging technical concept applicable to compatibility standards used between programmable systems. The Open Interface requirement supports compatibility to previous systems (backward compatibility) and to future systems (forward compatibility) that share the same interface. The idea that open standards should embody such a principle is new. But interest in Open Interfaces is increasing due to the considerable success of Open Interfaces in facsimile (T.30), telephone modems (V.8 and V.32 auto baud procedures), and Digital Subscriber Line transceivers (G.994.1 handshaking).

One way of achieving Open Interfaces is to implement a fairly new technique called an etiquette (Krechmer, 2000). Etiquettes are a mechanism to negotiate protocols. While a protocol terminates an X.200 (OSI) layer, an etiquette, which may negotiate multiple OSI layer protocols, does not terminate (replace) any protocol layer function. An etiquette is used only for negotiating which protocol, options, or features to employ. The purpose of etiquettes is connectivity and expandability. Proper etiquettes provide:

- Connectivity, negotiating between two devices in different spatial locations to determine compatible protocols.
- Means to allow both proprietary and public enhancements to the interface that do not impact backward or forward compatibility.
- Adaptability, so that a communications system can become compatible with a different communications system.
- Easier system troubleshooting by identifying specific incompatibilities.

As long as the etiquette is common between the equipment at both ends, it is possible to receive the code identifying each protocol supported by the equipment at a remote site. Checking this code against a database of such codes on the Web or in a manual, the user can determine what change is necessary in his or her system or in the remote system to enable compatibility.

One of the earliest etiquettes is ITU Recommendation T.30, which is used in all Group 3 facsimile machines. Part of its function includes mechanisms to interoperate with previous Group 2 facsimile machines, while allowing new features (public as well as proprietary) to be added to the system without the possibility of losing backward compatibility. Another etiquette is the ITU standard V.8, which is used to select among the V.34 and higher modem modulations. More recently, ITU G.994.1 provides a similar function in Digital Subscriber Line (DSL) equipment.

As an example of the usefulness of Open Interfaces, consider Microsoft APIs. Assume that a standard based upon the Microsoft Windows API is created. Then, any vendor could create an operating system (OS) to work with Microsoft's applications or to create applications to work with Microsoft's OS. If any vendor (including Microsoft) identified a new function, such as short message service or video conferencing, that was not supported across the basic API, that vendor then could offer the new function as an identified proprietary feature across the API to users that purchase that vendor's OS and applications. Since an Open Interface supports proprietary extensions (Krechmer, 2000), each vendor controls the way the new function is accessed across the API but does not change the basic compatibility of the API. In this manner, a vendor is able to maintain control and add value, based on the desirability of the new function.

Some aspects of the issue of open interfaces were explored in technical detail in 1995 (Clark, 1995). Since then, seven technical aspects of open interfaces have been identified (Krechmer, 2000). Currently, open interfaces only have been addressed at the standard committee level and not addressed at the SSO level, so no detailed quantification is offered here.

9. Open Use

Open Use describes the assurance a user requires to use an implementation. Open Use identifies the importance to users of known reliable standardized implementations. Often, a user will trust an implementation based on previous performance, its brand, or simply familiarity with the requirements. When such trust is not reliable, then compliance, conformance, and/or certification mechanisms for implementation testing, user evaluation, and identification may be necessary. This more exact form of Open Use is termed *conformity assessment*. The Open Use requirement is supported by ANEC, a European organization that focuses on consumer issues associated with standardization (ANEC, n.d.). ISO/IEC 17000 defines conformity assessment as the "demonstration that specific requirements relating to a product, process, system, person or body are fulfilled" (ISO/IEC 17000). Conformity assessment procedures such as testing, inspection, and certification offer assurance that products fulfill the requirements specified in the appropriate regulations or standards.

Open Use covers all possible parameters that may need to be identified as conforming to a standard for accurate, safe, and/or proper use. Such parameters could include physical access (e.g., access by people with disabilities), safety (e.g., CE or UL mark, the European and US indications that equipment is designed safely), and correct weights and measures (e.g., certification of scales and gasoline pumps). To achieve Open Use may require testing by implementers, regulators, users, or their testing agencies as well as known and controlled identification marks (e.g., UL, CE) to indicate conformity to certain requirements.

Open Use may represent requirements on the standardization process as well as requirements on implementations that use the standard to identify and assure compliance and, if necessary, conformance. For a manufacturer of a scale to measure weight, a self-certification process traceable to national standards may be required. For a communications equipment or communications software manufacturer, an interoperability event may be needed (often termed a *plug-fest*) to test whether different implementations interoperate. For the user, a simpler mark of conformity is often desirable. For example, in the European Union (EU), the CE marking is the manufacturer's indication that the product meets the essential (mostly safety) requirements of all relevant EU Directives. This specific marking indicating compliance reduces the user's safety concerns. Many consortia support plug-fests and compliance testing as part of their members' desire to promote associated products. Two levels of Open Use are quantified:

1. Open Use via plug-fests or over the Internet testing (implementer).
2. Open Use via conformance marking (user). This may include the first level of Open Use.

10. Ongoing Support

In Ongoing Support, standards should be supported until user interest ceases rather than when implementer interest declines. Ongoing Support of standards is of specific interest to standards users, as it may increase the life of their capital investment in equipment or software with standard interfaces. The user's desire for implementer-independent ongoing support is noted by Perens (1999) as one of the desirable aspects of open source software. The support of an existing standard consists of four distinct phases after the standard is created (Table 2).

This list may be used to quantify the ongoing support that a specific SSO provides by identifying which steps of the ongoing support process are widely announced by a specific SSO. This is a difficult requirement to quantify, as different SSOs have different procedures for making this process public, and many older SSOs may not make good use of the Internet to distribute such information to users.

It is difficult to interest users in the first phase of standards development (creation) shown in Table 2 (Naemura, 1995). Even the second phase—fixes—may be of more interest to the developers and implementers than to the users. The next three phases, however, are where users have an interest in maintaining their investment. Possibly greater user involvement in the ongoing support of standards would be practical by taking advantage of the Internet to distribute standards and allow users to keep abreast of the work in standards meetings. Increasing the users' involvement with these aspects of the standardization process also may represent new economic opportunities for SSOs. The ITU-T Telecommunications Standardization Bureau Director's Ad Hoc IPR Group report released in May 2005 includes "On-going support—maintained and supported over a long period of time" as one element of its Open Standards definition.[13]

Table 2. The SSOs' phases of support during a standard's lifetime

Phase	Activity	Description	Major Interest Group
0.	Create standard	The initial task of SSOs	creators
1.	Fixes (changes)	Rectify problems identified in initial implementations	implementers
2	Maintenance (changes)	Add new features and keep the standard up to date with related standards work	users
3.	Availability (no changes)	Continue to publish, without continuing maintenance	users
4.	Rescission	Removal of the published standard from distribution	users

COMPARING THESE 10 REQUIREMENTS WITH OTHER DEFINITIONS OF OPEN STANDARDS

Standards are a multi-disciplinary field. An academic view of the requirements of open standards should address each of the related disciplines—economics, law, engineering, social science, and political science. From the legal perspective, each of these 10 requirements may be a legal right of a specific group. As West (2004) notes, each of these requirements has an economic cost and a benefit to specific stakeholders. From an engineering perspective, two of these requirements (6 and 8) directly impact communications equipment compatibility and design. From a social science perspective, the dynamics of different stakeholders may be examined in terms of each requirement. From a political science perspective, the first three requirements are basic to any fair political process, including standardization.

Table 3. Rating "openness" at different SSOs

Reqmts.	Consortium (note 1)	ITU (note 2)	IEEE (note 2)	ATIS T1 (note 2)	ETSI (note 2)	W3C	IETF
6.1 OM	1	1	2	1	1	1	2
6.2 Con	0	1	1	1	1	1	1
6.3 DP	0	1	1	1	1	0	0
6.4 OW	1	1	0	0	0	1	1
6.5 OIPR	0	2	2	2	2	4	3 (note 3)
6.7 OC	0	1	1	1	1	1	1
6.8 OD	2	1	2	3	3	1	3
6.9 OI							
6.10 OU	2	0	0	0	1	1	1
6.11 OS	1	2	2	2	4	4	4
Score	7	10	11	11	14	14	16

Note 1: This hypothetical consortium is modeled on the description found at ConsortiumInfo.org[14].
Note 2: The ITU, ETSI, IEEE, and ATIS are recognized SSOs.
Note 3: The IETF IPR policy desires a royalty-free model but is flexible.

West (2004) defines "'open' for a standard as meaning rights to the standard are made available to economic actors other than the sponsor." This definition offers a succinct economic view of open standards. But economic rights cannot be maintained without supporting political rights such as balance, consensus, and due process. In order for the economic rights associated with compatibility standards to be available, some technical process (change control) and technical functionality (open interfaces) also are required. In order for specific economic rights associated with Intellectual Property Rights (IPR) to be available, specific SSO procedures must be defined.

Perens (n.d.) offers a software engineering perspective of Open Standards. He presents six requirements and related practices. The requirements proposed are (1) availability, (2) maximize end-user choice, (3) no royalty, (4) no discrimination, (5) extension or subset, and (6) no predatory practices. The 10 requirements proposed can be compared to the six principles proposed by Perens (n.d.):

- *Availability* is addressed by Open Documents.
- *Maximum end-user choice* is addressed by Open Use.
- *No royalty* is addressed under Open IPR.
- *No discrimination* is addressed by Open Meeting, Consensus, and Due Process.
- *Means to create extension or subset* is addressed by Open Interface.
- *Means to prevent predatory practices* is addressed by Open Change.

The six principles proposed by Perens (n.d.) map fully onto eight of the 10 requirements of Open Standards proposed. Perens (n.d.) does not address directly in the six principles the desires for or against Open World or the end user desire for Ongoing Support.

HOW OPEN ARE DIFFERENT SSOs?

Table 3 offers the author's quantification based on a review of the SSO's documentation (as of September 2004) of the specific requirements these SSOs support. By quantifying nine of the 10 requirements of Open Standards, it is possible to examine any SSO to determine what requirements are supported and what requirements are not, and how this impacts standards creators, implementers, and users. Then, the political, social, economic, technical, and practical implications of the standardization process and the machinations of all the stakeholders may be more rigorously analyzed and understood.

CONCLUSION

Microsoft is an example of a commercial organization that uses the term *open standards* when referring to its products (Gates, 1998). In fact, Microsoft's products are considered by many (including the European Commission) to be examples of non-openness. It seems clear that defining all the requirements of openness is necessary to avoid such misuse and the confusion it causes. Table 3 identifies that no SSO discussed meets all of the 10 requirements described, and each SSO differs significantly in which requirements they meet. It should not be surprising that implementers and users consider all SSOs with a jaundiced eye. This attempt to rate how well SSOs support open standards suggests that this implementer's and user's view is a wise one.

The 10 basic requirements presented here are the broadest possible views of the meaning of Open Standards. Are fewer requirements sufficient? That question only can be answered when stakeholders understand the consequences of what they may be giving up. The comprehension of the requirements that are supported by each SSO usually is buried in the fine print of the procedures of each SSO. Until each SSO clearly indicates which requirements of Open Standards it supports and at what level, Open Standards will be just another marketing slogan.

The author would like to thank Joel West and Henk de Vries for their continuing efforts to improve this article.

REFERENCES

American National Standards Institute. (1998). *Procedures for the development and coordination of American national standards*. Washington, DC: American National Standards Institute.

ANEC: The European consumer voice in standardization. (n.d.). Retrieved from http://www.anec.org/

Band, J. (1995). Competing definitions of "openness" on the NII. In B. Kahin & J. Abbate (Eds.), *Standards policy for information infrastructure* (pp. 351-367). Cambridge, MA: MIT Press.

Brady, R. A. (1929). *Industrial standardization*. New York: National Industrial Conference Board.

Cargill, C. (1989). *Information technology standardization*. Bedford, MA: Digital Press.

Cargill, C., & Bolin, S. (2004). Standardization: A failing paradigm. *Proceedings of the Standards and Public Policy Conference*, Chicago.

Clark, D. C. (1995). Interoperation, open interfaces and protocol architecture. Retrieved from http://www.csd.uch.gr/~hy490-05/lectures/Clark_interoperation.htm

Critchley, T. A., & Batty, K. C. (1993). *Open systems: The reality*. Hertfordshire, UK: Prentice Hall.

de Vries, H., Feilzer, A., & Verheul, H. (2004). Removing barriers for participation in formal standardization. In F. Bousquet, Y. Buntzly, H. Coenen, & K. Jakobs (Eds.), *EURAS Proceedings 2004* (pp. 171-176). Aachen, Germany: Aachener Beiträge zur Informatik.

Foray, D. (1995). Coalitions and committees: How users get involved in information technology (IT) standardization. In R. Hawkins, R. Mansell, & J. Skea (Eds.), *Standards, innovation and competitiveness* (pp. 192-212). Hants, UK: Edward Elgar Publishing.

Gabel, H. L. (1987). Open standards in the European computer industry: The case of X/Open. In H. L. Gabel (Ed.), *Product standardization and competitive strategy.* Amsterdam, North-Holland.

Gates, B. (1998). Compete, don't delete. *The Economist* (p. 19). Retrieved from http://www.economist.com

IETF working group guidelines and procedures, RFC 2418. (1998). Retrieved from http://www.ietf.org/rfc/rfc2418. txt

IFAN strategies and policies for 2000-2005. (2000). Retrieved from http://www.ifan-online.org/

ISO/IEC 17000. (2004). *Conformity assessment—Vocabulary and general principles.* Geneva: International Organization for Standardization (ISO).

Krechmer, K. (1998). The principles of open standards. *Standards Engineering, 50*(6).

Krechmer, K. (2000). The fundamental nature of standards: Technical perspective. *IEEE Communications Magazine, 38*(6), 70.

Naemura, K. (1995). User involvement in the life cycles of information technology and telecommunications standards. In R. Hawkins, R. Mansell, & J. Skea (Eds.), *Standards, innovation and competitiveness.* Hants, UK: Edward Elgar Publishing.

NRENAISSANCE Committee. (1994). *Realizing the information future.* Washington, DC: National Academy Press.

Perens, B. (1999). The open source definition. In C. DiBona, S. Ockman, & M. Stone (Eds.), *OpenSources voices from the open source revolution* (pp. 171-189). Sebastopol, CA: O'Reilly & Associates.

Perens, B. (n.d.). Open standards principles and practice. Retrieved from *http://xml.coverpages.org/openStandards. html*

Raymond, E.S. (2000, August 25). Homesteading the noosphere (Section 2). Retrieved from http://www.csaszar.org/index.php/csaszar/interesting/the_open_source_reader

Shapiro, S. (2001). Setting compatibility standards cooperation or collusion. In R. C. Dreyfuss, D. L. Zimmerman, & H. First (Eds.), *Expanding the boundaries of intellectual property.* Oxford, UK: Oxford University Press.

United States District Court for the District of Columbia Civil Action No. 98-1232 (TPJ).

Updegrove, A. (1995). Consortia and the role of the government in standards setting. In B. Kahin & J. Abbate (Eds.), *Standards policy for the information infrastructure* (pp. 321-348). Cambridge, MA: MIT Press.

Updegrove, A. (2004). Best practices and standard setting (how the "pros" do it). In S. Bolin (Ed.), *The standards edge, dynamic tension* (pp. 209-214). Ann Arbor, MI: Bolin Communications.

Updegrove, G. (2005). China, the United States and standards. *Consortium Standards Bulletin, IV*(4). Retrieved from http://www.consortiuminfo.org/bulletins/apr05.php#editorsnote

West, J. (2004). What are open standards? Implications for adoption, competition and policy. *Proceedings of the Standards and Public Policy Conference*, Chicago.

ENDNOTES

1 This article is a significant revision of the paper published in the Proceedings of the 38th Annual Hawaii International Conference on System Sciences (HICSS), January, 2005. In turn, the HICSS paper is a major revision of Krechmer (1998).

2 Wikipedia, http://en.wikipedia.org/wiki/X/Open

3 Wikipedia, http://en.wikipedia.org/wiki/Free_software

4 Wikipedia, http://en.wikipedia.org/wiki/The_Open_Group

5 IEEE, http://standards.ieee.org/sa/sa-view.html

6 ETSI, http://www.etsi.org/%40lis/background.htm

7 ANSI, http://www.ansi.org/standards_activities/overview/overview. aspx?menuid=3

8 The principle of open standards is formulated in European Interoperability Framework section 1.3, Underlying principles, derived from the eEurope Action Plan 2005 as well as the Decisions of the European Parliament, November, 2004, http://xml.coverpages.org/IDA-EIF-Final10.pdf

9 WTO, http://www.wto.org/english/tratop_e/tbt_e/tbtagr_e.htm #Annex%203

10 http://xml.coverpages.org/openStandards.html, May 2005

11 European Union, http://www.eurunion.org/news/press/2004/20040045.htm

12 European Union, http://www. eurunion.org/index.htm see Microsoft EU Tests Proposals

13 http://www.itu.int/ITU-T/othergroups/ipr-adhoc/openstandards.html

14 http://www.consortiuminfo.org/what, September, 2004

This work was previously published in International Journal of IT Standards and Standardization Research, 4(1)*, edited by K. Jakobs, pp. 43-61, copyright 2006 by IGI Publishing, formerly known as Idea Group Publishing (an imprint of IGI Global).*

Chapter V
The Role of Individuals and Social Capital in POSIX Standardization

Jim Isaak
New Hampshire University, USA

ABSTRACT

While standards are issued by organizations, individuals do the actual work, with significant collaboration required to agree on a common standard. This article explores the role of individuals in standards setting as well as the way these individuals are connected to one another through trusting networks and common values. This issue is studied in the context of the IEEE POSIX set of standards, for which the author was actively involved for more than 15 years. This case study demonstrates that the goals and influence of individual participants are not just that of their respective employers but may follow the individual through changes of employment. It also highlights changes in the relative importance of individual and corporate influence in UNIX-related standardization over time. Better understanding of the interaction between individuals and organizations in the context of social capital and standardization can provide both a foundation for related research and more productive participation in these types of forums.

INTRODUCTION

Standards development organizations (SDOs) provide a forum for interaction among individuals, firms, governments, and other stakeholders. These are an example of cooperative technical organizations (Rosenkopf, Metinu & George, 2001; Rosenkopf & Tushman, 1998) that bring together boundary-spanning individuals (Dokko & Rosenkopf, 2004; Tushman, 1977). This context should increase innovation for the firms involved (Rao & Drazin, 2002; Tushman, 1977), facilitate the creation of alliances (Rosenkopf, Metiu, & George, 2001), and promote a technological bandwagon (Wade, 1995). These forums also are breeding grounds for social capital (Adler & Kwon, 2002; Coleman, 1988; Portes, 1998; Putnum, 2000). There are a number of perspectives on social capital; this article adopts Putnum's (2000) definition: "Individuals connected to one another through trusting networks and common values" (p. 312).

This article will explore the roles of individuals and social capital in standards development using a case-study approach. The article will also look at the effects of the transition from individual to corporate engagement in the activities. The case considers POSIX standardization. The POSIX standards activity focuses on network externalities, specifically the applications programming interface spanning multiple vendors. This is a critical control point for operating systems and, therefore, for the heart and soul of the computer industry (West & Dedrick, 2000). The impact of the activity approaches the $100 billion mark (Hurd & Isaak, 2005; Unter, 1996) and created a foundation for the Linux system (Linux International, 2005), the most recent front on a continuing battle for de facto operating system standards control.

POSIX was a volunteer consensus standardization activity (Isaak, 1998; National Technology Transfer and Advancement Act, 1995; OMB, 1998) as opposed to a single vendor-controlled standard. It had roots in a UNIX industry association initially called /usr/group, now called Uniforum, IEEE (Institute of Electrical and Electronic Engineers), and more recently the Open Group (a merger of two consortia: X/Open and the Open Software Foundation (OSF)). Uniforum and IEEE both operated with individual memberships, so any number of participants from a single company or other organization could join and vote; the Open Group was a corporate consortium with limited membership but coordinated with IEEE to provide a forum where any interested party could vote or comment. The results were forwarded for adoption as international standards by ISO/IEC JTC1, where voting is done by national standardization bodies. Over time, POSIX standardization was less about individual discretion and more about corporate interests. The X/Open corporate consortium eventually supplanted the individual-centered IEEE process.

The POSIX API evolution is documented in both concurrent publications (Cargill, 1994, 1997; Jespersen, 1995; PASC, 2005; Walli, 1995) and retrospective publications (Isaak, 2005; Takahashi & Tojo, 1993). The history spans from the initial concepts of the early 1980s, although there was dramatic growth into the 1990s, finally settling into a mature maintained state at the end of that decade (Table 1). This also parallels the growth of the DOS (later Windows) proprietary platform standards. Findings from the literature are complemented with the author's own observations. The author served as chair of the IEEE POSIX work for 15 years, convener of the ISO working group for a similar period, the Digital Equipment Corp. (DEC) member of the X/Open Board of Directors, and part of the U.S. Delegation to the ISO/IEC JTC1 Technical Advisory Group on Applications Portability (see Table 2).

The case analysis provides insight on the role of individuals and the development of social capital in the various stages of development of this set of standards.

FORMATION: INDIVIDUAL INITIATIVE

Individuals, primarily from small companies, were the driving force behind the early UNIX standards work. This expanded on the anti-establishment alignment that was a strong thread in the UNIX community in the 1980s[1]. The driving force of these individuals was to establish a technology bandwagon, as suggested by Wade (1995), with a specific focus on the minicomputer market, a step above the PC capabilities of that era. The existence of two primary versions (Version 7 and BSD) as well as a few "compatible" systems created a need to agree on standard interfaces. Microprocessor technology had reached the point where it could support a viable standard operating environment that might move control from the then dominate mid-range players (IBM, DEC, etc.) to a new standard.

Table 1. POSIX timeline

1981 & 1983	/usr/group IEEE initiate individual expert based standards forums for developing a UNIX based operating system API.
1984	This work is consolidated into IEEE, and embraces the concept of "implementation independence" (acknowledging Version 6/7 AT&T UNIX, BSD, and independently developed compatible products.)
1985	IEEE publishes a trial use standard, and NIST takes a formal interest in testing with the intent to create a government procurement guideline.
1988	IEEE generates a Full Use standard NIST delivers the FIPS procurement guideline and test suite, Industry forums emerge for vendor driven standardization and user procurement Vendor divergence on windowing interfaces, the "GUI Wars" delay essential open systems capability for five years.
1990	Vendor initiated "guidelines" established to limit growth of POSIX.
1991	GNU/Linux started as POSIX based initiative (Linux International, 2005)
1995	VMS, MVS and Windows NT delivered POSIX conforming interfaces.
1998	Austin Group formed to manage evolution of POSIX in collaboration with the Open Group.

Table 2. Volunteer consensus SDO relationships

Acronym	Name and relation to other organizations
IEEE	Institute of Electronic and Electrical Engineers, a professional society which has been doing standards since 1890s; a founding member of ANSI. IEEE is accredited by ANSI and forwards its approved standards for adoption as ANSI standards. IEEE also forwards work for adoption by ISO and IEC. IEEE provides for individual membership, and more recently corporate membership forums
ANSI	American National Standards Institute—accredits U.S. volunteer consensus SDOs, adopts approved documents as American National Standards, and provides for representation of the U.S. in ISO and IEC activities.
POSIX	a family of portable operating system interface standards that originated with UNIX application programming interfaces.
PASC	Portable Applications Standards Committee, an IEEE forum which sponsored the 40 project working groups listed in Table 3 and forwarded draft standards to the IEEE for adoption
Computer Society	IEEE Society hosting the POSIX and other computer related standards activities. Authorizes POSIX and a number of other standards committees
ISO	International Organization for Standardization—membership by national bodies; national positions established by technical panels within each county. In the U.S. these are called TAGs (Technical Activities Groups), there is one for each ISO activity where the U.S. chooses to participate. PASC hosted the WG15 TAG for the U.S.
IEC	International Electro technical Commission (parallels ISO for these areas of work) (Note: the U.N. authorized ITC handles telecommunications standards, and is not tied to the POSIX work)
JTC1	Joint Technical Committee 1 of ISO and IEC—covers the areas of information technology
SC22	Subcommittee 22 of JTC 1—covers the areas of computer languages and applications portability
WG15	Working Group 15 of SC22—covers Operating System Interfaces (POSIX)
X3	Corporate membership SDO (renamed to INCITS) hosted by the Information Technology Industry Council
X3J11	X3 Committee responsible for the C Language
MAP	Manufacturing Applications Protocol—an industry effort lead by General Motors to encourage widespread adoption of a well defined set of OSI communications protocols for manufacturing
TOP	Technical/Office Protocols—a effort parallel to MAP lead by Boeing to encourage adoption of OSI communications for non-manufacturing applications—done with user driven consortia in collaboration with MAP
OSI	Open System Interconnection—the major international (ISO) communications standardization effort from 1973 to 1990 or so. Displaced by the Internet Protocols (which were available in public domain software) POSIX got started at the 'peak' of the OSI activities and then spanned into the displacement of that by the Internet.

The /usr/group committee specifically selected IEEE for formal standards development because it permitted individual voting. A large number of individuals from a single firm could all join and dominate a balloting process, but this did not occur in the many POSIX ballots. As voting blocks emerged, sometimes from a single company, they were encouraged to designate a single point of contact for resolution. This quickly migrated into reference ballots that could be cited by any voter and often spanned single company participation.

The establishment of community values for POSIX started with guidelines developed by the chair of the software engineering standards committee. These called for building consensus though a four-step process: (1) identify possible options; (2) select what appears to be the preferred approach (or some integration of options); (3) ask for objections (this step was a precursor of the formal balloting model), "without objection" being an approved resolution, but also assure that individuals' concerns are heard; (4) document any remaining objection and the rationale for not following that course. This last step prepares for the final approval process that IEEE uses, where the entire document is balloted and all objections must be addressed with either changes or rationale for not making changes (IEEE, 2005; PASC, 2005). This also avoided a win-lose interaction and encouraged understanding and mutual respect.

The collaborative decision making reflected in this four-step process is one of the building blocks associated with developing social capital in POSIX. /usr/group used this consensus process in making the decision to move the activity to IEEE, which established norms for future decision making. Hearing the options and objections put forward by individuals, addressing these, and then moving forward with general agreement rather than simple majority votes developed the respect and trust underlying the development of social capital. In each working group, starting with the /usr/group forum, the chairperson used this process for early, simple decisions, developing a comfort level with the process and a minimal level of social capital so that the group could move into more controversial decision making.

The group focused on the applications program interface for source code portability with the intent to avoid any dependence on implementation characteristics (IEEE, 1986). Given the makeup of the group, some building on traditional computers from DEC, DG, IBM, HP, others on Motorola 68000, and some on Intel 286 or other platforms (using BSD, ATT, or independently developed source code), this was the clear common ground. An early quid pro quo for the activity became "don't torpedo someone else's platform." Embracing the decision model and acknowledging the diversity of platforms and implementations, the POSIX group established the common values and norms of respect that are the foundation for social capital development.

Keypoint #1: Consensus standards building organizations, particularly individual membership forums, provide an environment that fosters social capital development. This builds on repeated collaborative problem solving over an extended period of time, building respect and trust, and reinforcing common values.

Individuals, even from larger vendors, joined the activity. Some may have had corporate sanction; others were affiliated with secondary product lines, reflecting the bottom-up perspective suggested by Rosenkopf, Metiu, and George (2001). These individuals, however, were both boundary-spanners and often stars, as described by Tushman (1977), within their corporations. Tushman (1977) uses boundary spanning to describe individuals who facilitate communications between different groups; for example, marketing and engineering. In the POSIX context, participants were inherently boundary spanners operating within the culture and norms of their employer and also within the culture and norms of the POSIX community.

The act of engaging in the SDO activities place them in the boundary-spanning role, and most served as primary cross-functional UNIX-Gurus within their firms. Participants needed to serve in a boundary-spanning role, since information and influence had to be exchanged between these two contexts.

Some of the individuals also were recognized as stars (Tushman, 1977), an expert resource that others draw upon for information about a specific topic. One of the co-chairs of P1003.2 was also the chief guru in technical editing of standards (after on-the-job training with POSIX documents); a few other working group participants gained notoriety as process experts. One individual was recognized as a star due to his clarity of vision and depth of understanding in writing balloting comments (I cannot recall a case where that individual's proposed changes in balloting were not solid improvements in whatever document he chose to address). Some individuals also spanned the boundaries between the POSIX community and other forums, such as IEEE, ISO, X/Open, and so forth. Leadership in this context required learning the ways of the community, the requirements of the standardization process, a sufficient understanding of the technology, and ability to help groups to bring these together into a coherent standard document. Individuals demonstrating skills in this area became visible stars, who were then appointed or elected to take on leadership roles.

Every meeting of the POSIX working group entailed discussions of the technical details and wording of the hundreds of pages of the document. These interactions from the trivial to the technically challenging built a momentum of respect, success, trust, and reinforced values. Notable examples include sidestepping the BSD vs. AT&T variation issues, reorganizing the document into functional vs. alphabetical sequence, and ultimately relinquishing control of a number of interface definitions to the C-Language committee (X3J11). In this situation, there was a quid pro quo compromise, with X3J11 agreeing to not cast ballots against POSIX due to any residual overlap, and POSIX giving up most of these interface specifications. Each minor discussion and resolution increased the level of social capital among the participants through the use of the four-step consensus process as well as ongoing dialogue outside of the meetings. Individuals used social capital resulting from one interaction in the next discussion with a typically monotonically increasing value. Individuals were emerging as star experts and collaborators with increasing influence as a result of these interactions.

Keypoint #2: Individuals use social capital to initiate activities that build additional social capital, whereas financial capital is expended when used.

The POSIX effort reflects a few significant differences from de facto standards battles such as described by West and Dedrick (2000). It is a formal standards process, exposing much of the processes to participants and, in many cases, in public documents (resulting standards, minutes, and balloting records). Focusing on the portability of applications source code caused POSIX standardization to emphasize applications development. This is one significant part of the cospecialization of assets described by West and Dedrick (2000). POSIX could not retain its platform independence and source code focus and fulfill the single-image-runs-everywhere standard that delivered the full positive network externalities of the Wintel standard.

Intellectual property, a more dominate factor in today's standardization, did not have a significant impact on POSIX. The only identified patent was relinquished by AT&T (for the set user ID bit). Focusing on interfaces and operating in a public forum precluded nondisclosures and exposure of trade secret information. AT&T did relinquish rights to derive the specifications from its documentation. Any number of these concessions might not have been available a few years later. All of these played into the transfer of control of the standard into the committee processes.

Around 1985, corporate users started to take a stronger interest in the UNIX world. In the data communications world, GM and Boeing were driving for OSI adoption via the MAP-TOP initiatives. One of these user interests was the U.S. government, represented by the National Institute for Standards and Technology (NIST) (previously the National Bureau of Standards). In the summer of 1985, NIST put out a call for UNIX testing, creating a clear message to industry that government procurement guidelines could be expected (NIST, 1986; Isaak, 1998). A common individual between NIST management and ISO/IEC JTC1/SC22 (recently formed) set the stage for the formation of SC22/WG15 (Table 2 provides a taxonomy of the SDO structure involved). This is an informative example of social capital established by that individual and her firm, providing influence that put the POSIX activity into the track for U.S. Government procurement and international standardization.

Finally, the X/Open group (initially called BISON for founders Bull, ICL, Siemens, Olivetti, and Nixdorf) was formed in Europe to promote AT&T UNIX as a European common ground to counter the domination of that market by U.S. manufacturers. Portes (1998) identifies negative aspects of social capital, including closed/exclusionary groups and downward leveling norms. X/Open was a specific effort to leverage the negative social capital potential for exclusion of interested stakeholders, such as users or non-European manufacturers.

THE BANDWAGON

With the November 1985 ballot completed, the activity started to spawn additional groups to focus on other areas of systems standardization. The POSIX working group divided into the P1003.2 Shell and Tools working group, P1003.3 for Test Methods and ongoing P1003.1 API groups. Each of the individuals appointed to head up these working groups distinguished themselves in their technical and collaborative contributions in the initial work, and all facilitated the creation of the management and procedural structure for the committees. Here, individuals who spanned the boundaries between a technical group and the administrative group and were sought out for advice (stars) with acquired social capital, increased their standards influence by taking on leadership roles.

The POSIX working groups met four times a year for a week in diverse locations, with the first international meeting in the spring of 1986 in Florence, Italy. By this time, a core group of active individuals had spent 600 hours together, eating, traveling, socializing, arguing, collaborating, and successfully solving nontrivial problems. This established levels of social capital that quite possibly exceeded their loyalty to their employers. One indication of this is that the chairs of all of these committees changed companies at least once between 1984 and 1995; none gave up their participation or their leadership roles. The right to continued participation and retention of leadership roles is another characteristic of an individual membership model such as IEEE's (Table 3).

Keypoint #3: The value of the POSIX community and effort exceeded corporate loyalty in many situations. After the first handful of core players changed companies but not standards committees, it became evident that individuals were seeking opportunities to remain active in the work by changing employment or even self-funding when employers reduced support.

The proliferation of POSIX groups did not stop with test methods. P1003.4 focused on real-time interfaces (inherited from /usr/group), P1003.6; security also transferred from /usr/group but with a change of leadership to individuals from IBM and NIST. This change reflected the lack of established social capital on the part of the imported leadership (from/usr/group) and, quite likely, some level of distrust of an AT&T advocate/employee as a leader of this activity. Dokko and Rosenkopf (2004) find that centralization/influence is transferable between corporate representatives in SDOs. This case of P1003.6 indicates that a lack of this influence also may be transferable. The POSIX committees shifted away from having star experts chair a group for two reasons. If they expressed their opinions, it would create a possible conflict of interest, and they deprived the group of their expertise if they remained quiet.

P1003.5 focused on the Ada programming language with a U.S. Army major taking the lead. This level of procurement influencing user investment provided a critical credibility for the effort. However, the military used two-year duty rotations at the time, which limited the ability of the individuals to develop and apply social capital. Later, DoD put forward individuals drawn from civilian staff and often from the Pentagon, retaining the visibility of commitment while increasing the time frame for individuals to acquire and apply influence.

This is the point where organizational engagement in the work becomes evident through firms' willingness to sponsor leadership roles. Note that corporations might have different motivations for supporting individuals in leadership roles. Some may have a strong interest in promoting the work or their corporate image, others in influencing it in a specific direction, or others to delay progress in order to minimize possible conflict with their business strategies. In all cases, a combination of individual human capital, social capital, star status, and some inherited value of organizational affiliation factored into the leadership selections. To encourage corporate commitment, balance interests, and advance the work, organizational affiliation assumed an increasing influence in the appointment of committee officers, as the market impact of the work became apparent.

POSIX allowed other SDOs, trade associations, and professional organizations to participate though a concept of Institutional Representatives. These individuals often developed significant influence. This stemmed from their individual expertise and interactions in the activities, their organizational tie to significant expert communities, and, in some cases, the substantial technical contributions from these communities.

In 1986-1987, AT&T started to roll out their system V interface definitions (SVID)[2] and test suites. This created a tension between the private specifications of AT&T—SVID and ownership of the UNIX trademark—and the question of what the reference specification should be. DEC, with their BSD-based Ultrix, challenged a U.S. Air Force Command and Control (AFCAC) procurement for requiring SVID-based systems. While the challenge was not successful, it became clear that the stakes were becoming significant and that the government procurement process would benefit from adoption of procurement guidelines. NIST set the stage for a Federal Information Processing Standard (NIST, 1986), hoping to target it on a 1988 version of POSIX 1003.1.

In Europe, the X/Open group started to formalize their organization, with DEC as the first U.S.-headquartered company to join. The orientation of this group toward the AT&T implementation and DEC toward the BSD version was an additional deviation from the X/Open group norms. X/Open was able to obtain concurrence from AT&T to build their specifications on SVID and proceeded with that publication. X/Open used a high annual fee, along with concurrence of the current members as a means to maintain an exclusive membership, forming a major-vendor-only fraternity[3]. With the introduction of the 3B2 computer systems from AT&T, they also joined X/Open (as did HP, Sun, and IBM). While this matches the concept of participation in SDO/CTOs leading to corporate alliances (Rosenkopf, Metiu, & George, 2001), it is significant to note that the individuals participating in IEEE vs. X/Open tended to be quite different. The POSIX activity

Table 3. PASC/POSIX projects 1983-2000 (pasc.org, 2005)

PASC Project Number/year	Topic area	Initial Chair/Corporate relation	Results
P1003.0—1987 [Also ISO/IEC 14252]	Open Systems Guide	Al Hankinson, NIST Kevin Lewis, DEC	Published 1995
P1003.1—1983 [also ISO/IEC 9945-1]	C Language API	Jim Isaak, Charles River Data Systems; DEC	Trial Use 1986 Multiple Revisions
P1003.2-1986 [Also ISO/IEC 9945-2]	Shell and Tools Environment	Hal Jespersen, Unisoft; Amdahl, Sun Don Cragun; Microport; Sun	Full Use: 1991 Multiple Revisions
P1003.3—1986	Test methods for .1	Roger Martin, NIST; Sun	Full Use: 1996 (as 2003.1)
P1003.4—1987	Real Time API	Susan Corwin, Intel	Integrated into 1003.1 as 1003.1b [1993]
P1003.5—1987	Ada Bindings	Terry Fong, US Army	Full use 1992
P1003.6—1987	Security API	Jeanne Baccash, AT&T	Mirved[7] into P1003.1e and .2c
P1003.7—1989	Administration	Steve Carter, Belcore	Mirved into P1387 (3 parts)
P1003.8—1989	Networking API	Dave Dodge, Oracle	Mirved into P1003.1f
P1003.9—1989	Fortran Binding	Dan Magenheimer, & John Mc-Grory, HP	Full Use 1992
P1003.10—1989 [Also ISO/IEC 15287]	SuperComputing Profile	Karen Sheaffer, Sandia	Full use 1995
P1003.11—1989	Transaction Processing Profile	Elliot Brebner, Unisys	Withdrawn Draft available
P1003.12—1990	Network API	Les Wibberley,	Full use 2000
P1003.13—1990	Real Time Profile	Susan Corwin, Intel	Full use 1998 Multiple revisions
P1003.14—1990	Multiprocessor Profile	Bob Knighten, Encore	withdrawn
P1003.15—1990	Batch Services	Karen Sheaffer, Sandia	Full use 1995
P1003.16—1991	C Language Bind	Donn Terry, HP	Reverted into .1
P1003.17—1991	Directory Services	Robert Spade, Motorola	To 1224.2
P1003.18—1991	Traditional UNIX profile	Donn Terry, HP	withdrawn
P1003.19—1993	Fortran 90 binding		withdrawn
P1003.20—1993	Ada real time		Mirved into .5
P1003.21—1993	Real Time Comm.	Craig Meyers, SEI/CMU	withdrawn
P1003.22—1993	Security Framework	Lynne Ambuel, DoD	Withdrawn
P1003.23—1997	OSE Guide	Sandra Swearingen, USAF	1998
P1003.24—1996	Ada binding for X/Windows		withdrawn
P1003.25—1998	Fault tolerance	Helmet Roth, DISA	Mirved out of P1003.1h, withdrawn
P1003.25—2000	Device API's		
P1201.1—1989	Windowing	Sunil Metha, Convergent Tech-nologies	withdrawn
P1201.2—1989	GUI Usability	Lin Brown, Sun	withdrawn
P1224—1991	X400 API	Steve Trus, NIST	Full Use 1993
P1237—1990	RPC API	Ken Holiday	Moved to X3

Table 3. Continued

PASC Project Number/year	Topic area	Initial Chair/Corporate relation	Results
P1238.1—1990	OSI API	Kester Fong, NIST	Full Use 1994
P1238.2—1990	FTAM API		
P1251—1994	ACSE LIS		Full Use
P1253—1994	ACSC C language		Full Use
P1295.1—1992	MOTIF/GUI	John Hurd, DEC	Full use 1993
P1295.2—1992	Open Look/GUI	Dave Bealby, Sun	withdrawn
P1372—1994	Language Independent API	Jay Meyer, Unisys	Withdrawn
P1387—1992	Administration	Martin Kirk, X/Open	1995 (part 2)
P1494	National Profiles	Keld Simonsen, Danish Standards Expert	Withdrawn
P2000.1—1997	Y2k Terms	Kevin Lewis, DEC	Full Use 1998
P2003.2—1990	.2 Test methods	Ray Wilkes, Unisys	Full Use: 1996

built on a more renegade group of evangelists, where even X/Open's technical work tended to draw on mainstream engineering managers from corporate UNIX development groups.

The user community was taking a stronger interest and becoming concerned over the apparent divergence of the major vendors, in which one camp was moving ahead with SVID, while another was promoting BSD and standardization through POSIX. Note that users were not at the table in X/Open, and while they could be equal partners in the IEEE industry, users rarely invested their resources to build sufficient influence. U.S. government agencies were active in POSIX, particularly NIST and DoD. Individuals from user organizations had significant sway when they chose to engage. An individual from a user organization advocated the requirements for application and implementation conformance. He did not hold a leadership role or significant social capital within the group, but his perspective as a user was a significant factor in the adoption of those concepts. However, major users created their own organization in 1991—User Alliance for Open Systems (UAOS, 1991)—and few participated in POSIX on an individual level.

Keypoint #4: Many IT user organizations tended to relinquish formal standards participation to vendors or to "representative" users including government. At this time users created their own consortia rather than collaborating with vendors. The result was a lack of individuals in user organizations spanning the boundary into the SDOs and building the social capital needed to exert influence in the standards process.

The POSIX group sought compromise in two areas: elimination of differences between SVID and POSIX, and elimination of duplication between the X3 C language committee and the POSIX API. The final resolution was to eliminate as many SVID deviations as possible from POSIX and then introduce options that would permit SVID or BSD conformance (some options were mutually exclusive).

One place where divergence was not an option was in a data interchange format. SVID had a utility and related format called *cpio*, and BSD had a different one called *tar*. Each had legitimate advantages, and the distribution of software and data dictated a common facility. The Institutional Representative from Usenix resolved the Tar

Wars problem by promising to develop a public domain piece of software that would combine the functionality of both; in return, the divergent groups agreed to abandon their insistence on their existing format. The resulting utility/format was appropriately called *pax*. This change required a change in the distribution process for suppliers of software or data for use on POSIX-conforming systems, impacting the software duplication/manufacturing process, not just the suite of utilities provided with the system. Some individuals in this case put the community goal of interchange interoperability ahead of corporate interests. This is a level of individual discretion that would not have been feasible 10 years later.

ENTRANCE OF THE CORPORATE PLAYERS

The 1988 approval of the full use version of POSIX corresponded with FIPS 151 so that the U.S. Government procurement guidelines were firmly aligned with the standard. The FIPS selected from the POSIX options, reducing the viable range of variations. Harmonization activities within the WG15 group resulted in concurrent progression to draft international standard status. The model adopted at that level called for resolution of any significant international objections by means of a rapid amendment to the IEEE documents. To avoid divergence between the IEEE and ISO published documents, IEEE adopted the (European) A4 page size for the IEEE version. ISO typically reprinted IEEE standards in this format, creating the perception of divergence. IEEE built on its relationship with ISO, making this concession, and became the publisher for the standard for ISO.

At the international level, in theory, participation in technical committees and subcommittees is by nation state. Also, in theory, working group participants are chosen because of their expertise. Some countries, such as the UK and Japan, selected individuals from a university as their delegation chairs, often with corporate individuals as participants. The international span of POSIX resulted in significant overlap between the IEEE participation and the WG15 participation. The common core of individuals participating in the international travel for WG15 meetings as well as the POSIX meetings experienced an extended series of interactions, solving problems and building social capital. This became important in having national bodies defer some of their expectations for future action, trusting IEEE and the individuals involved to deliver these. The language-independent specification was a key example of such a deferral. Unlike most other transactions in POSIX, which involved individuals banking social capital by agreeing to the requests of others, here there existed a clear quid pro quo. U.S. participants agreed to replace the international versions of the C-language-specific POSIX documents in the future by a programming language-independent version. Ultimately, this obligation passed from the ISO national body level interactions through government/corporate levels down to individuals, who assumed the responsibility for pursuing this in IEEE.

X/Open started to participate formally in the POSIX process in 1988. At their first meeting, their lead individual called for a meeting of all of the employees of the X/Open firms and enjoined them to "vote the party line" on issues where X/Open had established a position. Those present soundly rejected this. It countered the individual nature of IEEE participation. From another perspective, this can be seen as X/Open and/or their spokesperson having not acquired sufficient social capital among that set of participants to carry out this course of action. In this same time frame, X/Open reopened a previous decision that the committee thought was resolved. They brought a clear technical argument for their recommendation, and the prior decision was reversed. This demonstrates how expertise was more important in POSIX decision making than formal title or authority.

Over the years, this changed, with X/Open becoming the primary venue for POSIX evolution. This was partially a result of X/Open establishing its reputation and that of the representative individuals within POSIX. It also reflected declining corporate support for POSIX, as firms stopped financing the participation of individual employees. In some cases, no individual participated; in others, corporate standards experts replaced previously active individuals; and in yet other cases, individuals continued by taking vacation and funding their own expenses. This personal funding situation was sufficiently common so that the SEC established a policy to waive the registration fee for such individuals.

Three major challenges facing the POSIX movement started to emerge. First, an API specification, even in combination with the C language standard, was not sufficiently rich to establish a viable operating environment for complex applications software. Second, the Macintosh, fledgling Microsoft Windows, and X-Windows system pointed to the value of a windowing-based environment. Third, the limitations of source code portability started to emerge; applications vendors did not want to implement, test, and support software on an unbounded set of platforms.

In 1987, Sun and AT&T announced a partnership (formalized later as Unix International), which raised significant concerns among other players about how the next set of reference interfaces might be manipulated to specific corporate advantage (or disadvantage.) HP and DEC initiated a counter-movement in 1988—the Open Software Foundation (OSF). The author was part of the OSF formation process. OSF sought to address the evident concerns of the user and applications community by both endorsing a specific suite of standards (or profile, as they were becoming known, a concept evolved out of the OSI world), including Motif for windowing; and also collaborating on engineering a specific implementation that would eliminate options and divergences in the code base. It is important to realize here that this was only source code compatibility, not binary compatibility, because OSF would be implemented on multiple incompatible processors.

To provide a windowing environment for applications, Sun started to promote Open Look. DEC and HP joined forces with their related work and proposed the Motif graphical user interface (GUI) environment. Both were higher-level sets of widgets based on the X-Windows system. Corporations started to take sides, splitting both the X/Open community (vendor-based) and the POSIX community (individuals, but with strong vendor alignment). For example, IBM and Siemens joined OSF, whereas ICL and Fujitsu joined Unix International. These GUI Wars strained the goodwill of participants, resulting in some meetings that relied more on *Roberts Rules* than consensus. Cargill (1997) suggests that this was common for the POSIX activities, but the norm was consensus through a process of "no objections." Standards creation calls for seeking consensus, which is a positive factor in developing social capital, resorting to *Roberts Rules* runs counter to that trust. Consortia such as X/Open and OSF could avoid either situation by selecting only participants who agreed prior to joining, using consensus by exclusion as a decision-making model. Ultimately, POSIX could not come to a resolution of the GUI problem within their accepted values, and the executive committee refused to adopt any resolution in spite of strong employer lobbying in many directions.

The OSF was both divisive and ultimately decisive in bringing together the diverging windowing approaches. IBM joined just prior to announcement, with the caveat that OSF adopt its operating system kernel. Others from the X/Open community joined, as well. The users who had viewed the BSD/SVID compromise as a frustration now faced an OSF/Sun-AT&T supplier split with the same concerns. Applications vendors found little solace in the hope that they might find the world reduced to two operating system bases on however-many processors.

X/Open sought a common-ground role and considered the promotion of a set of binary platform interfaces. Essentially, the selection was a group of specific processors and API profiles that would be used to define a limited number of binary targets. In theory, an applications supplier could provide a single product with multiple binary images, one for each in the set it chose to support. To solve this problem, a UK group proposed ANDF (Architecture Neutral Distribution Format) to define an intermediate code that could be targeted by all vendors; this processor-independent approach was similar to that of UCSD Pascal in the 1970s, Java in the 1990s (Egyedi, 2001), and .NET (ECMA, 2002) in 2002. ANDF lacked both a strong corporate sponsor and individuals with sufficiently established relationships to gain traction in either X/Open or POSIX.

These divisions and battle lines emerged five years after the standardization efforts started and concurrently with the establishment of major user demand (U.S. Government FIPS, GM, Boeing, et al.). Vendors had identified user interest in this open-systems concept but not really embraced the level of interoperability that the users sought. The IEEE and/usr/group models of individual participation were influenced strongly by the employers funding the individuals. However, it is significant that the individual relationships built up over the prior years often formed substantial ties that spanned vendor differences, holding the momentum of the activity through this era of discontent.

Chairs of the 1003 committees sat on the Sponsor Executive Committee (SEC)[4]; by 1990, the SEC became the focal point for areas of corporate engagement. Part of the challenge of managing the SEC, particularly in bringing new members to the table, was identifying a reasonable balance between various camps while assuring the SEC had the significant players at the table. A simple example was the appointment of administrative roles to engage targeted players, such as appointing an individual from IBM as logistics chair, at a time when an IBM employee did not hold another seat.

One drawback of the individual level commitment to the POSIX work is that these individuals tended to become evangelists in their own corporations. This could backfire in areas where other operating systems were considered a proprietary advantage. DEC is one example where this internal tension was fairly evident; although DEC did eventually provide a POSIX-compliant VMS, the Microsoft team (of ex-DEC engineers) that ultimately developed Windows NT rejected this path when developing a similar operating system at DEC. Where individual interests diverged strongly from vested corporate interests, the boundary-spanning stars from the SDOs may have alienated internal groups rather than influenced their innovation to align with the SDO's goals. At the same time, the UNIX commitment by individuals often was stronger than corporate loyalty, with individuals moving to new companies or starting up their own firms in this area. Later, Microsoft acquired one of these startups—Softway—to support NT with POSIX compliance.

THE TRANSITION TO CORPORATE ORIENTATION

The April 1990 meeting was the azimuth of POSIX's independence. Vendors realized that new committees formed to address niche markets could create specifications that users might blindly require. A group of vendors proposed a motion that sought to curtail the potentially unbounded expansion of POSIX projects. It established criteria for new projects and formalized concepts, such as existing practice, a base document, a realistic scope, a comparable level of acceptance, and an integration plan. These had previously been considerations but not constraints. This was an explicit outgrowth of corporate engagement in influence roles and the formation of alliances emerging from the participation. Groups like X/Open, OSF, and Unix International sought control over the evolution of the standards, and the open membership requirement for POSIX as well as its ANSI accredited standing interfered with this objective. Anybody could join and bring forward new projects that might be mandated by buyers who preferred accredited standards, such as the U.S. Government. This can be viewed as a formalization of downward leveling norms or exclusion, two of Portes' (1998) negative aspects of social capital.

The GUI War (Motif/Open-Look) reflected many of these elements. At the time of Jespersen's (1995) coverage of this, the war was barely over, but he does raise many of the key points. This was actually a battle of three parties, not two: the Motif camp, the Open Look camp, and the users. The strategy in the Motif camp was to advocate for multiple standards—define them, and let the market choose. In reality, there were users who asked for that option and implied they would select Motif. The apparent strategy of the Open Look camp was to delay standardization, often citing the concern of premature standardization or existing practice. It was my perception that the intent was to develop greater market share (on both sides) so the final standard would be an endorsement of the winner. However, user applications developers strongly urged the vendors to deliver a single target environment.

At the beginning of this battle, Windows was not a viable competitor in either terms of maturity or the capability of the PC platforms on which it ran; by 1995, this picture had changed, and UNIX vendors started to recognize the real competition. If POSIX had won even a small percentage of the hundreds of millions of windowing PCs sold after 1990, this would have expanded the UNIX market significantly. This could only have happened if POSIX systems had captured the high-end windowing applications before the Windows on Intel-compatible devices (Wintel) systems caught up. This parallels in time and antagonist the fall of NEC's PC-98 standard (West & Dedrick, 2000); it also reflects the same ultimate causes—a sufficient solution from Wintel and failure of the standards owner to respond in a timely manner. It is notable that no amount of technical rationale, market analysis, clearly articulated user demand, or vendor collaboration (much less, social capital) could bring these divergent camps together. This only happened after Windows 95 made the competitive threat undeniable.

In 1991, the SEC established two additional criteria for 1003 projects: Test Methods and Language Independent Specifications (LIS). NIST was the driving force behind testing. They were funding POSIX validation suites and chairing the system interface test methods work (P1003.3). The test method work identified ambiguities in the POSIX specifications as well as provided guidance for conformance testing that increased user confidence and facilitated vendor branding.

Academic interests involved in the ISO work led the language-independence effort. In part, this was a compromise in response to a competing effort in ISO (Open Distributed Processing [ODP]) as well as partly a level of concern about Japanese activities on the Systems Software Interface (SSI) (Takahashi & Tojo, 1993), which sought to span the full gambit of interfaces needed for applications (operating system, windowing, database, etc.). A few years later, the SEC abandoned these requirements, but not before a systems interface document was established in this form and an effort at a more formal "Z" specification was attempted.

Takahashi and Tojo (1993) document the type of pressure brought on the SSI work around 1988-1991. This visionary work sought to address the broad range of portability issues beyond the operating system interface. U.S. and European organizations were concerned that Ken Sakamura's TRON work might give Japan a technical lead in this area. Also, the Open Systems Interconnection (OSI) experience with academic (as opposed to existing implementation) interfaces and the overall scope of addressing a full range of interfaces raised questions. If users did not broadly adopt SSI, it would be a major waste of corporate resources in the standardization process; if it were successful, it would threaten the existing product differentiation. Major U.S. and European vendors were not prepared to launch a comprehensive drive for applications portability, either building on the POSIX activity or following the broader SSI perspective. Just considering the POSIX vendor costs provides some insight. At the April 1990 meeting, POSIX leadership tried to gauge the overall annual cost of the standards effort—an initial estimate of $15 million a year was doubled when an individual from IBM indicated that they were expending that much alone.

In June 1991, a community of users referred to as SOS had their CEOs send letters to their suppliers outlining the kind of open systems environment they expected to have available. Needless to say, when the CEO of DEC got 10 identical letters from CEOs of Fortune 500 companies that indicated a preference for a given computing environment, it got corporate attention. This environment paralleled the OSF suite, the collection of standards identified in 1003.0, the X/Open Open System Environment documents, and so forth (UAOS, 1991). A question asked cynically at the time (not to the user CEOs, as far as I know) was, "That's nice, but if I sell it for 20% less, will you abandon your request for standards conformance?" There is every indication that this question was answered in the affirmative, as those companies abandoned their formal standards-based ideals for PCs that sold for less, purchased directly by individual departments. The combined social capital of users and vendors alike, even with significant strategic incentives, were not enough to overcome the lure of lower list price systems with a sufficient core of applications.

The GUI War within POSIX came to an abrupt end when Sun announced that they would support Motif and join OSF (Cargill, 1997). Sun's leadership was probably painfully aware that the battle had only led to user alienation, and perhaps it could see the specter of Microsoft on the horizon, as well.

One project lost in this battle was an effort to improve drivability, applying human factor analysis to improve the consistency between Graphical User Interfaces, particularly to provide a bridge between Motif and Open Look. Ironically, if this effort had completed with the appropriate adoption and mandates from organizations like NIST and the European Union akin to those for POSIX, it would have provided a basis for ease-of-use spanning UNIX, Mac, and Windows systems, greatly reducing today's user frustration with divergent applications interface conventions But by this point, the GUI Wars had so dissipated the trust between POSIX standardization participants that this effort never received enough time or energy to be successful. Also, the human factors expertise to define GUI drivability differed from that of the mainline POSIX participants, which also reduced their level of commitment to champion the work and attract a broad base of experts.

The P1003.0 project, open systems environment, reflected the need for a more comprehensive suite of capabilities than just the operating system. The focus was on both the question of what defines an open system environment and a guide for users on how to describe it. This group worked for some time on a definition, which continues to show up beyond the POSIX arena. The group also tried to develop a reference model that followed the OSI concept (and was influenced in part by the Japanese SSI effort). Perhaps a major value of this effort was to provide a forum for dialog among representatives of major user organizations, although some of the engineer participants suggested that it provided a place for the "suits" to go and get them out of the technical meetings. This referred to both the marketing people and the end users—again, a form of social capital's potential for exclusion/segregation and a failure of these engineers to recognize the need for boundary-spanning individuals.

MIGRATION TO CORPORATE FOCUS

A gradual decline in POSIX participation became evident, as vendors consolidated and X/Open started to become the forum for the limited technical evolution of the standards (Walli, 1995). POSIX retained an ongoing role in complementary work: threads, the Ada, FORTRAN bindings, and even the language-independent specification. The SEC denied new projects the 1003 and POSIX designations. The hurdles established by the SEC for POSIX projects, such as language independence and "comparable level of acceptance," discouraged individuals from bringing forward new work.

NIST abandoned its role in IT standards rather abruptly with a change of management. This affected leadership involvement in POSIX but also the perception (and perhaps reality) of a declining U.S. Government commitment to open systems. The new government mantra was "commercial off the shelf" (COTS), which, in most cases, amounted to a policy to buy Wintel systems. DoD remained the strongest user or government voice in POSIX. While DoD was an advocate of COTS, it also recognized that mission-critical systems, secure systems, and other objectives were better suited to the POSIX environments. In terms of social capital, DoD retained strong connections to the activity and strong influence, as well. NIST pulled back on its engagement and lost much of its influence after many of its key representatives took jobs in private industry, taking their social capital along with them. Within the POSIX community, NIST lost significant trust and respect as result of its withdrawal, something that may have reduced its influence in overlapping forums.

The Layered POSIX implementations emerged in this time frame, as well. DEC announced OpenVMS with a POSIX-conforming implementation. IBM provided a similar capability in MVS. Even Microsoft delivered a level of POSIX conformance with Windows NT. This version of NT did not support multiple user IDs, which triggered the POSIX interpretation process, a formal request to determine what the standard required. NIST tried to find out if a single-user system was a legitimate implementation after finding out that they were unable to run their validation tests on NT for interfaces that expected multiple IDs. The POSIX interpretations process evaluated the standard and determined that it did not require support for multiple user Ids, and NT was certified. It would have been straightforward to amend the standard to require some minimum number of users. This minimally would have added some additional documentation for each implementation. The vendors' desire to not change even documentation outweighed the value of establishing an advantage over Microsoft.[5]

Collaboration between POSIX and X/Open led to an agreement to develop the POSIX component of X/Open via a joint and, eventually, X/Open-hosted forum. The elements of POSIX that were not in X/Open's environment remained in IEEE but with limited support. The one exception in part was the DoD support for real-time-related extensions that it targeted for mission-critical applications. DoD ran an extensive evaluation and decision process to select between real-time POSIX and other systems for the Joint Strike Fighter project. The network connections of individual participants involved in both the POSIX and DoD decision process were a key factor in this decision.

As POSIX-based products were maturing, vendors moved from "no additions" to "no substantive changes" to "we don't want to even change the documentation" constraints on the evolution of the work. Clearly, this put a cap on changes of the core standards to respond to new requirements. The tsunami of Microsoft Windows swamped much of the low-end market and washed right up into the mid-range systems. At the end of the 1990s, the traditional vendors were cutting back with major layoffs and cost reductions. Standards activities were cut back in IBM, DEC, HP, and

NIST. Bucking the trend were Sun and Microsoft, who hired experienced standard individuals, but they often were working outside of the POSIX forums. In 1998, vendors formed the Austin Group to coordinate the evolution of the Single UNIX Specification, providing a forum for the individual (POSIX) and Corporate (X/Open) involvement and publishing a single resulting document.

The other significant events impacting the viability of the POSIX work that emerged in this time frame were viable windows systems from Microsoft and the GNU/Linux[6] open-source implementation of POSIX (Linux International, 2005). The failure to deliver windowing with POSIX in the late 1980s drove applications suppliers to the single implementation platform (operating system and microprocessor) Microsoft environment. Lower prices and applications availability also shifted demand to Windows. This triggered a downward spiral in the demand for and evolution of POSIX. The 1990s were the infancy for Linux with limited commercial visibility, but the sizable market share this UNIX clone gained in the Web server segment reflects advantages that POSIX held over Windows NT in the 1990s in terms of multitasking, multi-user security concepts, and communications interfaces.

APPLICATION PORTABILITY STANDARDS: THE REAL SOCIAL BENEFIT

The opportunity to meet the POSIX objective of applications portability surfaced again, as Sun put forward their Java language. As noted before, real portability needed to address binary compatibility. Prior attempts—UCSD Pascal and ANDF—did not have the requisite processor power. POSIX refused to pursue Java standardization without Sun advocating for the project. Sun chose a more closely controlled model for promoting that work (Egyedi, 2001), a methodology that might not match guidelines for volunteer consensus standards as specified by U.S. Government guidelines. Java was unrivaled for six years, until Microsoft brought forward .NET. However, Microsoft concurrently introduced .NET and formal standards for the C# language and the .NET common execution environment through ECMA and ISO (ECMA, 2002). As suggested by Rao and Drazin (2002), recruiting talent from rivals can yield innovation and also the transfer of social capital influence (Dokko & Rosenkopf, 2004). Microsoft's acquisition of a significant number of DEC's standards participants demonstrates this transfer of capital. These individuals facilitated the adoption of .NET by ECMA. At this point, it would appear that the .NET environment provided the standards that may have been vendor-independent and processor-independent and would provide applications portability at a binary-like level.

At the same time, Linux, with its open source advantage and wide use in key markets, is now the flagship for the evolution of UNIX-based applications portability. There is little interest in using Linux to drive the evolution of the formal POSIX specifications. One reason is that the Linux community is organized around individual experts, while POSIX has become a corporate-dominated activity.

CONCLUSION AND AREAS FOR FUTURE WORK

The POSIX experience appears to confirm a number of the analytical findings related to individuals and cooperative technical organizations. The leadership and major players in the POSIX activity were boundary-spanning stars spanning both the corporate/SDO worlds and the technical/administrative ones. These individuals acquired significant social capital. This contributed substantially to their influence within this activity and, in some cases, spanned into related activities (X/Open, ISO, and other forums). The IEEE Computer Society Hans Karlsson award for outstanding leadership and achievement through collaboration was given to two POSIX leaders—Jim Isaak and Roger Martin—acknowledging their success in building and applying social capital to achieve standardization goals.

Individual committee members often changed jobs from one employer to the next. In most cases, as anticipated by Dokko and Rosenkopf (2004), this benefited the hiring firm but did not negatively impact the prior employer. There is some indication that mobility did benefit innovation, as predicted by Rosenkopf, Metiu, and George (2001) and Tushman (1977). Expanding the metrics on innovation to include spinout companies and new products and not just patent applications could make this more apparent, although such data are harder to obtain.

The role of these individuals in terms of both individual and firm influence and centrality, as predicted by Dokko and Rosenkopf (2004) and Rosenkopf, Metiu, and George (2001) appears to be very strong. User firms and other organizations seeking to influence the standards processes need to recognize this and invest the appropriate person-hours in the SDO efforts needed to gain the desired influence.

Some analysis may overlook key influence roles within the standards process. For responsibilities such as interpretations, ballot resolution, chairing a project, or serving as a technical editor, the discretionary decisions made by key individuals shaped the outcome of the standardization process, and thus, it mattered who served in those roles. In IEEE, the final approval voting is independent of meeting participation, which also affects evaluation of centrality and influence within this type of forum, since effective influence must touch both on those attending and the potentially much larger group of individuals who vote but never attend a meeting.

The POSIX perspective shows the following:

- Collaborative problem solving and developing respect as well as trust are factors in social capital formation.
- Long-term, repeated, individual problem-solving interactions facilitate the creation of significant social capital.
- Drawing on social capital to initiate the resolution of a subsequent project of common interest expands the level of respect and trust, creating a positive feedback loop for social capital development.
- Individual roles (customer, government) and expertise can have a significant influence on technical standards decisions independent from social capital effects.
- Individuals resist violating the trust/respect norms and depleting social capital within a community, as with X/Open's suggestion that participants vote along party lines or get directly involved in the GUI Wars.

Opportunities for further research include the following:

- What is the impact of consensus decision-making methods on social capital development?
- Is the high level of social capital in the POSIX seen in other SDOs that use individual membership? Would we see the same level of social capital in SDOs organized around corporate membership? What are the similarities and differences between these two types of organizations in how social capital is created and used?
- Does the ability to build social capital diminish as technology-driven standards organizations mature into more stable,

corporate-/market-dominated activities? Or does the shift to market-driven standardization reflect a permanent change in how industry views standardization?

- Are firms actually able to retain social capital after losing key boundary-spanning stars? Further evaluation of individual membership SDOs could show the ways in which social capital is retained by individuals who remain active compared to their prior employer.
- How do individuals use the social capital developed through SDO participation in other contexts? For example, were the POSIX veterans more effective in their work in Internet-related SDOs than those without a common background?
- It has been suggested that social capital and human capital reduce firm dissolution, but a significant number of the firms involved with POSIX seem to have undergone dissolution. Was such dissolution in spite of their social capital or due to effects of negative social capital?
- Future researchers might examine the role of social capital in comparable standardization efforts addressing portability, which did not complete the adoption process such as the Japanese SSI and UK ODP initiatives. Rejection of these has at least the appearance of exclusion and downward leveling norms in combination with a lack of established individual influence in key forums. At the same time, any such study must consider not only the organization of these standardization efforts but also the economic interests of corporations who fund participation at the international standards level and who would have faced the costs of implementation, should these efforts have become economically significant

REFERENCES

Adler, P., & Kwon, S. (2002). Social capital: Prospects for a new concept. *Academy of Management Review, 27*(1) 17-40.

Cargill, C. (1994). Evolution and revolution in open systems. *StandardsView, 2*(1), 3-13.

Cargill, C. (1997). Section 2: Sun and standardization wars. *StandardsView, 5*(4), 133-135.

Coleman, J. (1988). Social capital in the creation of human capital. *The American Journal of Sociology, 94*, S95-s120.

Dokko, G., & Rosenkopf, L. (2004). *Social capital for hire? Mobility of technical professionals and firm influence in wireless standards committees* [work in process].

ECMA. (2002a). ECMA 334 C# language specification. Retrieved September 19, 2005, from *http://www.ecma-international.org*

ECMA. (2002b). ECMA 335 common language infrastructure (CLI). Retrieved September 19, 2005, from *http://www.ecma-international. org*

Egyedi, T. M. (2001). Why Java was—not—standardized twice. *Computer Standards & Interfaces, 23,* 253-265.

Hurd, J., & Isaak, J. (2005). IT standardization: The billion dollar strategy. *International Journal of IT Standards and Standardization Research, 3*(1), 68-74.

IEEE. (1986). *IEEE 1003 trial use standard—Portable operating system for computer environments.* New York: IEEE.

IEEE. (2005). *IEEE procedures, project status and history.* Retrieved September 19, 2005, from http://standards.IEEE.org

Isaak, J. (1998). The role of government in IT standards. *IEEE Computer, 31*(12), 129,132.

Isaak, J. (2005). POSIX—Inside: A case study. *Proceedings of the 38th Annual Hawaii International Conference on System Sciences*, Waikoloa, Hawaii.

Jespersen, H. (1995). POSIX retrospective. *StandardView, 3*(1), 2-10.

Krechmer, K. (2001). The need for openness in standards. *IEEE Computer, 34*(6), 100-101.

Linux International. (2005). *Linux history.* Retrieved September 19, 2005, from http://www.li.org/linuxhistory.php

National technology transfer and advancement act of 1995. (1996). Pub. L. No. 104-113, 110 Stat. 775. Retrieved September 19, 2005, from http://standards.gov/standards_gov/index.cfm?do=documents.NTTAA

NIST. (1986). FIPS 151: Proposed federal information processing standard for UNIX operating system derived environments. *Federal Register, 51*(168), 90896-30897.

OMB. (1998). *OMB circular A-119.* Federal participation in the development and use of voluntary consensus standards and in conformity assessment activities. Retrieved September 19, 2005, from http://www.eh.doe.gov/techstds.overview/omba119.pdf

PASC. (2005). *PASC operating procedures and history.* Retrieved September 19, 2005, from http://www.pasc.org/plato

Portes, A. (1998). Social capital: Its origins and applications in modern sociology. *Annual Review of Sociology, 24*, 1-24.

Putnum, R. (2000). *Bowling alone.* New York: Simon & Schuster.

Rao, H., & Drazin, R. (2002). Overcoming resource constraints on product innovation by recruiting talent from rivals: A study of the mutual fund industry 1986-94. *Academy of Management Journal, 43*(3), 491-507.

Rosenkopf, L, Metiu. A., & George, V. P. (2001). From the bottom up? Technical committee activity and alliance formation. *Administrative Science Quarterly, 46*, 748-772.

Rosenkopf, L., & Tushman, M. L. (1998). The co-evolution of community networks and technology: Lessons from the flight simulation industry. *Industry & Corporate Change, 7*(2), 311-346.

Takahashi, S., & Tojo, A. (1993). The SSI story: What it is and how it was stalled and eliminated in the international standards arena. *Computer Standards and Interfaces, 15*(3), 523-538.

Tushman, M. (1977). Special boundary roles in the innovation process. *Administrative Science Quarterly, 22*(4), 587-605.

UAOS. (1991, January 27). User alliance for open systems. *Overcoming Barrier to Open Systems Information Technology.* McLean, VA.

Unter, B. (1996). The importance of standards to HP's competitive business strategy. *ASTM Standardization News, 24*(12), 13-17.

Wade, J. (1995). Dynamics of organizational communities and technological bandwagons: An empirical investigation of community evolution in the microprocessor market. *Strategic Management Journal, 16*, 111-133.

Walli, S. (1995). The POSIX family of standards. *StandardView, 3*(1), 11-17.

West, J., & Dedrick, J. (2000). Innovation and control in standards architectures: The rise and fall of Japan's PC-98. *Information Systems Research, 11*(2), 197-216.

ENDNOTES

1. A lapel button of this era was "Sex, Drugs and UNIX. "A few years later, it was replaced by "Condoms, Aspirin and POSIX."

2 The numbering sequence for UNIX is Version 6, Version 7, System III, System V (and internally, the next one was called Plan 9). Indications were that Microsoft adoption of Version 7 included a constraint that there would not be a Version 8, a limitation resolved by changing the numbering system.

3 A four-hour, six-course lunch at the Henry VI pavilion near Paris or a few days in a villa in Tuscany was a sharp contrast in meeting venues and style between the X/Open Board and the IEEE's "yet another Marriott Buffet" operating models. Most of the X/Open Board members had titles like "director" or "Vice President" and tended to be from marketing, not technical, organizations, and to my recollection, all were male. Venue selection was a deliberate affectation to match the desired level of titled participants. The technical substance of X/Open was handled by the Technical Managers group, not the Board, and they held quite different meetings and activities, such as whitewater rafting under the midnight sun.

4 The group developing POSIX standards was actually called the Portable Applications Standards Committee (PASC). To simplify the discussion, I refer to it as the "POSIX committee" throughout the article. This was formalized when the SEC was formed in 1986 and additional working groups were formed.

5 NT demonstrated an instance of the "wierdnix" model, a concept developed in 1986 by the co-chair of the P1003.1. A wierdnix contest was actually held in 1986 for the most pathological conforming implementation concept as a way to identify gaps in the specifications.

6 GNU is the environment developed by the Free Software Foundation (FSF) that provided both a foundation for the creation of Linux and the relevant software license. Individuals from the FSF participated in the POSIX effort early on, and later, FSF acquired formal standing in the process as an Institutional Representative.

7 "Mirved" is a term suggesting the concept of nuclear missiles with multiple warheads in a single nose cone; the term was introduced as we tried to combine multiple standards projects and groups back into their logical documents. In many cases, this consolidated things; in some cases, it generated new project numbers.

This work was previously published in The Journal of IT Standards and Standardization Research, 4(1), edited by K. Jakobs, pp. 1-23, copyright 2006 by IGI Publishing, formerly known as Idea Group Publishing (an imprint of IGI Global).

Chapter VI
Linguistic Qualities of International Standards

Hans Teichmann
Erasmus University, The Netherlands

Henk J. de Vries
Erasmus University, The Netherlands

Albert Feilzer
Academic Centre for Dentistry, The Netherlands

ABSTRACT

Linguistic qualities are essential for the fitness for use of every standard. The intentions of the standards developers should become perfectly clear to those who will finally use the documents, but language barriers at several project stages may hinder this. This chapter addresses the topic for standards at the global and regional levels using a case study about the linguistic qualities of the standards published by the IEC (International Electrotechnical Commission). Most IEC standards are bilingual (English and French), and they are frequently translated into national languages. Feedback on standards use, translation practices, and user satisfaction has been obtained by means of two questionnaires sent to the IEC national committees (NCs). These data are assessed with respect to the language skills of the technical experts concerned, the particular linguistic aspects of the standards, the process of standards development, national translating practices, and standards user satisfaction. Standards development in two languages adds to their fitness for use, but this advantage should be balanced against the cost of bilingualism. The current practice satisfies more or less all parties involved; nevertheless, some improvements can be suggested. The issue of bilingualism vs. unilingualism also has an important cultural and political dimension.

INTRODUCTION

Obviously, standards are developed in order to be read and used. They should therefore be fit for use. The main element of this fitness for use concerns the technical contents: Do the specified provisions meet the user's requirements? As a technical standard is necessarily a written document, its linguistic qualities are directly associated with its fitness for use. In the case of international standards (ISs), the number of user complaints concerning language equals the number of complaints about technical content (Aben, 2002). Standards, but also other technical documents such as specifications and instructions for use, are often available in more than one language. In this context, the quality of the translation is also of importance for the evaluation of fitness for use.

The linguistic qualities of technical and scientific texts are hardly studied in a systematic manner (Nuffield Languages Inquiry, 2000). Among the rare exceptions are Kocourek (1982), Sager, Dungworth, and McDonald (1980), and Von Hahn (1983). The equally limited literature on technical and scientific translating includes Jumpelt (1961), Maillot (1981), Pinchuck (1977), and Wright (2004). On the other hand, the literature on standardization in general does not cover issues of linguistic qualities (e.g., Ailleret, 1982; de Vries, 1999; Diesken & Hoffmann, 1992; Egyedi, 1996; Fomin, 2001; Higgins, 2005; International Electrotechnical Commission [IEC], 2006; Winckler, 1994). Similarly, the general literature on applied linguistics (which includes translating) appears to neglect the field of international standardization (e.g., Beardsmore, 1986; Catford, 1980; Edwards, 1994; Hatim, 2001; Newmark, 1986; Nida, 2001; Nord, 2005; Vinay & Darbelnet, 1995; Wilss, 1982).

Technical and scientific translating involves much more than the mechanical looking up of equivalents of special terms in dictionaries. It requires a good knowledge of the source language (SL) and the target language (TL), a good general education, the knowledge of any text-specific conventions if applicable, and a solid technical and scientific background. The inappropriate use of calques should be avoided (A.6 in Appendix A). According to Maillot (1981), the difficulties of technical translating are frequently underestimated. Kocourek (1982), for instance, claims that technical translating is not a very difficult task. A study on language issues specific for the domain of standardization is the publication of Teichmann (2003b) on translating texts of the IEC; we shall refer to this publication in the present study.

This chapter concerns the fitness for use of standards as far as their linguistic aspects are concerned. In its most basic form, a standard is developed and subsequently used (see Figure 1). The standard's developers have to use both natural language and nonlinguistic means (e.g., formulae, drawings, and figures) to express the intended technical contents (International Organization for Standardization/International Electrotechnical Commission [ISO/IEC], 2004c; Sager et al., 1980). The user has to interpret the final text and hopefully fully understands what the standard's writers had intended.

By far, most standards are developed in an international setting where the English language dominates, though many of the participants will not have that language as their mother tongue. However, English is not the only major world

Figure 1. Untranslated monolingual standards

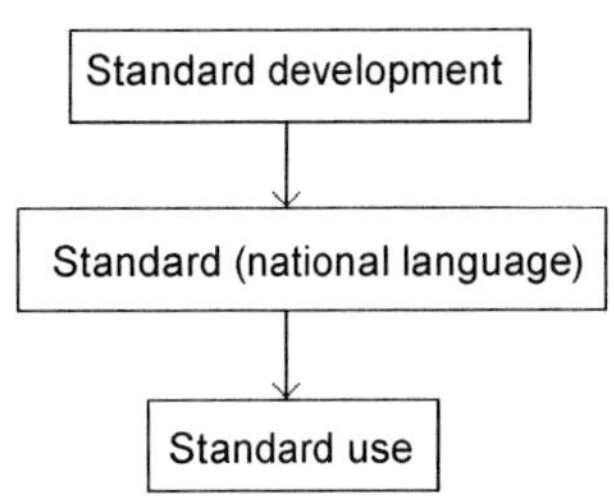

Figure 2. Translated monolingual standards

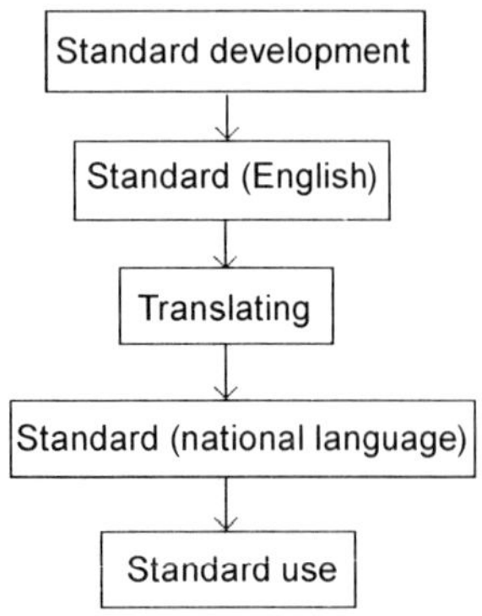

Figure 3. Translated multilingual standards

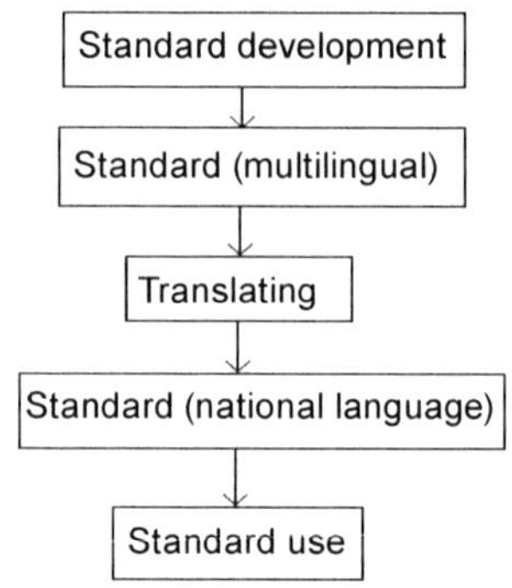

language (Nuffield Languages Inquiry, 2000). Some figures, in millions of first-language speakers, are 1,113 for Chinese; 372 for English; 316 for Hindi or Urdu; 304 for Spanish; and 201 for Arabic. It is therefore not self-evident that the standard users will have sufficient language skills in English (or another official language) to be able to interpret the standards fully and correctly. In such cases, national translations of the standards may be required (see Figure 2).

The main standardization organizations at the global level are the ISO, the IEC, and the Standardization Sector ITU-T of the International Telecommunication Union (ITU). ISO and IEC standards are developed in English and French. Russian is the third official language of both organizations, but in practice, the Russian versions are prepared only after the English and French versions have been published. IEC and ISO use English (standard British English) and French (standard French), while the ITU has a different language policy (A.9 in Appendix A). Of the three main standardization bodies at the European level, the European Telecommunication Standards Institute (ETSI) uses English only. The other European organizations, that is, the Comité Européen de Normalisation (European Committee for Standardization, CEN) and the Comité Européen de Normalisation Electrotechnique (European Committee for Electrotechnical Standardization, CENELEC), publish their standards in three languages: English, French, and German. Again, translations into other national languages may be prepared as required (see Figure 3). This applies in particular to those ISO, IEC, CEN, and CENELEC standards that are implemented as national standards.

In this chapter, we intend to identify possible gaps between the standards developers' intentions and the final versions of the international standards, and to investigate their origins. Therefore, we have to take into account the language skills of technical experts, the process of standards development, the most important linguistic aspects of the standards (including major potential shortcomings), the process of translating them into a national language (including certain potential problems), and the perceptions of the standards users. The aim of the chapter is to discover the essential elements affecting the linguistic qualities of the ISs by examining the chain from standards development (see Figure 4) to standards use.

International standards should define clear and unambiguous provisions in order to solve matching problems for the business community (De Vries, 1999). To achieve this objective, the standards shall be:

- Consistent, clear, and accurate
- Comprehensible also to qualified persons who have not participated in their preparation
- Fit for application and adoption without change as regional or national standards
- Fit for use also by experts with mother tongues other than English or French (or, in the case of European standards, also German)

Adequate language quality does not necessarily mean perfection. In fact, adequacy is determined by three mutually conflicting factors:

- Precision of the texts
- Intelligibility of the texts
- Timely development of the standards

RESEARCH DESIGN

Because the practices of the different standardization organizations vary only to some degree, one of them can be taken as a case. IEC was chosen for the following reasons.

- An organization working at the global (rather than only the European) level is preferred in order to avoid typical European restrictions concerning, for instance, politics (European Union) or culture.
- It provides us with the opportunity to better build on our own previous research, which also had the IEC as a case (Teichmann, 2000a; 2000b; 2003a; 2003b, 2006).
- The first author has many years of working experience at the IEC central office (CO).
- IEC standards are prepared and used mainly by technical experts. Their language skills vary within a wide range and constitute the starting point for this study. In the first section, different categories of foreign language capabilities are distinguished.

In the second section, specific linguistic aspects of the IEC standards themselves are discussed. As this topic has already been addressed in a previous study (Teichmann, 2003b), some of its main findings are summarized.

Subsequently, the procedure of how the IEC develops its standards (usually bilingual) is described. The contribution of its CO to their linguistic qualities is presented. This section is based on the ISO/IEC directives (ISO/IEC, 2004b), an internal IEC document (IEC, 2003), and on personal experience of the first author.

The next step concerns optional translations into national languages. As this practice depends on the individual NCs, a questionnaire to the 63 IEC NCs was circulated in order to gather relevant data. These national organizations are in most countries also the ISO member bodies (De Vries, 1999).

Thereafter, the other side of the process is investigated: the users' satisfaction with the standards. For this purpose, preferably the users themselves should be consulted. The International Federation for the Application of Standards (IFAN) has already carried out such a survey among standards users (Aben, 2002). For this purpose, a Web questionnaire was filled out by 288 respondents from 52 countries, 94% of which use ISO standards, 45% IEC standards, and 11% other international standards. 188 replies (65%) came from 25 European countries; the remaining 100 replies (35%) came from 3 African, 8 American, and 14 Asian countries, as well as from 2 countries of the South Pacific region. On the question "What are your greatest problems in using international standards?" 89 respondents (30%) mentioned the technical content and 82 (28%) the language. As there was no further specification of the term *language*, it is not clear to which extent the language quality of the international standards, the quality of national translations, or the lack of such a translation was concerned. There may be millions of standard users all over the world, in industry, scientific institutions, and many other types of organizations, making it almost impossible to create a representative random sample of users that are willing and able to answer specific questions on linguistic issues. Therefore, it was decided to investigate the users' experience in an indirect way by putting the questions to the NCs that represent them (first questionnaire). A second questionnaire was sent only to 18 NCs that had indicated their interest in the project. The use of two separate questionnaires made it possible to have a short and straightforward first questionnaire, thereby enhancing the response rate, which turned out to be 49% (31 of the 63 NCs). The second questionnaire was more complex and sent only to the NCs with a certain interest and expertise in specific issues of linguistic nature. Moreover, it was possible to streamline the second questionnaire in the light of the first batch of responses received. In this case, we received 14 filled-in questionnaires out of 18 (78%).

The questionnaires were also used for a third category of questions concerning the perceived language quality of bilingual IEC standards based on the feedback by translators and end users.

Finally, a number of results from the case study are listed together with recommendations aiming at the improvement of the linguistic qualities of international standards.

THE IEC CASE

Language Skills of Technical Experts

IEC standards are prepared by technical experts, and the main user group also consists of technical experts. Other users of IEC standards include executives who refer to them in contracts, governments that refer to them in legislation, lawyers who have to settle disputes in commerce and industry, and NGOs (nongovernmental organizations). The concept of functional bilingualism covers both the productive and the receptive language skills of a language user (A.7; Beardsmore, 1986). Based on this principle, we distinguish in the context of IEC standards five particularly relevant types of users and contributors (Teichmann, 2000a).

1. People without receptive skills (reading and listening comprehension) and without productive skills (oral and written production capability) in English or French. They have to rely on translated IEC texts. Participation in IEC work on the international level, that is, at committee meetings and in project teams (PTs) is not possible.
2. People with reading comprehension of English or French (but without listening comprehension in English and without productive skills in English). These experts are able use the English and French IEC standards, but active participation in IEC work on the international level is still not possible.
3. People with full receptive skills in English (that is, with both listening and reading comprehension), but without productive skills in English. These engineers are able to use the standards and may attend committee meetings for information. However, they may not serve as PT members. In other words, active participation in IEC work on the international level is not yet possible.
4. People with full receptive skills in English (that is, with both listening and reading comprehension) and with written (but without oral) productive skills in English. They are able to use the standards. They also may attend committee meetings for information and make written contributions. Participation in PTs as a corresponding member is also possible.

5. People with full receptive skills in English (that is, with both listening and reading comprehension) and with full productive skills in English (that is, for both oral and written production). These experts are the most suited contributors to IEC work on several levels: technical committees (TCs), subcommittees (SCs), PTs, and editing committees (A.8). This group includes, but is of course not limited to, the native English speakers and bilingual native French speakers. It also includes the secretaries of the TCs and SCs as well as those experts whose mother tongue is other than English but who are qualified for translating IEC standards into their national language.

Linguistic Aspects of IEC Standards

In the context of IEC standards, besides the application of the rules for drafting (ISO/IEC, 2004c), five linguistic aspects are particularly important (Teichmann, 2000b).

Contributors with Native Language other than English

In the context of IEC texts, the starting point is in most cases a new work item proposal (NP) written in English only by people whose mother tongue is frequently other than English and who are technical experts rather than language experts.

Where IEC documents were prepared by technical experts with mother tongue other than English (French), it sometimes requires effort on the part of the reader to understand the text properly. Individual intrusions may be responsible for the lack of intelligibility. In particular, errors related to polysemy (A.1), faux amis (A.2), paronyms (A.3), calques (A.6), incorrect language structure, and incorrect style may occur. Users with limited receptive language skills in English or French (A.7) may find it particularly difficult to deal with problems like ambiguous sentences, very formal style, and overly long sentences. Similar facts were reported by Daoud (1991).

IEC's English and French Bilingualism

Most ISs are bilingual, written in English and French. Both versions have equal legal status (ISO/IEC, 2004b).

Modal Auxiliaries

Modal auxiliaries (A.5) have to be used in a correct and consistent manner. Otherwise, the English and the French texts will not be equivalent. The same principle applies, of course, also to the ISs translated into other national languages.

Situational Dimensions of IEC Standards

House (1977) has developed a model for assessing translation quality. In two previous papers (Teichmann, 2000b, 2003b), House's concept of "situational dimensions" was applied to the textual characteristics of IEC standards. In the present context of standard development and use, three of these dimensions are particularly relevant: social attitude, province, and modality.

The dimension of social attitude describes the degree of social distance or proximity of the text. It was found that, in particular for technical experts working in a foreign language, it can be very difficult to apply the formal style appropriately as educated native speakers do. The most common problem involves formulating sentences with the wrong level of abstraction or the wrong style. Incorrect language structure may also be a problem.

The dimension of province reflects an occupational and professional activity, as well as the topic of the text in its sense of the area of operation. Relevant shortcomings may concern, in particular, the use of standardized technical terminology; technical quantities, units, and their symbols; standardized graphical symbols; mathematical formulae; as well as technical diagrams, tables, and figures.

Modality is an additional situational dimension (not used by House) that refers to differences in the form and medium of communication such as the differences between a report, an essay, or a letter. In the case of IEC standards, numerous conventions are prescribed in Part 2 of the directives (ISO/IEC, 2004c). They have to be implemented in a systematic manner. On the other hand, translators of IEC standards into languages other than English or French face a special problem if Part 2 is not available in their target language: Without a sanctioned terminology, it is very difficult to obtain consistency within a given text, and even more so between several texts of the same series.

Figure 4. Development stages of IEC standards

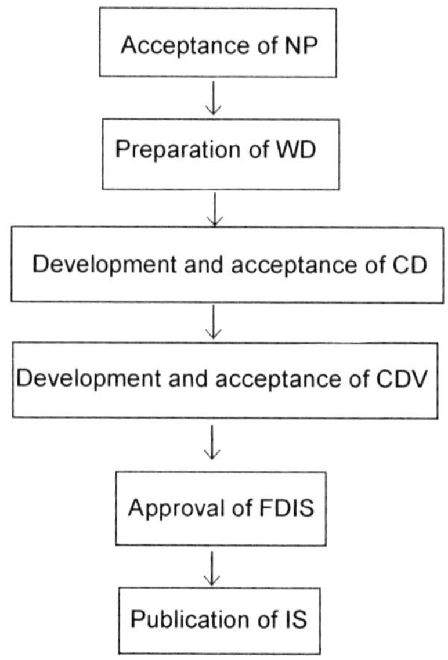

Horizontal Standards (A.4)

For the users of IEC standards, the consistent application of a unified terminology (as well as unified quantities and units, and of unified graphical symbols) is an important criterion by which even the overall quality of the standards may be judged.

Language Issues During the Development Process of IEC Standards

In IEC's decentralized system, most TC and SC secretaries and chairmen are technical experts from industry and scientific institutions; the others are mostly staff members of the NCs. The NCs appoint the secretaries. The IEC CO provides general managerial support to these committees, as well as the editing of the final draft of the international standards (FDIS) during the approval stage. The rules for drafting are always applicable. Editing by the CO is carried out according to its list for checking and editing (IEC, 2003). A close look at the numerous elements of the language used shows, however, that some of these elements are covered neither by the rules for drafting nor by the list for checking and editing.

The NCs that participate in a TC or SC are requested to submit their comments: technical comments on the committee drafts (CDs) and the enquiry drafts (CDVs), as well as editorial comments on the CDs, CDVs, and FDIS. These contributions commonly contribute considerably to the technical and linguistic improvement of the projects as they move through the subsequent stages. Editing committees are an efficient means for achieving adequate language quality (A.8).

The development of IEC standards uses a project approach that stipulates that all projects move through six project stages (ISO/IEC, 2004b; see Figure 4).

Proposal Stage

Proposed new work generally originates from industry via an NC. It is communicated to the members of the appropriate TC or SC accompanied by a voting form. The linguistic quality of the original NPs (always in English) is sometimes rather poor. The CO engineer in charge should make sure that no confusing NPs are circulated.

Preparatory Stage

During this stage, a working draft (WD, in English only) is prepared, generally by the project leader within a project team. Sometimes, the linguistic quality of the WD is rather poor.

Committee Stage

At this point, the document is submitted to the NCs as a CD (in English only) for comments. These comments are usually both technical and editorial. In most cases, more than one CD is required before the fourth stage can be reached (for instance, CD 1, CD 2, etc.). It is good practice to set up an editing committee in order to deal with all editorial matters.

Enquiry Stage

Before passing to the approval stage, the CDV (usually in English only, sometimes in English and French) is submitted to all participating NCs for a 5-month voting period. Technical and editorial comments are requested. It is the last stage at which technical comments can be taken into consideration. The linguistic quality of the CDVs is usually better than the quality of the preceding CDs (in the meantime, the inputs by the NCs have been taken into account by the editing committee or the secretary).

Approval Stage

The FDIS is circulated to the relevant NCs for voting. Editorial comments on the FDIS (at present, approximately 65% are bilingual in English and French, and 35% are English only) may still be submitted. The FDIS is edited by the IEC CO (IEC, 2003).

Publication Stage

The IEC CO publishes the IS in English and French or, exceptionally, English only.

The language used in the PTs, TCs, and SCs is almost exclusively English. The French member body of the IEC, UTE (Union Technique de l'Electricité et de la Communication), is responsible for the preparation of the French version of the texts. It decides in consultation with the relevant TC or SC whether or not a French version of the forthcoming FDIS will be required. If the French NC undertakes to prepare the translation, this task has to be completed within 60 days. Otherwise, the FDIS will be circulated without delay and the standard will be published in English only. In such a case, the UTE may provide a French version at a later time and then request the publication of a bilingual version that supersedes the existing monolingual one (at present, this practice concerns approximately 5% of all projects). As a result, IEC's bilingualism will no longer be responsible for any significant delays. On the other hand, the preparation of the French versions may bring to light errors and cases of ambiguity in the English versions that would otherwise not have been noticed. Improvements of the English texts are common if appropriate feedback is provided by competent translators. It is important that the English version and the French version have an equal official status.

The ISO/IEC Joint Technical Committee 1 (JTC 1) on IT works in English only. In fact, this committee applies the ISO/IEC *Directives for the Technical Work of ISO/IEC JTC 1* (ISO/IEC, 2004a).

National Translations

Many ISs are translated into languages like Arabic, German, Russian, and Spanish. The preparation of such translations is not covered by the directives; it is the responsibility of the NCs concerned.

By using a questionnaire, we have investigated the translation practices among the NCs of the IEC. The questionnaire was sent to all 63 NCs, and 31 (49%) responded. Their geographical distribution was as follows: 2 from Africa (out of 4), 3 from North and South America (out 5), 6 from Asia (out of 19), and 20 from Europe (out of 33).

The NCs use the English and/or the French version of the IEC standards in their country, or they decide to prepare a national translation. The application of the English version of the IEC standards (71%) is far more common than the application of the French version (6%) or the application of both the English and the French version (10%).

Of the 31 NCs considered, 21 or 68% prepare translations into their national language (sometimes into one of several national languages). Whilst IEC deals with the English and French versions of the ISs, the NCs are responsible for their own translating activities. Depending on the country, the percentage of IEC standards for which a national translation is prepared differs from very few percent to almost 100%.

In most cases, the choice of whether or not to translate an IEC standard into the national language is a common decision of the national standards body and responsible experts (57%). Alternatives are decisions made by the standards body (19%), by the government (10%), or by others (19%).

There is usually a combination of several criteria for the decision to prepare a national translation:

- Expected number of standards users (71%)
- Need for absolutely correct interpretation (71%)
- Potential users who lack knowledge of English and French (57%)
- Expected sales revenues (33%)
- Others (19%)

In most cases, the translations are prepared by technical experts who are active in their NCs. Other options are a member of the NC staff or a professional translator. Usually, the initial translation is checked by one or more technical experts and/or a member of the NC staff.

In 43% of the NCs concerned, the rules for drafting are available in the national language. The availability of the most important horizontal standards (A.4) prepared by TC 1, TC 3, and TC 25 in the national language is even more common (62%).

National translations are based almost exclusively on the English versions of the IS. However, 64% of the NCs reported to use the French version for the clarification of English texts as the need arises. Several countries pointed out that their translators sometimes discover contradictions between the English and the French versions.

User Satisfaction

The second questionnaire deals with user satisfaction. We received 14 questionnaires (out of 18; 78%). The geographical distribution of the respondents to the second questionnaire was as follows: two from Africa (out of two), zero from North and South America (out of one), five from Asia (out of six), and seven from Europe (out of nine).

In the perception of the IEC NCs, users are in general satisfied with the language quality of the English version of the IEC standards (77%). Nineteen percent report that their users are sometimes not satisfied and 10% report occasional complaints. Most NCs (81%) do not keep any records on customer complaints concerning the linguistic qualities. One third of the relatively small number of countries using the French versions report occasional complaints on language quality. Shortcomings in IEC standards that may reduce their fitness for use concern mostly paronyms (A.3), incorrect language structure, incorrect style, polysemy (A.1), and faux amis (A.2). Improvements are also requested with regard to the ambiguity of sentences, very formal style, and very long sentences. Further shortcomings are too many corrigenda and errors that are not covered by corrigenda.

The issue of conflicting meanings between the English and the French versions of IEC standards is not perceived as a major problem. The items covered by both the ISO/IEC rules for drafting (ISO/IEC, 2004c) and IEC's list for checking and editing (IEC, 2003) do not present major problems, except for mathematical formulae (Teichmann, 2004). A recent study (Teichmann, 2006) has confirmed Daoud's (1991) findings that readers of adequate technical texts written in a foreign language may interpret them incorrectly.

CASE DISCUSSION AND CONCLUSION

Technical and scientific experts are not necessarily good technical writers. However, the efficiency level of technical experts participating in international standards development depends to a considerable degree on their receptive and productive skills in English as the main working language. Similarly, good receptive skills in English (French) are essential for the efficient use of the ISs. Lacking linguistic abilities and other shortcomings in the chain from the NP to the publication (Figure 4) and application of the ISs may create a gap between the intentions of the standards writers and the interpretation by the final user. If the NCs are not aware of the potential problems, they may appoint inefficient TC and SC secretaries, select inactive project team members, or send incompetent national delegates to meetings. They also may provide insufficient or inadequate inputs to the TC or SC work (failure to provide technical and editorial comments, or a lack of support of editing committees) and take wrong decisions concerning translating CDVs or ISs. All these shortcomings tend to reduce the linguistic qualities of the standards.

In IEC, it is mainly the task of the editing committees to solve problems of poor linguistic quality. In the different standards-setting organizations, the average degree of receptive and productive capabilities in English may vary: The IT community, for instance, is probably more accustomed to using the English language than experts in other more conventional technical areas. It is also obvious that the percentage of people with French language skills in Africa, Europe, and Canada exceeds the relevant figures in the other American countries, and in Asia and Australia. Therefore, the increasing involvement of Asian and Latin American nations in international standardization may tend to reduce the relative importance of the French language.

The domination of English as first IEC language is very pronounced, but experience shows that the use of bilingualism from an early stage of drafting can be of great assistance in the preparation of clear and unambiguous texts. IEC's use of bilingualism enhances the correctness and unambiguity not only of the English versions of the standards, but also the translatability of the English versions into other languages. Another reason for having a French version is that it enhances the fitness for use in countries where the French language is important, for example, in many African countries. These advantages should be balanced against the investment in money (carried by the French NC UTE) and the disadvantage of delays of up to 60 days. Furthermore, it should not be overlooked that the issue of bilingualism vs. unilingualism has an important cultural and political dimension that has also to be taken into account (Teichmann, 2005).

At the national level, most NCs use the English version, but some prefer the French one or use both versions. National translations are based almost exclusively on the English version, but many translators use the French version for clarification when necessary. In general, NCs are satisfied with the language quality of IEC standards, though several categories of mistakes were found fault with. Three situational dimensions are of immediate concern to standard writers and users: social attitude, province, and modality. The remaining six dimensions are intrinsic to all international standards (House, 1977; Teichmann, 2000b). Though the contributions of the technical department of IEC in principle were rated satisfactory, more attention should be paid to the following occasional problems: polysemy, faux amis, ambiguity of sentences, sentences that are too long, and especially incorrect mathematical formulae. Many standard users considered the effect of incorrect formulae as unacceptable. Furthermore, there was criticism that errors are not always covered by corrigenda.

Hardly any NCs keep records on customer complaints concerning the language quality of the international standards. However, a systematic feedback on such problems would be most important with a view to taking corrective action. It goes without saying that systematic feedback regarding technical errors is even more urgent.

GENERALIZATION

The present case concerns IEC standards. Whether or not the linguistic qualities achieved by other standards-setting organizations and the relevant user satisfaction are comparable to the case of IEC could not be investigated, but since their processes are comparable, we expect similar results. This might be investigated by using a questionnaire with a few essential questions. The standards development process of ISO is equal, and ISO also operates at an international scale and uses French as a second official language. Therefore, we expect that findings from the IEC case to a large extent also apply to ISO. In ITU, the processes are slightly different and the language policy is adapted to ITU's particular environment, where political issues tend to be more frequent and important than in ISO and IEC. As a result, ITU has six official languages: French, English, Spanish, Arabic, Chinese, and Russian. In the case of CEN and CENELEC, three instead of two languages are concerned (also German), and there is a limitation to Europe; however, the standards development process is almost identical to IEC's process. Therefore, we assume that most findings will also apply to CEN and CENELEC. It would also be interesting to compare the perceived linguistic qualities of the monolingual IT standards developed by ISO/IEC JTC 1 and ETSI with the perceived qualities of the English version of bilingual IEC standards.

RECOMMENDATIONS

Our case study results concern six major players who are involved in the development and use of international standards: the management of the IEC, the NCs, the TCs and SCs (including their chairmen, secretaries, project teams, and editing committees), the users of the IEC standards, translators of IEC texts (in particular, of bilingual and monolingual standards), as well as the other international and regional standards-setting organizations. A given recommendation may apply to several stakeholders.

International Level

Both the TC and SC secretaries and editing committees should use standard British English as well as standard French, and apply the horizontal standards in a systematic manner. Product committees should not compete with the horizontal committees. The secretaries should also be aware of the difficulties that standard writers and standard users who are not of English or French mother tongue may experience.

As regards the IEC, by far most translators use as source text only the English version of the ISs. They may discover errors in these texts. If translators use both the English and the French versions as source text, they may also discover contradictions between these two versions. Systematic feedback on all shortcomings (shortcomings in English texts or in French texts, or conflicting meanings) would be most important. ISO, ITU, CEN, and CENELEC should also support such a policy.

It would probably be worthwhile preparing a guide titled *Translating ISO/IEC Standards into a National Language*. Related work has been reported by the German standards institute Deutsches Institut für Normung (DIN, 2004).

National Level

When sending delegates to participate in international standardization, the national standardization organizations should pay attention to their language abilities.

The decision on whether or not to translate an international or regional standard will depend mostly on the scope of the standard. Though international and regional standardization organizations are neither involved in, nor responsible for, translations into a national language, it is in the interest of the entire standardization community that these translations are of adequate quality. The availability of Part 2 of the directives and of the most important horizontal standards in the national language is essential for this.

Translations of international and regional standards are prepared preferably by persons with a technical and scientific education rather than professional translators. The translators should be familiar with the directives and use the relevant horizontal standards whenever applicable.

To a certain extent, the quality of national translations may be enhanced if the IS is bilingual: If required, the French version may serve as backup.

The national bodies should keep records on customer complaints concerning the technical and linguistic qualities of the ISs and report them to the TC secretariats. In the case of mistakes in the French version, the French national standards body (AFNOR in the case of ISO and CEN; UTE in the case of IEC and CENELEC) is responsible for the linguistic quality.

ACKNOWLEDGMENT

This study was prepared in cooperation with the IEC central office, Geneva, Switzerland. Thanks are also due to the IEC national committees that completed our questionnaires and offered the wealth of their experience, as well as to several unknown reviewers.

REFERENCES

Aben, G. (2002). *Report on the survey regarding the use of international standards.* Geneva, Switzerland: International Federation of Standards Users.

Ailleret, P. (1982). *Essai de la théorie de la normalisation.* Paris: Eyrolles.

Beardsmore, H. (1986). *Bilingualism: Basic principles.* Clevedon, UK: Multilingual Matters.

Catford, J. (1980). *A linguistic theory of translation.* Oxford: Oxford University Press.

Daoud, M. (1991). *The processing of EST discourse: Arabic and French native speakers' recognition of rhetorical relationships in engineering texts.* Los Angeles: University of California.

De Vries, H. J. (1999). *Standardization: A business approach to the role of national standardization organizations.* Boston: Kluwer Academic Publishers.

Diesken, M., & Hoffmann, U. (1992). *New technology at the outset.* Frankfurt, Germany: Campus.

Deutsches Institut für Normung (DIN). (2004). *Translator's manual I, (DIN-Normen), (German-English).* Unpublished manuscript.

Edwards, J. (1994). *Multilingualism.* London: Penguin.

Egyedi, T. (1996). *Shaping standardization.* Delft, the Netherlands: Delft University Press.

Fédération des Institutions Internationales Semi-Officielles et Privées Établies à Genève (FIIG). (1979). *The use of languages in organizations and in meetings.* Geneva, Switzerland: Author.

Fomin, V. (2001). *The process of standards making: The case of the cellular mobile telephony.* Jyväskylä, Finland: University of Jyväskylä.

Gove, Ph. B. (1993). *Webster's third new international dictionary of the English language unabridged.* Cologne, Germany: Könemann Verlagsgesellschaft.

Hatim, B. (2001). *Communication across cultures: Translation theory and contrastive text linguistics.* Exeter, United Kingdom: University of Exeter Press.

Higgins, W. (2005). *Engine of change: Standards Australia since 1922.* Blackheath, Australia: Brandl & Schlesinger.

House, J. (1977). *A model for translation quality assessment.* Tübingen, Germany: Gunter Narr.

International Electrotechnical Commission (IEC). (2003). *Cahier des charges et types d' editing.* Unpublished manuscript.

International Electrotechnical Commission (IEC). (2006). *The electric century.* Geneva, Switzerland: Author.

International Organization for Standardization/International Electrotechnical Commission (ISO/IEC). (2004a). *Directives for the technical work of ISO/IEC JTC 1 on information technology.* Geneva, Switzerland: Author.

International Organization for Standardization/International Electrotechnical Commission (ISO/IEC). (2004b). *Part 1: Procedures for the technical work.* Geneva, Switzerland: Author.

International Organization for Standardization/International Electrotechnical Commission (ISO/IEC). (2004c). *Part 2: Rules for the structure and drafting of international standards.* Geneva, Switzerland: Author.

Jumpelt, R. (1961). *Die Uebersetzung Naturwissenschaftlicher Und Technischer Literatur.* Berlin, Germany: Langenscheidt.

Kocourek, R. (1982). *La langue française de la technique et de la science.* Wiesbaden, Germany: Brandstetter.

Maillot, J. (1981). *La traduction scientifique & technique.* Paris: Technique & Documentation.

Newmark, P. (1986). *Approaches to translation.* Oxford: Pergamon Press.

Nida, E. (2001). *Contexts in translating.* Amsterdam: Benjamins.

Nord, C. (2005). *Text analysis in translation: Theory, methodology, and didactic application of a model for translation-oriented text analysis.* Amsterdam: Rodopi.

Nuffield Languages Inquiry. (2000). *Languages: The next generation.* London: The Nuffield Foundation.

Pinchuck, I. (1977). *Scientific and technical translation.* London: André Deutsch.

Quirk, R., & Greenbaum, S. (1973). *A university grammar of English.* London: Longman.

Sager, J., Dungworth, D., & McDonald, F. (1980). *English special languages: Principles and practice in science and technology.* Wiesbaden, Germany: Brandstetter.

Teichmann, H. (2000a). IEC work on terminology and fundamental concepts (IEC/TC 1 and 25). *Conférence pour une Infrastructure Terminologique en Europe* (pp. 43-46).

Teichmann, H. (2000b). Situational dimensions of IEC standards. *TermNet News, 67*, 6-12.

Teichmann, H. (2003a). Efficiency of IEC technical committee secretaries. *EURAS Yearbook of Standardization, 4*, 125-145.

Teichmann, H. (2003b). Translating IEC texts. *Journal of the International Institute for Terminology Research, 11*(1), 1-135.

Teichmann, H. (2004). Market study on adequate language quality of IEC international standards. *Aachener Beiträge zur Informatik, 36*, 53-61.

Teichmann, H. (2005). Ethnolinguistic aspects of international standardization. *Aachener Beiträge zur Informatik, 37*, 71-78.

Teichmann, H. (2006). Linguistic shortcomings in international standards. *International Conference on Terminology, Standardization and Technology Transfer* (pp. 46-58).

Vinay, J.-P., & Darbelnet, J. (1995). *Comparative stylistics of French and English.* Amsterdam: Benjamins.

Von Hahn, W. (1983). *Fachkommunikation.* Berlin, Germany: de Gruyter.

Wilss, W. (1982). *The science of translation.* Tübingen, Germany: Gunther Narr.

Winckler, R. (1994). *Electrotechnical standardization in Europe.* Brussels, Belgium: CENELEC.

Wright, S. (2004). *Standards for translation assessment and quality assurance.* Kent State University.

APPENDIX A

A.1 Polysemy (Maillot, 1981)

A term is monosemic if it is used for a single notion. Polysemy means that a language uses a given term for several distinct notions. Translating from monosemy in the SL to polysemy in the TL is straightforward. Examples are the following.

Heating (qualitative aspect of the phenomenon): Echauffement
Temperature rise (quantitative term in degrees Celsius): Echauffement

Translating from polysemy in the SL to monosemy in the TL is more demanding: For several distinct notions, the SL has one term and the TL has different terms. Therefore, the context has to be taken into account. Examples are as follows.

Manufacturer: Fabricant (de câbles)
Manufacturer: Constructeur (de transformateurs)
Conductivity (electrical): Conductibilité (propriété de conduire le courant électrique)
Conductivity (electrical): Conductivité (grandeur qui mesure cette propriété)

A.2 Faux Amis (Maillot, 1981)

Faux amis are terms of different languages and of common origin that have different meanings but may be taken for equivalents. This problem is especially frequent where there has been a great deal of borrowing between the two languages concerned, for example, between French and English. Examples are the following.

Conductor in a cable is not *le conducteur* (isolé), but *l'âme* (conductrice).
Isolator is not *un isolateur*, but a synonym of *isolating switch* (sectionneur).
Commutator is not *un commutateur*, but *le collecteur* (d'une machine tournante).

A.3 Paronyms (Maillot, 1981)

Unlike the faux amis, paronyms do not belong to different languages, but to the same language. They are terms based upon the same root, that is, having the same derivation. Paronyms are sometimes taken as equivalents although they have different meanings. Some English, French, and German examples follow.

Resistance and resistor, axle and axis, reactance and reactor
Condensateur and condenseur; déchiffrement and déchiffrage
Rational and rationell; formal and formell

A.4 Horizontal Standards (Teichmann, 2000a)

IEC's (and ISO's) product-oriented committees would not be able to function without the horizontal standards prepared by the horizontal committees. Product committees are responsible for a specific product or a group of related products. They are far more numerous than the horizontal committees and should not do horizontal work. Examples of product committees are as follows.

TC 2: Rotating Machinery, TC 4: Hydraulic Turbines

Horizontal committees are responsible for subjects such as fundamental principles, concepts, terminology, or technical characteristics that are relevant to a number of product-oriented TCs or SCs. Their projects should serve the purposes of the product committees. Examples of horizontal committees are the following.

TC 1: Terminology; TC 3: Information Structures, Documentation, and Graphical Symbols; TC 25: Quantities and Units, and Their Letter Symbols

Horizontal committees obtain usually less support by industry than product committees because their fields of activity are not closely related to financial rewards.

A.5 Modal Auxiliaries (ISO/IEC, 2004c)

A standard does not in itself impose any obligation upon anyone to apply it. However, such an obligation may be imposed, for example, by legislation or by contract. In order to be able to claim compliance with a standard, the user needs to be able to identify the requirements he or she is obliged to satisfy. The user needs also to be able to distinguish these requirements from other provisions where there is a certain freedom of choice. According to the directives, the following modal auxiliaries have to be used for the English and the French versions.

Requirement: shall, shall not; doit, ne doit pas

Recommendation: should, should not; il convient de, il convient de ne pas

Permission: may, may not; peut, peut ne pas être

Possibility and capability: can, cannot; peut, ne peut pas

A.6 Calques

Calques concern linguistic borrowing that consists of the imitation in one language of some part of the peculiar range of meaning of a particular word in another language (Gove, 1993). In languages other than English, there is a trend toward a mid-Atlantic jargon. This is unfortunately the impression often given by texts published by international organizations (Vinay & Darbelnet, 1995). Their members, either through ignorance or because of a mistaken insistence on literalness, seem to demand translations that are largely based on calques. The unacceptable style has its roots in ill-digested translations of Anglo-American originals. Such texts can be identified by their use of false comparative, artificial or prestigious allusion, and an unusual verbosity. Examples of calques include the following:

a. Standard English: unless otherwise specified
   Calque: s'il n'en est autrement spécifié
   Standard French: sauf spécification contraire
b. Standard French: remplacer (un document)
   Calque: to replace
   Standard English: to delete and substitute

A.7 FUNCTIONAL BILINGUALISM

The practice of alternatively using two languages is called bilingualism, and the persons involved are bilinguals. There are very different degrees of bilingualism, and a notion of relativism must be introduced whereby the individual degrees can be ascertained (Beardsmore, 1986).

Functional bilingualism means that a speaker is able to conduct his or her activities in a dual linguistic environment satisfactorily. Such a speaker may well use patterns that are alien to the monoglot reference group (e.g., native English speakers) and show heavy signs of interference in phonology, lexis, and syntax. However, to the extent that these do not impede communication between the speaker and the listener, they do not get in the way of functional bilingualism.

The relevant language skills are either receptive (listening comprehension and reading comprehension) or productive (oral production and written production).

The person who must use a foreign language to communicate can feel himself or herself to be in an inferior position in relation to those who express themselves in their mother tongue (Fédération des Institutions Internationales Semi-Officielles et Privées Établies à Genève [FIIG], 1979).

A.8 Editing Committees

The tasks of editing committees are to update and edit committee drafts, enquiry drafts, and final drafts of international standards (ISO/IEC, 2004b). If the documents are bilingual, the editing committee should comprise at least one technical expert of English mother tongue with an adequate knowledge of French, one technical expert of French mother tongue with an adequate knowledge of English, and the project leader. The project leader and/or secretary may take direct responsibility for one of the language versions concerned. Standard British English and standard French are the official language varieties (ISO/IEC, 2004c; Quirk & Greenbaum, 1973).

A.9 ITU's Language Policy and Procedures

The official working languages of ITU are French, English, Spanish, Arabic, Chinese, and Russian on an equal footing. These languages may all be used at major conferences. ITU has three substantive sectors plus the general secretariat. The three sectors deal with standards, development, and radio communications.

As far as the standards sector (ITU-T) is concerned, all study groups with the exception of Study Group 3 work in English only. They use the alternative approval procedure to approve the recommendations (standards) in English, and the recommendations are subsequently translated into all other official languages. (Until recently, they were only translated into Spanish and French since Arabic, Chinese, and Russian were used far less within ITU.)

Study Group 3 is a fairly political study group dealing with tariff and accounting principles including related telecommunication economic and political issues. It is the only ITU-T study group to have interpretation, and it is the only one to have contributions to it translated, along with its reports, into all other official languages. Its standards are approved by the traditional approval procedure.

Abbreviations

- **AFNOR:** Association Française de Normalisation (French Standards Association)
- **CD:** Committee Draft
- **CDV:** Enquiry Draft
- **CEN:** Comité Européen de Normalisation (European Committee for Standardization)
- **CENELEC:** Comité Européen de Normalisation Electrotechnique (European Committee for Electrotechnical Standardization)
- **CO:** Central Office (of IEC)
- **DIN:** Deutsches Institut für Normung (German Standards Institute)
- **ETSI:** European Telecommunication Standards Institute
- **FDIS:** Final Draft International Standard
- **FIIG:** Fédération des Institutions Internationales Semi-Officielles et Privées Établies à Genève (Federation of Semi-Official and Private International Institutions Established in Geneva)
- **IEC:** International Electrotechnical Commission
- **IFAN:** International Federation of Standards Users
- **IS:** International Standard
- **ISO:** International Organization for Standardization
- **IT:** Information Technology
- **ITU:** International Telecommunication Union
- **ITU-T:** International Telecommunication Union-Standardization Sector
- **JTC:** Joint Technical Committee
- **NC:** National Committee
- **NGO:** Nongovernmental Organization
- **NP:** New Work Item Proposal
- **PT:** Project Team
- **SC:** Subcommittee
- **SL:** Source Language
- **TC:** Technical Committee
- **TL:** Target Language
- **UTE:** Union Technique de l'Electricité et de la Communication (Technical Union for Electricity and Communication)
- **WD:** Working Draft

Section III
Diffusion and Adoption of Standards

Chapter VII
A Diffusion Model for Communication Standards in Supply Networks

Michael Schwind
University of Kaiserslautern, Germany

Tim Stockheim
Frankfurt University, Germany

Kilian Weiss
Frankfurt University, Germany

ABSTRACT

This chapter presents a simulation framework that analyzes the diffusion of communication standards in different supply networks. We show that agents' decisions depend on potential cost reduction, pressure from members of their communication network, and implementation costs of their communication standards. Besides focusing on process-specific market power distributions, the impact of relationship stability and process connectivity are analyzed as determinants of the diffusion of communication standards within different supply network topologies. In this context, two real-world scenarios, from the automotive and paper and publishing industries, are used as examples for different network topologies. The results support the thesis that increasing relationship dynamics and process connectivity lead to decreasing competition of communication standards. In certain circumstances, local communication clusters appear along the value chain, enabling these clusters to preserve their globally inferior standardization decision.

INTRODUCTION

During the last decade, the importance of communication standards has increased due to an intensification of the relationships between supply chain companies. The ensuing demand for intra-organizational process integration and advanced communication technologies in a broad range of software and hardware, for example, XML (extensible markup language) interfaces in standard software, Web electronic data interchange (WebEDI), and so forth, has made the standardization of communication protocols a crucial task. As a result, a discipline within standardization research has emerged that is dedicated to analyzing the impact of network topology on the diffusion of communication standards and to improving the understanding of diffusion processes in generic topological structures (Economides & Salop, 1992; Weitzel, Wendt, Westarp, & König, 2003). The objective of the work presented here is to analyze how the underlying process topology determines the dynamics of the communication infrastructure within supply networks. Based on Weitzel et al., our central hypothesis states that network topology and market power strongly influence the diffusion process of communication standards in supply networks. In particular, we pose the following questions.[1]

1. Does market power influence the diffusion of communication standards along the value chain?
2. How does relationship stability between supply network organizations affect the diffusion process?
3. What effect does connectivity between the particular value chains have on the tendency toward the concentration of standards in supply networks?
4. How do communication standards propagate along value chains and what are the consequences for the diffusion within the whole network?

This chapter introduces a simulation framework for the diffusion of communication standards in supply networks, focusing on the impact of the underlying topologies. On the assumption that the communication relationships result from supply network structures, we employ empirical data sets to model industry-specific supply networks. These topologies are the basis for simulations that analyze diffusion dynamics in supply chain settings under varying levels of flexibility, market power, and interconnectivity. However, how will the results of this analysis help when struggling to determine the right communication standard? This chapter shows what impact different topologies and process structures in supply networks have on the diffusion of communication standards. Decision makers should not only consider the installed base[2] and the cost of standards, but should also carefully take into account the following aspects.

- How flexible is the choice of suppliers in the value chain?
- How connected are the value chains within the supply network?
- Who are the central players?
- How is the market power distributed?

In the following sections, we proceed as follows. First we provide an overview of the three most relevant areas of research concerning communication standard application; next we systematically identify relevant determinants of diffusion processes within supply networks. Based on these determinants, we then explain the simulation model and its implementation. After describing the supply web scenarios from two specific industries to analyze the diffusion of communication standards, we present and analyze the results of the simulations. Finally, we conclude with an evaluation of our findings.

LITERATURE REVIEW

Network Effects

The adoption of standards could create positive and/or negative network externalities called network effects (NEs) in macroeconomics. Katz and Shapiro (1985, p. 424) provide the following definition of a positive NE: "[A NE occurs in a situation where] the utility that a given user derives from the good depends upon the number of users who are in the same 'network' as he or she." A first general classification of positive NEs concerns their impact on returns. Decreasing costs of communication are one example of a positive NE, and increasing returns through the usage of a larger distribution network are another. Changing the standards creates one-time switching costs that contribute to the lock-in effect (Arthur, 1989; Liebowitz & Margolis, 1995). The lock-in effect is considered to be an example of a negative NE due to the costs created by the introduction of a new standard that promises future advantages. The evolution of various standards cannot be explained solely by a sequence of rational decisions taken in a deterministic manner, but must also be attributed to the impact of small exogenous incidents that occur in an economic development process. This phenomenon is called path dependency (David, 1985). The evolution of communication standards can thus be seen to contain a random element, one that requires nondeterministic models instead of deterministic ones (David & Greenstein, 1990).

The Standardization Problem

The standardization problem[3] (SP) refers to the dilemma encountered by agents as they attempt to choose their communication standards in a network not knowing which standards other players will adopt. Unfortunately, the utility of the selected standard depends on the adoption decisions of the other players in the network (Buxmann, Weitzel, Westarp, & König, 1999). Stockheim, Schwind, and König (2003) identify three basic classes of methodological approaches employing NE theory to describe and analyze the SP.

The first class concentrates on macroeconomic considerations, explaining the evolution of network monopolies and oligopolies under NEs. Of the vast number of publications in this class, some of the most prominent should be cited. Economides (1996), using welfare analysis, shows how competition can fail due to the impact of NEs. Economides and Himmelberg (1995) consider the case of sponsored standards, that is, those standards where the beneficiary of a particular standard promotes the goods, generating an externality that establishes the sought-for de facto standard (Shurmer & Swann, 1995) and supplies the sponsor with exploitable monopoly power in the case of a successful lock-in. Katz and Shapiro (1985, 1986) show that, in the case of an oligopoly with two competing incompatible standards, the level of industry output is greater than in any other equilibrium of incompatible firms. In addition, Saloner and Farrell (1985) demonstrate that the Pareto-superior technology will always be adopted for the SP when information is complete, whereas with incomplete information, inferior standards may be chosen.

The second approach to formulate and evaluate the SP is based on game theory. Starting with the individual utility and costs function, equations for the participants' behavior are formulated and described as an equilibrium model, which is mostly solved in a closed-form expression. The marketing related "mix-and-match" approach by Matutes and Regibeau (1998), as well as the models of Economides (1989), Economides and Salop (1992), and Economides and Lehr (1994), all characterized by a lack of a priori assumptions in respect to the NE's existence, are prominent models constructed in the game theory domain.

The third, most recent approach for analyzing diffusion dynamics is agent-based computational economics (ACE; Tesfatsion, 2002). ACE methods simulate the behavior of individual network participants and observe the equilibrium outcome. This is an advantage insofar as the description of the agent's behavior does not have to be compatible with the mathematical equilibrium formulation, which leads to fewer constraints on the action space of the agent. Wendt and Westarp (2000) introduce a topological structure into the network effect problem, which they formulate as an ACE problem, and analyze the dynamics of diffusion processes under different topologies. Similar ACE diffusion models have been introduced by Arthur (1989) and Axelrod (1997). Based on these two models, Leydesdorff (2001) develops an ACE model for cultural dissemination where NEs and lock-in effects can be combined with local learning.

Supply Networks

Supply networks can be defined as "a group of buyers, sellers and the pattern of links that connect them" (Kranton & Minehart, 2001, p. 485). A supply network consists of different organizations with specific roles along the value chain. The network topology is strongly influenced by the underlying processes of the supply chain, with relationships ranging from very little to almost complete integration and collaboration. Another important aspect is the power exerted by the different organizations within the supply network. The asymmetric distribution of power within the business relationships, that is, the existence of few strong manufacturers in the automotive industry (Benton & Maloni, 2000) is important when considering diffusion dynamics. Such asymmetric distribution of power influences the diffusion and thus the resulting concentration of standards within the network. Settings with two or three strong standards can be stable equilibria, obviously being the result of subgroups with strong intragroup communication and fewer links to other groups, collectively resisting external pressure to switch their selected standard (Wendt & Westarp, 2000). Different determinants define the type of relationships between the organizations: formalization, intensity, frequency, standardization, and reciprocity (Chow, Heaver, & Henriksson, 1995). Recently, research has shown how organizations' power influences the emergence of different supply network topologies (Benton & Maloni; Cox, 2001).

DETERMINANTS OF STANDARDIZATION DYNAMICS IN SUPPLY NETWORKS

The benefits of communication technology stem from the possibility of exchanging data with other organizations. For participants of communication networks, it is crucial to use compatible communication standards and therefore coordinate the use of their individual technology (Buxmann et al., 1999). While this coordination is relatively easy for a single organization, coordination becomes much more difficult in an interorganizational context. Our multiagent model simulates diffusion effects as a series of individual decisions conducted by participants of a relational diffusion network. The decision to adopt a communication standard is not based on the IB,[4] but by the adoption decisions within the personal (stochastic) communication network of each agent.

Wendt and Westarp (2000) identify the following determinants of diffusion dynamics, using sociological and geographical concepts from research on innovation diffusion in relational networks (Valente, 1995).

- Intensity of communication
- Costs
- Personal network exposure
- Opinion leadership (power)
- Opinion leadership (location)
- Intragroup pressure toward conformity
- Stand-alone utility (functionality) of the products

THE SIMULATION PROCESS

To gain insight into the diffusion of standards in the supply networks of different industries, we use a Monte Carlo simulation, applying the outlined decision functions to a multiperiod decision model. The agents repeat their standardization decisions until equilibrium is reached. The deciding agent has complete information about the standards adopted by the other agents during the last period. At the end of the period, all agent standards are updated and a new period starts.

During the simulation, the occurrence of oscillations must be taken into account. A simple example for such a behavior can be shown in a setting where two agents belong to two different clusters. If one partner estimates that he or she will achieve cost reduction by adopting the other standardization decision while the other drops its standard, in the next round both agents recalculate their decision and swap their adopted standards. They could go on like this and never reach the same standardization decision as long as they decide simultaneously. To avoid such behavior, which seems rather unrealistic in a real-world scenario, we decreased the probability of adopting another standard from 100% to 90%.

FORMAL DESCRIPTION OF THE SIMULATION MODEL

Based on the previously listed determinants, Table 1 defines the underlying parameters of the supply chain model specified in the next section. This parameter setting can be used to describe a broad range of supply networks typical for different industries and to analyze the diffusion process of communication standards in these networks. By adjusting these parameters, the process structure of two different industry sectors can be mapped onto the simulation model.

In our simulation model, the agents choose between three homogeneous standards $s_i \in \{1, 2, 3\}$ and no implementation at all $s_i=0$. The resulting benefits from implementing a standard depend on

Table 1. Determinants and parameters simulating diffusion dynamics in supply networks

ID	Determinants	Parameters	Description
1	Communication intensity	Size of enterprise q_i	Size of an organization (define size distribution for each type)
2		Number of sourcing relationships g_i	Sourced parts of agent i from different tier
3	Costs	Implementation costs $K(q_i)$	Costs to implement a standard
4		Maximum communication costs $C(q_i)$	Maximum savings per relationship
5	Communication partner influence	Communication cost savings $C(q_i, q_j)$	Reduction of communication costs for partners implementing the same standards
6	Personal network exposure	Relationship stability between tiers p_c^{main}	Probability of transaction with main supplier
7		Number of suppliers n_c	Number of potential suppliers per sourced part
8	Opinion leadership (power)	Economic base power b_i^{eco}	Power sources other than size-based technological and economic power
9		Technological knowledge b_i^{tech}	Network participant's technological knowledge
10	Opinion leadership (location)	Set of agents in tier $O_c \subset Agents$	Position (tier c) within the supply network
11	Intragroup pressure toward conformity	Market pressure functions	Pressure that partner agents can exert on the deciding agent to implement their standard

Table 2. Relative power from technology

Size of Enterprise	$C(q_i)$	$K(q_i)$:
Small ($q_i = 1$)	2	2
Medium ($q_i = 2$)	3	4
Large ($q_i = 3$)	4	6

the size of the enterprise and are characterized by the following properties.

- Implementation costs $K(q_i)$: The implementation costs describe the one-time costs that occur if a new standard s is implemented; implementation costs will depend on the agent's size.
- Savings $C(q_i, q_j) = \min(C(q_i), C(q_j))$: Communication cost savings can be achieved when agent i and j implement a standard s at the same time. We suggest that the amount of communication—and the amount of possible savings—depends on the smaller partner. This stems from the observation that the amount of material flow is directly connected with the amount and complexity of the data flow.

On the assumption that costs and potential benefits of the communication standards increase with the size of the adopters, the values in Table 2 were used as estimates, although other variants could be justified.

Supply Network Initialization

This section describes the initialization process that generates a network topology of organizations and their relationships. Table 3 shows the parameters used for initializing the different process roles.

In the context of supply chains, often one supplier has an outstanding position, meaning that most of the supplies are delivered by this main supplier. We include this observation into our model

Table 3. Initialization parameters

Parameter	Description
Distribution of sizes	How are the sizes (1, 2, 3) distributed within each cluster?
Distribution of economic power	How is the economic power distributed?
Distribution of technological power	How is the technological power distributed?
Size of tiers	How many agents belong to each tier?
Number of tiers	How many tiers belong to the supply network?

by introducing a probability p_c^{main} that specifies how likely an agent (at tier c) receives the required materials from its main supplier. Because we add the supplier relationships in a random order, the first supplier is chosen as the main supplier. The probability of each remaining supplier's delivery $p_{j,i}$ is calculated accordingly. Based on these parameters, the initialization procedure illustrated in Figure 1 can be formulated.

The first step of the initialization process is the generation of the different agents, each assigned to a cluster (in the supply web context, tiers).

Figure 1. Initialization of the supply network

```
For each tier (cluster) c
  For each agent i in tier c (a_i ∈ O_c)

    Initialize a_i = ⟨q_i, b_i^eco, b_i^tech⟩
    Set g_i := g_c and randomly set s_i ∈ {0, 1, 2, 3}
For each tier c
  For each agent i in tier c
    For each part (g_i)
      Set counter := 0
      While counter < n_c
        Select random supplier j from tier O_{c-1}

        If (Rand(0,1,2,3) < 2^{q_j - 1}), then
          Add communication relationship
          Increment counter

          If main supplier not selected, then p_{i,j} := p_c^main

          Else p_{i,j} := (1 - p_c^main) / (n_c - 1)
```

$$a_i = \langle q_i, b_i^{eco}, b_i^{tech} \rangle \text{ where } q_i, b_i^{eco}, b_i^{tech} \in \{1, 2, 3\} \quad (1)$$

Properties are assigned to each agent based on data about different tiers. Each initialization process has a stochastic component producing diverse topologies that may cause changing simulation results and lead to the necessity of repeated Monte Carlo simulations.

After the agents have been initialized, their relationship structure is created. The number of clients possessed by an agent of a certain cluster is not evenly distributed as the relationships are initialized semirandomly in conjunction with the influence of the potential sourcing partner's size. The probability that agent i chooses agent j as a sourcing partner increases with its size q_i (see Figure 1).

Agents' Decisions

Every period, the agents within the communication network decide which standard they want to implement: $s_i = s^*$ if agent i implements standard s^*, or else $s_i = 0$. This decision is based on a first-choice model and relies on a utility function that generates individual utility values in money units for each standard.

For more clarity, we have assumed risk-neutral agents. Although the assumption of risk-adverse agents is more realistic, the impact on the outcome concentration from such an assumption will be small. Since the uncertainty is equally distributed between the different standards, the introduction of risk-adverse agents has the same effect as reducing the potential savings of communication costs. Depending on the relative implementation costs, this would lead to less implemented standards, but should not have a high impact on the concentration rate.

The utility of an implemented standard $U_{i,s}$ is calculated as the sum of three components:

$$U_{i,s} = V_{i,s} + M_{i,s}^{adjusted} - K(q_i) \quad (2)$$

where $V_{i,s}$ is the expected savings of communication costs for agent i when implementing standard s and $M_{i,s}^{adjusted}$ is the utility from market pressure in money units for the decision of agent i implementing standard s.

In addition, implementation costs $K(q_i)$ (see Table 2) pose a negative impact on the agent's utility. The expected sum of the savings for an agent with the implemented standard s is calculated as follows:

$$V_{i,s} = \sum_{j=1}^{J_i} \left(C(q_i, q_j) \cdot p_{i,j} \cdot \hat{x}_{j,s} \right) \quad (3)$$

where $C(q_i, q_j)$ is the savings of agent i if partner j has implemented the same standard, J_i is the number of communication partners for agent i, and $\hat{x}_{j,s}$ is the estimator that agent j will adopt communication standard s.

In our simulations $\hat{x}_{j,s} = 1$, if $s = s_j$ of the last period and $\hat{x}_{j,s} = 0$ for all other cases. We have modeled the ability of other agents to influence the deciding agent to implement their preferred standards in terms of market pressure ($M_{i,s}$). The partner agents offer incentives for certain standards, that is, by assuring revenues or threatening to end the business relationship, which leads to a new decision base for the decider. To calculate the utility, we adjusted the market pressure using the following function:

$$M_{i,s}^{adjusted} = M_{i,s} - \frac{1}{S} \cdot \sum_{s'=1}^{S} M_{i,s'} \quad (4)$$

We based market pressure on the following equation:

$$M_{i,s} = \sum_{j=1}^{J_i} (C(q_i, q_j) \cdot r_{j,i}^{adj} \cdot p_{i,j} \cdot \hat{x}_{j,s}) \qquad (5)$$

where $M_{i,s}$ is the utility from market pressure in money units for the decision of agent i to implement standard s and $r_{j,i}^{adj}$ is the relative power of partner agents j on agent i.

The basic thought behind equation 5 is that the partner agent's pressure depends on the possible savings of a_j, the relative power of agent a_j to influence the decision of a_i, and the probability of the communication relationship ($p_{i,j}$). The first determinant of maximal pressure (utility in money units) the partner agent is able to put on the deciding agent is savings. This is based on the assumption that pressure above the partner's savings would lead to a compensation payment for the deciding agent. The second determinant is the probability of a future communication relationship stored within the network structure. If, on the one hand, an agent is highly certain that a communication relationship will exist in the future, he or she will put more pressure on the partner. If, on the other hand, the probability of a future relationship is very small, he or she will put less effort on influencing this partner. Since we postulated a minimum function for the maximum pressure, a linear relationship of market pressure and probability is problematic. The question is how relevant the communication costs are in relation to the value of the business relationship. To simplify the analysis, we have assumed a rather strong relevance of the communication costs, an assumption that is realistic in many cases, for example, EDI solutions (Buxmann et al., 1999). This enables us to use a proportional relationship between the market pressure and the probability of the business relationship. The third determinant for the pressure the partner agent can put on the deciding agent is relative market power between the two agents. To derive an equation for the relative power of one agent toward another in the context of supply networks, we postulate the following assumptions.

- Market pressure on a deciding agent depends on the partners' relative power and their previous savings from a particular implementation decision. The pressure is restricted by the lost savings of the partners as higher market pressure would be counteracted with settlement payments for the partners' lost savings.
- The economic power of the decider depends on size and an exogenously provided base power component, which aggregates different aspects, that is, the legal situation and transaction costs.
- The agent's technological power describes the level of technological knowledge, and influences its possibility to push partner agents to implement a preferred standard.
- The relative market power of an agent is influenced by his or her buyer-supplier role. The buyer gains power whereas the supplier suffers a loss.

The relative power $r_{j,i}$ of agent j toward agent j has been calculated using the following logistic function:

$$r_{j,i}^{adj} = \frac{1}{1 + e^{-15 \cdot (r_{j,i} - 0.5)}} . \qquad (6)$$

The adjustment leads to an amplification of the relative power of two communicating agents. The reason behind this is that the upper boundary of the possible market pressure restricts the linear equation and therefore leads to decreasing marginal pressure with increased relative power. This also preserves the symmetry of $r_{j,i}$ and guarantees that $r_{j,i}^{adj} \in [0; 1]$.

Formalization of Market Power

Based on the aforementioned assumptions, we calculate the relative power between two agents as the weighted sum of the economic market power and the technology-related market power of the partner agent (equation 7):

$$r_{j,i} = \alpha \cdot r_{j,i}^{eco} + (1 - \alpha) \cdot r_{j,i}^{tech} \quad (7)$$

where $r_{j,i}^{eco}$ is the economic market power of the partner agent and $r_{j,i}^{tech}$ is the technology-related market power of the partner agent.

Equation 8 shows the mapping of the agent's economic market power from size $r_{j,i}^{size}$ and the relative power based on unconsidered economic factors $r_{j,i}^{base}$, which are then aggregated to create relative economic market power:

$$r_{j,i}^{eco} = \beta \cdot r_{j,i}^{size} + (1 - \beta) \cdot r_{j,i}^{base} \quad (8)$$

where $r_{j,i}^{size}$ is the market power of the partner agent based on size and $r_{j,i}^{base}$ is the power of the partner agent based on economic factors.

Within our implementation we set α and β to 0.5, which ensures that both factors possess influence identically. Equations 9, 10, and 11 show the mapping of the agent's basic attributes (size q_i, technological knowledge b_i^{tech}, and economic base power b_i^{eco}) on the relative power (based on size, technology, and economic market power).

$$r_{j,i}^{size} = f(q_j, q_i) \text{ with } f \in \Re \text{ and } 0 \leq f \leq 1 \quad (9)$$

$$r_{j,i}^{tech} = g(b_j^{tech}, b_i^{tech}) \text{ with } g \in \Re \text{ and } 0 \leq g \leq 1 \quad (10)$$

$$r_{j,i}^{base} = h(b_j^{eco}, b_i^{eco}) \text{ with } h \in \Re \text{ and } 0 \leq h \leq 1 \quad (11)$$

The relative impact of the determinants of market power varies within different industries, a circumstance that makes different concrete instantiation necessary depending on the analyzed industry. Thus, for reasons of comparability, we use the following instantiations even though other variants might be more realistic in our specific markets.

$$r_{j,i}^{size} = (2 + q_j - q_i)/4 \quad (12)$$

$$r_{j,i}^{tech} = (2 + b_j^{tech} - b_i^{tech})/4 \quad (13)$$

$$r_{j,i}^{base} = (2 + b_j^{eco} - b_i^{eco})/4 \quad (14)$$

The next section outlines our example scenarios and the simulation results.

EXAMPLE SUPPLY WEB SCENARIOS

Typical examples for supply networks are the automotive and publishing industries. They will be used to simulate the diffusion processes of communication standards.

Automotive Industry (Germany)

After going through many changes, from a job-shop-driven industry toward mass customization, today's automotive supply chain can be described as an *n*-tier supply network. Carmakers and their suppliers have gone through a considerable consolidation process, as is the case with Daimler Chrysler's merger and General Motors' equity alliances with at least four other manufacturers (Biller, Bish, & Muriel, 2001). Additionally, the efforts of carmakers to reduce their suppliers and to source more complex components have led to rather intensive partnerships with their direct suppliers (Caprile, CIREM Foundation, & Llorens, 2000).

Another specific aspect of Germany's automotive supply chain is that the distribution of new cars is dominated from dealerships carrying only brands from one carmaker, which leads to very static business relationships. Based on Biller et al. (2001), Crain Communications (2002), Fricke (2004), Hertwig, Mühge, and Tackenberg (2002), and Verband der Automobilindustrie e.V. (2003), the following (simplified) process roles and tiers within the automotive supply chain are mapped onto the simulation.

- **Retailer:** The retailers sell cars to customers and provide services. Most of the organizations are small to medium sized, do not have much technological knowledge, and hold medium economic power. They are also confronted with very high transaction costs if they change their (single) manufacturer.
- **Carmakers:** There are relatively few manufacturers within the automotive industry. They are large companies possessing much economic power and technological knowledge, although the technological competence is shifting toward Tier 1 suppliers, which provide increasingly complex components.
- **Supplier (Tier 1):** The group of Tier 1 suppliers is dominated by medium-sized companies, which focus on different special areas, delivering components to one or more car manufacturers. Whereas the economic power relative to the large manufacturers is small, more pressure can be put on the Tier 2 suppliers. The technological knowledge has been increasing during the last decade, but is still not dominant relative to the carmakers. Because of the component strategies of the carmakers, business relationships remain stable.
- **Supplier (Tier 2):** The Tier 2 organizations supply products of little complexity primarily to the Tier 1 suppliers but also to the carmakers. Their economical and technological power is rather small.

Paper and Publishing Industry

According to the statistical systematic of economic branches in the European community, Class DE (21, 22), the paper and publishing industry can be broken down into the following units: pulp manufacturing, paper and paper products, and publishing and printing. The paper and publishing industry supply chain is simple compared to the automotive supply chain. The main difference is the higher level of relationship dynamics, especially between the printers and the producers of the paper. Moreover, since its organizations are much more interconnected, the paper and publishing supply chain differs in structure from the automotive supply chain. It is composed of the following organizations (Pyzdrowski & Donnelley, 2003; Röder, 2002).

- **Publishing house:** The publishing house creates publishing products and uses different printers and different kinds of paper.
- **Printer:** The printing industry provides mostly generic services, but has relatively intensive relationships with its clients.
- **Producer:** The producers of paper are dominated by medium to large companies, although specialized small firms coexist.

Simulation Scenarios

Figures 2 and 3 show unified modeling language (UML) diagrams of the settings derived for our simulation model. The base power distributions for size, technology, and economics are given in the form {*Smallest, medium, largest characteristic*}. The results gained by applying various parameter constellations to this setting are shown in the next section.

RESULTS AND CONCLUSION

The Herfindahl Index, applied to measure the relative concentration of communication standards, was used to analyze the outcome of the simulations[5]:

Figure 2. Supply Web abstractions of automotive industry

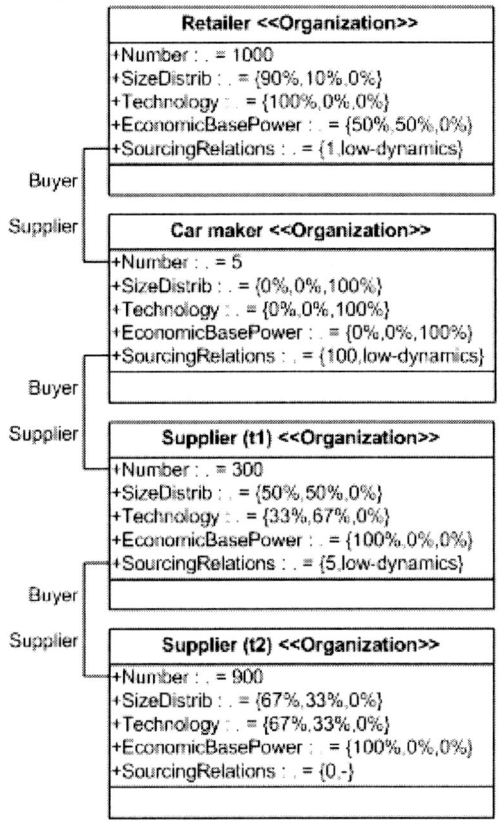

Figure 3. Supply Web abstraction of paper and publishing industry

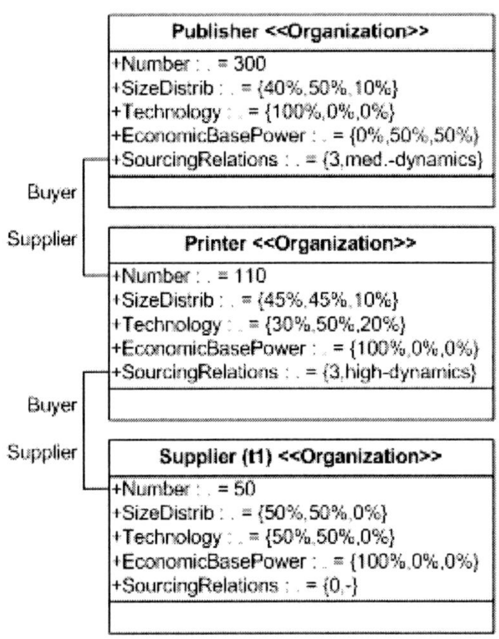

$$H = \sum_{s=1}^{|S|} sm^2_{s} \qquad (15)$$

where H is the Herfindahl Index and ms_s is the relative market share of standard s.

Impact of Supply Chain Connectivity

The aim of the first simulation was to assess the impact of connectivity between particular value chains in the automotive supply network on the diffusion of three homogeneous standards. Connectivity between the particular value chains describes the density of communication relationships within a supply network and is defined as the amount of sourcing relationships in the supply network. Table 4 shows the different connectivity scenarios, with scenario 1 representing the least connected network.

Table 5 shows the impact of increased process connectivity on the communication standards concentration in an automotive supply network with stable relationships. One can observe a positive correlation between concentration and process connectivity in all four tiers along the value chain. This outcome can be explained by looking at the decision function of the agents: Higher connectivity leads to an increase of com-

Table 4. Process connectivity scenarios (number of suppliers)

Scenario	Sourcing Rel. Retailer	Sourcing Rel. Manufacturer	Sourcing Rel. Supplier (t1)
1	1	100	5
2	1	105	5
3	1	110	6
4	1	115	6
5	1	120	7
6	2	125	7
7	2	130	8
8	2	135	8
9	2	140	9

Table 5. Impact of process connectivity

Scenario	Retailer	Manufacturer	Supplier (t1)	Supplier (t2)
1	0.481	0.485	0.443	0.424
2	0.470	0.475	0.463	0.438
3	0.502	0.488	0.494	0.471
4	0.544	0.558	0.573	0.542
5	0.487	0.480	0.594	0.564
6	0.521	0.506	0.655	0.625
7	0.668	0.715	0.756	0.737
8	0.653	0.688	0.791	0.765
9	0.735	0.782	0.928	0.909

munication relationships, resulting in less isolated groups of agents that are able to maintain their intragroup standards.

Impact of Relationship Stability

This section analyzes the impact of relationship stability, which, in our simulation, has been defined as the probability that an object is sourced from the main supplier. An example of low relationship stability is regular sourcing from the cheapest supplier. Table 6 shows a negative impact of the relationships' stability on the concentration of communication standards in the automotive supply chain with connectivity Scenario 1. This can be explained analogously to the impact of connectivity: The reduction of relationship stability leads to a higher connectivity and reduces the probability of the formation of subnetworks.

Table 6. Impact of relationship stability

Rel. Stability	Retailer	Manufacturer	Supplier (t1)	Supplier (t2)
100%	0.481	0.485	0.443	0.424
90 %	0.509	0.499	0.488	0.414
80 %	0.510	0.512	0.590	0.479
70 %	0.643	0.650	0.756	0.599
60 %	0.915	0.934	0.969	0.804
50 %	0.922	0.926	0.941	0.784
40 %	0.973	0.981	0.963	0.844
30 %	0.990	0.984	0.984	0.887
20 %	0.984	0.984	0.985	0.890

Generalization of the Analysis

We then simulated the multivariate impact of relationship stability and process connectivity on the concentration of communication standards. Figure 2 shows the impact of the two factors, revealing a negative correlation (0.1%) of relationship stability and a positive correlation (0.1%) of supply chain connectivity with the concentration of the communication clusters for low to moderate flexibility and connectivity (see Figure 2). This result allows the generalization of our results in the previous sections, except for highly connected and flexible settings, where we have observed a slight decrease of the Herfindahl Index. This effect could be explained by a decreasing diffusion of a dominating standard from the retailers to the suppliers. Our simulation data show that the backward propagation of the standards decreases due to a shift of power toward the suppliers, leading to a 7% decrease of the concentration in Tiers 1 and 2.

Diffusion of Standards along the Supply Chain

We simulated the impact of locally decreased relationship stability within the automotive supply network and monitored the propagation of the increased concentration along the value chain. Table 7 shows the relative increase of the Herfindahl Index within the different tiers, changing from a

Figure 4. Impact of increased relationship stability and connectivity on concentration

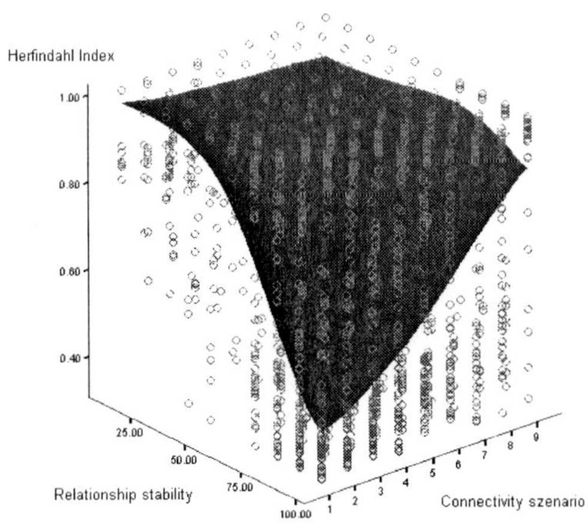

totally static communication scenario to different (locally restricted) dynamic communication relationships. The relationship stability (left row) describes the probability of communication with the main supplier.

The impact of the changes strongly differs in location and strength depending on the position of the initial increase of concentration. We can observe two main diffusion zones: retailer to manufacturer and supplier t1 to supplier t2. These clusters can be explained by the network topology and the agents' market power. Within a linear process structure, like the automotive supply chain, the standard has to promote from tier to tier. The diffusion process can be blocked by powerful, central agents, resulting in a lower standardization within the network.

Table 8 also underscores our thesis that diffusion clusters exist in the automotive supply network. It shows the correlations (0.1%) between the first standard's concentration rates in the different supply chain tiers. The correlations were derived from 500 simulations and vary significantly for the connected tiers. The correlation between the manufacturer and the t1 supplier is relatively low (0.654), whereas the correlation between the other directly connected tiers is higher (0.874 and 0.734). The low correlation between the manufacturer and the t1 supplier testifies to the existence of a diffusion barrier.

The Automotive and the Paper Industries in Comparison

Finally, we simulated the diffusion processes in two different network structures: the automotive

Table 7. Impact of increased dynamic within individual tiers

Scenario	Retailer	Manufacturer	Supplier (t1)	Supplier (t2)
100%, 100%, 100%	0%	0%	0%	0%
50%, 100%, 100%	25%	29%	9%	7%
100%, 50%, 100%	0%	0%	4%	0%
100%, 100%, 50%	0%	0%	102%	111%
50%, 50%, 50%	38%	48%	113%	120%

Table 8. Correlation of the distribution of the first communication standard

	Retailer	Manufacturer	Supplier (t1)	Supplier (t2)
Retailer	1	0.874	0.547	0.353
Manufacturer	0.874	1	0.654	0.437
Supplier (t1)	0.547	0.654	1	0.734
Supplier (t2)	0.353	0.437	0.734	1

and the paper supply network. A strong tendency toward concentration within the paper industry and the coexistence of different standards within the automotive industry could be observed (see Figure 5). This outcome matches empirical findings: The European automotive industry deploys four relatively homogeneous standards for the supplier-manufacturer relationship (Verband der Automobilindustrie [VDA], Odette, Edifact, intermediate document [IDOC]; Fricke, 2004) and standards for automotive retail (STARS) for the manufacturer-retailer relationship (Nelson, Shaw, & Qualls, 2005), whereas the paper and publishing industry globally deploys one standard (PapiNet[6]; Pyzdrowski & Donnelley, 2003). Using our findings about the influence of relationship dynamics on the concentration process, we can explain the very high concentration of standards within the paper and publishing industry. This process is relatively flexible, whereas the automotive supply chain mainly consists of static relationships; that is, many manufacturers source complex components specially developed for them. Another reason for the findings is the different network topology resulting from different underlying processes. Because the power concentration is higher in the automotive industry, the formation of clusters within the automotive network is more likely than within the paper supply network.

CONCLUSION

The results prove our central hypothesis that network topology and participants' market power within supply chains strongly influence the diffusion process of communication standards.

1. High relationship stability (inflexible supply chain) leads to a decrease of concentration.
2. Supply chain connectivity strongly influences the outcome in two ways. First, an increase leads to a higher tendency toward the standards' concentration since an increased connectivity reduces the probability of subnetworks. Second, an increase in connectivity often leads to a shift of market pressure along the supply chains, leading to changed relevance of the tier organizations.
3. Market power and network topology strongly influence the propagation of standards along the supply chains. These results are confirmed by a recent study of Gerst and Bunduchi (2005) who recognize a strong impact of powerful firms in the automotive supply chain, particularly practiced by using the influence of few manufacturers on the ISO (International Organization for Standardization) standardization committee. In certain constellations, however, the

Figure 5. Concentration of communication standards (automotive and publishing industry)

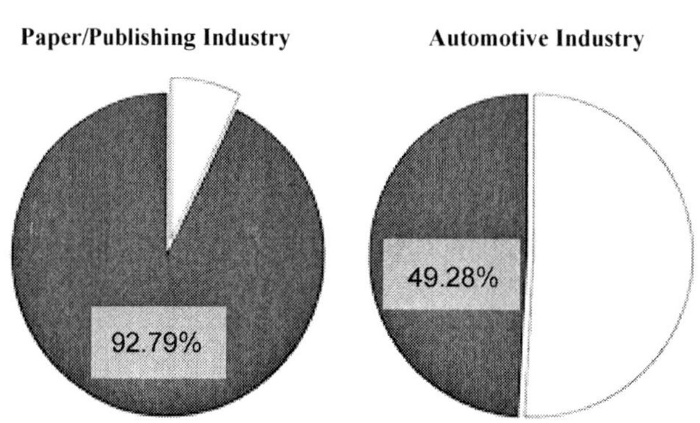

market pressure on organizations does not suffice to enforce a further propagation of the communication standard to the next tier. This leads to a low chance of a single standard's domination.

Based on our findings, the topological indicators for a tendency toward open standards can be summarized as follows.

- Low relationship stability
- Highly connected supply chains
- Low centrality of supply structures
- Homogeneous market power

The findings of this chapter provide a model for the impact of process structures on the diffusion of communication standards. Furthermore, they enable software manufacturers to develop better standardization strategies. The findings help to assess the risks of establishing propriety standards in process-based network structures (i.e., supply networks) and enable software providers to explicitly target diffusion barriers with differentiated pricing.

Another focus of future research should address latest findings for real-world topologies, like the discovery of the existence of small-world and scale-free properties in multiple occurrences of networks that are influenced by social structures (Hein, Schwind, & König, 2006). An empirical investigation of both supply chain structures regarded in this work with respect to the existence of such properties and their integration into our scenario might be a further step to more realistic simulation results.

REFERENCES

Arthur, B. (1989). Competing technologies, increasing returns, and lock-in by historical events. *The Economic Journal, 99*, 116-131.

Axelrod, R. (1997). The dissemination of culture: A model with local convergence and global polarization. *Journal of Conflict Resolution, 41*, 203-226.

Benton, W. C., & Maloni, M. (2000). Power influences in the supply chain. *Journal of Business Logistics, 21*, 49-74.

Biller, S., Bish, E. K., & Muriel, A. (2001). Impact of manufacturing flexibility on supply chain performance in the automotive industry. In J. Song & D. Yao (Eds.), *Supply chain structures: Coordination, information and optimization* (pp. 73-118). Boston: Kluwer Academic Publishers.

Buxmann, P. (1996). *Standardisierung betrieblicher Informationssysteme.* Wiesbaden, Germany: Gabler Verlag.

Buxmann, P., Weitzel, T., Westarp, F. von, & König, W. (1999). *The standardization problem: An economic analysis of standards in information systems.* Proceedings of the First IEEE Conference on Standardization and Innovation in Information Technology, SIIT '99, Aachen, Germany.

Caprile, M., CIREM Foundation, & Llorens, C. (2000). *Outsourcing und Arbeitsbeziehungen in der Automobilindustrie* (Tech. Rep.). Retrieved August 25, 2004, from http://www.eiro.eurofound.eu.int/2000/08/study/tn0008203s.html

Chow, G., Heaver, T. D., & Henriksson, L. E. (1995). Strategy, structure and performance: A framework for logistics research. *Logistics and Transportation Review, 31*(4), 285-308.

Cox, A. (2001). Managing with power: Strategies for improving value appropriation from supply relationships. *Journal of Supply Chain Management, 37*(2), 42-47.

Crain Communications. (2002, April). Car cutaways: An inside look at who supplies what on some of Europe's most important new models. *Automotive News Europe.*

Cubine, M., & Smith, K. (2001). Lack of communication, standards builds barriers to paper e-commerce. *Pulp & Paper Magazine, 75*(2), 32-35.

David, P. A. (1985). Clio and the economics of QWERTY. *The American Economic Review, 75*(2), 332-337.

David, P. A., & Greenstein, S. M. (1990). The economics of compatibility standards: An introduction to recent research. *Economics of Innovation and New Technology, 1*(1), 3-41.

Economides, N. (1989). Desirability of compatibility in the absence of network externalities. *The American Economic Review, 79*, 1165-1181.

Economides, N. (1996). The economics of networks. *International Journal of Industrial Organization, 14*, 673-699.

Economides, N., & Himmelberg, C. (1995). *Critical mass and network size with application to the US fax market* (Discussion Paper No. EC-95-11). NYU Stern School of Business.

Economides, N., & Lehr, L. (1994). The quality of complex systems and industry structure. In W. Lehr (Ed.), *Quality and reliability of telecommunications infrastructure.* Hillsdale, NJ: Lawrence Erlbaum.

Economides, N., & Salop, S. C. (1992). Competition and integration among complements, and network market structure. *The Journal of Industrial Economics, 40*(1), 105-123.

Farrell, J. G. (1984). Installed base and compatibility: Innovation, product preannouncements, and predation. *The American Economic Review, 76*, 940-955.

Fricke, M. (2004). *Information logistics in supply chain networks: Concept, empirical analysis, and design.* Stuttgart, Germany: Ibidem-Verlag.

Gerst, M., & Bunduchi, R. (2005). Shaping IT standardisation in the automotive industry: The role of power in driving portal standardisation. *Electronic Markets, 15*(4), 335-343

Hein, O., Schwind M., & König, W. (2006). Scale-free networks: The impact of fat tailed degree distribution on diffusion and communication processes. *Wirtschaftsinformatik, 48*(4), 267-275.

Hertwig, M., Mühge, G., & Tackenberg, H. (2002). The impact of e-business on the organization of the German automobile supply industry. *Zeitschrift für die Gesamte Wertschöpfungskette Automobilwirtschaft, 6*, 17-25.

Katz, M. L., & Shapiro, C. (1985). Network externalities, competition and compatibility. *The American Economic Review, 75*, 424-440.

Katz, M. L., & Shapiro, C. (1986). Technology adoption in the presence of network externalities. *The Journal of Political Economy, 94*, 822-841.

Kranton, R. E., & Minehart, D. F. (2001). A theory of buyer-seller networks. *American Economic Review, 91*(3), 485-508.

Leydesdorff, L. (2001). Technology and culture: The dissemination and the potential "lock-in" of new technologies. *Journal of Artificial Societies and Social Simulation, 4*(3).

Liebowitz, S., & Margolis, S. (1995). Path dependence, lock-in and history. *Journal of Law, Economics and Organization, 11*, 205-226.

Matutes, C., & Regibeau, P. (1998). "Mix and match": Product compatibility without network externalities. *The RAND Journal of Economics, 19*(2), 221-234.

Nelson, M., Shaw, M., & Qualls, W. (2005), Interorganizational system standards: Development in vertical industries. *Electronic Markets, 15*(4), 378-392.

Pyzdrowski, R., & Donnelley, R. (2003). *E-enabling an efficient supply chain.* Retrieved from http://www.papinet.org

Röder, H. (2002). *Studie: Struktur und Marktanalyse der Holz verbrauchenden Industrie in Nordrhein-Westfalen* (Tech. Rep.). Retrieved August 25, 2004, from http://www.forst.nrw.de/nutzung/cluster/6_1.Absatzstufe.pdf

Saloner, G., & Farrell, J. (1985). Standardization, compatibility and innovation. *RAND Journal of Economics, 16*(1), 70-83.

Shurmer, M., & Swann, P. (1995). An analysis of the process generating de facto standards in the PC spreadsheet software market. *Journal of Evolutionary Economics, 5*(2), 119-132.

Stockheim, T., Schwind, M., & König, W. (2003). A model for the emergence and diffusion of software standards. *Proceedings of the 36th Hawaii International Conference on System Sciences* (pp. 59-68).

Tesfatsion, L. (2002). Agent-based computational economics: Growing economies from the bottom up. *Artificial Life, 8*, 55-82.

Valente, T. W. (1995). *Network models of the diffusion of innovations.* Cresskill, NJ: Hampton Press.

Verband der Automobilindustrie e.V. (2003). *Auto jahresbericht.*

Weitzel, T., Wendt, O., Westarp, F. von, & König, W. (2003). Network effects and diffusion theory: Extending economic network analysis. *The International Journal of IT Standards & Standardization Research, 2*, 1-21.

Wendt, O., & Westarp, F. von. (2000). *Determinants of diffusion in network effect markets.* Proceedings of the IRMA International Conference, Anchorage, AK.

ENDNOTES

1 A supply networks is defined as a set of parallel, partially inter-connected value chains leading to particular products.

2 The number of systems sharing the same standard is called Installed Base (IB) (Saloner & Farrell, 1984).

3 The Standardization Problem was postulated by Buxmann (1996) in his PhD-thesis.

4 The installed-base refers to the customer base of digital products. Due to the fact that a communication standard is to a greater or lesser extent an interface to an existing ERP system the term installed base seem suitable.

5 Only implemented standards are considered ($s \neq 0$)

6 The paper and publishing industry shows a strong tendency towards a unified communication standard called PapiNet (http://www.papinet.org) which is growing heavily (50% new members in 2003). Since the value of a unified standard was recognized early in the paper and publishing industry, almost all known EDI-standards for this industry (PMLsm, GCA XML, WoodX, XBits) are integrated into PapiNet to this day (Cubine, 2001).

Chapter VIII
Scope and Timing of Deployment:
Moderators of Organizational Adoption of the Linux Server Platform

Joel West
San Jose State University, USA

Jason Dedrick
University of California—Irvine, USA

ABSTRACT

Here we present a qualitative study of how organizations do (or do not) adopt a new computer server platform standard; namely, Linux using PC-compatible hardware. While discussions of Linux typically focus on its open source origins, our respondents were interested primarily in low price. Despite this relative advantage in price, incumbent standards enjoyed other advantages identified by prior theory; namely, network effects and switching costs. We show when, how, and why such incumbent advantages are overcome by a new standard. We find that Linux adoption within organizations began for uses with a comparatively limited scope of deployment, thus minimizing network effect and switching costs disadvantages. We identify four attributes of information systems that potentially limit the scope of deployment: few links of the system to organizational processes, special-purpose computer systems, new uses, and replacement of obsolete systems. We also identify an organizational level variable—internal standardization—which increases scope of deployment and, thus, the attractiveness of the incumbent standard.

INTRODUCTION

Economic theories have suggested how individual consumers make decisions between two or more de facto product compatibility standards. Positive network effects, mediated by the supply of third-party complements, make the more popular standard more attractive to potential adopters (Katz & Shapiro, 1985). However, in many cases, adopters choose standards (or their associated products) for much simpler reasons: they have a relative advantage on some dimension of price, performance, or features (Liebowitz & Margolis, 1994).

The research on network effects generally examines choices between contemporaneous standards rather than successive generations (see Sheremata, 2004, for a rare exception). However, potential users often consider adoption of a new standard in light of an investment in a prior standard, with switching costs discouraging adoption of a new standard (Klemperer, 1987; Greenstein, 1993). Given the advantages that incumbent standards generally hold in network effects and prospective switching costs, a new standard must enjoy some relative advantage on another dimension in order to attract new adopters. For example, successive generations of videogame consoles displaced investments in earlier consoles and software libraries by offering superior graphics and realism of play (Gallagher & Park, 2002).

Within this accepted theory, there are some gaps in our knowledge. De facto competition models tend to cover aggregate decisions of rational individual adopters and do not suggest which adopters will be the first to adopt. Based on an empirical study of organizational information technology (IT) standards, we suggest that there are at least three ways in which organizational standards adoption differ from consumer ones. First, such organizational standards decisions involve multiple decision makers and perspectives. Also, a large organization typically will employ multiple standards simultaneously. Finally, there are important differences in the attractiveness of a new standard between systems within a single organization.

In order to try to explain such differences both between and within organizations, we focus on two research questions:

- How do organizations adopt new standards?
- How can a new standard can get adopted despite the network effects and switching costs that favor a successful incumbent technology?

As an example of an organizational standards decision in which a new standard recently has gained broad acceptance in the face of established incumbent technologies, we consider the selection of server platform standards. Using a qualitative study of management information systems (MIS) departments in 14 organizations, we look at the choice between three major server platforms: Windows, proprietary Unix, and Linux-based systems. Using inductive theory generation, we generate a set of propositions about factors that directly or indirectly influence organizational IT standards adoption.

We identify a new construct—scope of deployment—that considers the degree to which the adoption decision is coupled to other internal and external factors, including the organization's IT architecture, business processes, and supply of third-party complements. We show how the scope of deployment moderates the impact of external network effects and internal switching costs, in that new standards are most likely to be adopted for uses with a limited scope of deployment. We also suggest that the goals of internal technology standardization directly conflict with the opportunities offered by trial adoption of new standards.

THEORY

Research that directly considers how organizations adopt IT standards is comparatively rare (see West, 1999 and Hanseth & Braa, 2000, for notable exceptions). However, we can draw on two related literatures. The first considers the economic motivations for standards adoption, originally arrived from atomistic theories of individual consumers. The second focuses on organizational technology adoption but often ignores key factors such as the hardware-software paradigm that drive standards adoption in the economics literature.

Economics of Standards

In order to explain adoption decisions made between two or more competing technology standards, the most influential stream of research is that on the economics of standards. Such research identifies the roles of positive network effects and switching costs in cementing the lead of an established standard, with both effects mediated through the provision of specialized complementary assets (i.e., software libraries). Early and oft-cited examples of this stream include Katz and Shapiro (1985, 1986), Farrell and Saloner (1985), Teece (1986), and David (1987).

One benefit accruing to producers of established standards is the presence of switching costs. Among the first to consider such costs was von Weizsäcker (1984), who modeled how users would consider the net present value of anticipated future switching costs. Klemperer (1987) classified switching costs into three categories: transitory transaction costs, learning costs (e.g., IT worker skills), and contractual costs (e.g., contract termination penalties) deliberately introduced by vendors in order to build barriers to subsequent competitors. The exploitation of these costs by vendors has been referred to as *lock in* (Shapiro and Varian, 1999) or *groove in* (Arthur, 1996).

The other hypothesized factor in the economics of standards adoption is the role of positive network effects that accrue to all adopters of a popular standard.[1] Katz and Shapiro (1985) showed how an indirect network effect—the availability of software to support a given hardware standard—would make the more popular standard more attractive to future adopters. Both network effects and switching costs for a given standard are increased for products that require complementary assets (i.e., software) that must be adapted (or specialized) for that standard (Teece, 1986).

The empirical support for network effects is limited. Aggregate studies have imputed a network premium as the gap between price paid and that predicted from product features and other likely measures; if a higher market share led to a higher price, then researchers concluded that the higher price paid was accounted for by the positive-feedback value that accrued to buyers due to the larger network of adopters (Brynjolfsson & Kemerer 1996; Gallaugher & Wang, 2002). The importance of switching costs was supported by Greenstein (1993, 1997), who showed that U.S. federal government agencies preferred compatibility in their subsequent computer purchases of mainframe computers from 1971 to 1983. However, Liebowitz and Margolis (1994, 1999) disputed the empirical support for network effects and customer lock-in, attributing the success of various winning standards to relative advantage.

Finally, organizational standards decisions often focus on an architecture of related hardware, operating system, and middleware standards that form a computer platform (Morris & Ferguson, 1993; Bresnahan & Greenstein, 1999). Control of the value of the platform rests with the control of complementary assets, which, for a personal computer, means the programming interfaces for pre-packaged application software (West & Dedrick, 2000). Historically, vertically integrated computer companies controlled all layers of the platform, but with Unix (and later Linux), firms outsourced provision of the operating system, while Wintel PC makers delegated control of the entire platform to suppliers (West, 2003).

Organizational Technology Adoption

A variety of approaches has been used to consider how organizations adopt information technology standards for their own internal use. Generally, these studies fall into two camps: those that use network effects to predict aggregate market share and those that attempt to predict firm decisions without regard to network effects.

As an example of the former, Tam and Hui (2001) combined network effects with other product and manufacturer attributes to predict aggregate market share for mainframe and minicomputer vendors (which generally use a single de facto platform standard within a given product category). Similarly, Gallaugher and Wang (2002) used a combination of network effects and product attributes to estimate the hedonic (quality-adjusted) price for a variety of Web server packages.

At the other extreme, Chau and Tam (1997) treat the Unix-compatible operating systems standard (aka open systems) as an innovation in the context of Rogers (1983) and Tornastky and Fleischer (1990) but do not measure network effects or switching costs as a predictor of switching propensity. While not directly related to standards, the diffusion of innovation (DOI) framework of Rogers (1962, 1983) does offer insights into differences between adopters and, thus, potentially predicts who might be the earliest adopters. Also, Moore (1991) has argued that the DOI adopter categories can be used to predict whether a new technology will win widespread adoption.

Each approach has its limitations. The network effects models do not explain which firms are most likely to adopt a new standard (or when). The diffusion of innovations approach builds upon established theory in this regard but ignores the large body of work that establishes the importance of network effects in standards adoption. As a generalization of a person-centric communications model, it also has another limitation: it tends to assume the adoption decisions of individuals. However, within an organization, many technologies are "too big and complex to be grasped by a single person's cognitive power—or usually, to be acquired or deployed within the discretionary authority of any single organizational participant" (Eveland & Tornatzky, 1990, p. 133).

Also, both literatures tend to adopt a simplifying assumption that the decision process is based on either utility maximization or bounded rationality. Both, in turn, assume that organizations make rational decisions to best achieve an established objective such as profits or utility. However, there is a large body of research in MIS that shows that organizational technology decisions often are made based on factors such as the internal distribution of decision power and interorganizational politics (Dutton, 1981; Markus, 1983; Pinsonneault & Kraemer, 1993, 1997; Davis, 2000).

Finally, even knowing which organizations will adopt new IT standards does not explain how those standards diffuse within the organization. As modeled in the network effects literature, an individual consumer adopts a single VCR standard with an investment in hardware and content exclusively in that format. However, such an assumption is unrealistic for a large organization in which multiple competing standards may be utilized simultaneously for various purposes and organizational subunits.

Ideally, then, to explain organizational adoption of standards, we would want to be able to predict not only when a given class of organization will adopt a new standard but also where and why. By explaining inter- and intraorganization heterogeneity of adoption propensity, we would be able both to interpret the pace of standards adoption and also to offer managerially relevant predictions as to where adoption is most likely to happen next.

RESEARCH DESIGN

Platform Choices, Standards and Implementations

We chose to study the organizational choice of a computer platform as defined by Bresnahan and Greenstein (1999). Such a platform decision is a crucial standards-related decision for organizations, as it both constrains and is constrained by the choice of internal software systems, off-the-shelf application software, hardware peripherals, and related skills and services. At the same time, optimizing the platform decision is made more complex by the coupling of both hardware and operating system selections, since not all operating systems are available with all hardware systems.

Among such decisions, we chose to study server platforms for two reasons. First, at the time of our study, there was a wide range of economically viable platform standards. Unlike the single dominant desktop platform, for servers, there were three major categories: Unix, Wintel and Lintel (Table 1). Second, the three competing platforms reflect three distinct standards strategies: Unix servers using proprietary RISC-based processors, servers based on proprietary Microsoft Windows and commodity Intel-compatible hardware (Wintel), and those using the open source Linux operating system and the same commodity hardware (Lintel).

One theoretical confound is whether firms conceptualize platform adoption as adopting a set of standards or a specific implementation of those standards. The relationship of two constructs is clearly different between the case of an open multivendor standard and a proprietary de facto standard, and these differences were particularly salient for the operating system portion of the server platform.

Organizations often favor open standards, because they believe in the principle that the standard will allow them a choice of implementations, reducing lock-in; they also may believe such openness will attract adoption and, thus, a supply of complementary products (Gabel, 1987; West, 2006). Vendors thus must weigh appropriating rents through lock-in against the benefit that (perceived or actual) openness has in attracting adoption (West, 2003).

Table 1. Representative server platforms

	Proprietary		Open source
Platform	**Sun**	**"Wintel"***	**"Lintel"***
System			
Name	Sun Fire	PC-compatible	PC-compatible
Producer	Sun	commodity†	commodity†
Operating System			
Name	Solaris	Windows 2000	Linux/BSD
Producer	Sun	Microsoft	open source
APIs	Unix	Windows	Unix-derived
CPU			
Name	UltraSparc	Pentium	Pentium
Producer	Sun	Intel§	Intel§

* *"Wintel" = Windows/Intel; "Lintel" = Linux/Intel*

† *Available from both branded (Dell, HP, Gateway) and unbranded suppliers*

§ *Available from competing suppliers*

At the same time, organizations don't directly use open standards but rather the implementations of those standards. Adoption-related attributes such as support services actually are associated with a particular implementation (Krechmer, 2006). The lines between a standard and its implementation are further blurred in the case of a proprietary de facto standard.[2] If a for-profit entity has provided specifications for attaching third-party complements but has discouraged other firms from implementing those APIs in order to host the same complements (as with Windows), then attributes of the de facto specification are tantamount to those of its only implementation.[3]

Defining Linux as a standard or an implementation is especially problematic. Many open source projects (e.g., the sendmail mail transfer agent or the BIND domain name server) focus on providing a single implementation of external standards established through formal standardization efforts. For these Internet standards, the existence of an implementation aids the standardization process, but there is an assumption or even formal requirement for multiple implementations (West & Dedrick, 2001). However, the developers of Linux have eschewed more than 20 years of formal multivendor standardization efforts for Unix (Isaak, 2006) and seem intent on continuing Linux as an implementation-defined de facto standard akin to Windows.

However, one key aspect of Linux implementations is common across all open source projects. Firms have both the technical ability and the economic incentive to create their own variant implementations of Linux in a process called *forking* (West & Dedrick, 2001).[4] In this regard, the possibility of multiple (partially interoperable) implementations is always present. For Linux, these implementations are produced both by Linux distributors (e.g., Red Hat, Red Flag, SuSE) and non-profit projects (e.g., Slackware, Debian, Knoppix). In fact, the Linux Standard Base (Wu & Lin, 2001) has been established to provide formal Linux standardization in hopes of improving the application interoperability between these various de facto implementations.

Thus, Linux is like Windows in that the standard is defined as a de facto standard rather than the formal (IEEE-sponsored) standardization used for Unix. Linux, however, is like Unix in that it is a true multivendor standard, reducing lock-in; in fact, the barriers to creating new implementations (and thus, the threat of lock-in) is much lower than for Unix (West, 2003). Despite these differences, all three choices—Linux, Windows, and Unix—function as a platform. Switching costs (for training, commercial software, and internal software) are lower if firms make future computer choices within a given platform (Bresnahan & Greenstein, 1999). The organizations in our sample acted consistently with this view, in that selection of a platform appeared to be a long-term decision, but they more easily considered changing implementations within that platform.

Methods

While theoretical modeling is often used in the economics of standards literature, for our study of organizational standards adoption, we selected an empirical approach, specifically a comparative case study approach (Benbasat, Goldstein & Mead, 1987; Yin, 1994). We used established procedures for inductively generating theory from qualitative case study data (Glaser & Strauss, 1967; Eisenhardt, 1989), which can capture the complexity of an organizational IT adoption decision (Orlikowski, 1993).

We conducted a series of in-depth interviews from November 2002 through October 2004. We interviewed the CIO or other senior MIS executive, and, when possible, another person in the MIS department who was closer to the actual technical issues raised, such as a system administrator. We hoped both to develop a more complete picture and to provide a degree of data triangulation by comparing the responses of the two interviewees for consistency (Benbasat et al., 1987).

In order to draw inferences about a wide range of organizational adoption patterns, we used theoretical sampling (Glaser & Strauss, 1967) to capture a range of variation on key organizational variables, organizational size, and degree of adoption intensity. We continued to sample until we had interviewed organizations that were extensive users of each of the three major platforms and represented a different range of adoption of the new Lintel standard, from non-adoption through complete internal standardization on that platform (Table 2).

We studied standardization within the boundaries of a given MIS department, either for the entire organization or, in some cases, for an organizational subunit (e.g., a government research lab). In one case, we studied two subunits of a single organization: professional schools at separate campuses of the same public university. All 14 of our organizations were US-headquartered, and 11 of the organizations (or their subunits) were based in California.

After using the initial interview protocol in a pilot study at four sites, both the site selection criteria and protocol evolved during our study to capture new patterns identified during the research. The primary data consisted of semistructured interviews based on a common protocol. Interviews were conducted in person (or, in a few cases, by telephone) and typically lasted from 45 to 90 minutes. Field notes were taken, and the interviews were tape recorded and partially transcribed. Basic organizational data were collected via a questionnaire, with

Table 2. Characteristics of sample firms

Name	Business	Org. (unit) Size†	Primary Platform	Lintel Adoption	Informants
Beach Co.	Rec. equipment	80	Windows	Website only	1
Bio Branch	Pharmaceuticals	560 (150)	Linux	Predominant	1
Biotech	Pharmaceuticals	1,000	Unix	Internet and database applications	2
Dataco	Online data retrieval	2,700 (1,500)	Linux	Phasing out Unix	1
E-store	E-commerce	7,500	Unix	Shifting from Unix to Linux	1
FastFood	Restaurant chain	200,000	Mixed	None	1
FinCo	Financial services	130,000	Mixed	Partial adoption	1
NatLab	Government research lab	(8,000)	Unix	Phasing out Unix	2
ISP	Internet service provider	11	Linux	Since founding	1
NewMedia	Content provider	35	Unix	Partial transition	2
NorthU	Public university professional school	114,000 (325)	Mixed	Replacing Unix with Linux, while keeping Windows	3
Semico	Semiconductor design	2,500	Mixed	Limited; evaluating further use	2
SouthU	Public university professional school	114,000 (300)	Windows	Abandoned previous limited use	2
Travel Service	Travel-related reservations	6,000	Mainframe	Partial adoption	1

Total: 14 companies, 21 informants

† Size of parent organization (unit) in number of employees

background data for companies compiled from standard sources such as Hoovers and Dunn & Bradstreet. As needed, follow-up questions were asked by phone or by e-mail.

The final set of propositions was developed through interpretation, discussion, and anomaly resolution involving both authors, until we matched empirical patterns against multiple cases in the interview data against the propositions. Each proposition met the following criteria: it was based on comments of multiple respondents; it was emphasized strongly by at least one respondent; and there was a consistent pattern of responses, or conflicting evidence could be explained by differences in context. We also identified a logical chain of evidence from research questions to case study data to proposition, as recommended by Benbasat et al. (1987).

FINDINGS

Our findings address the question of when a new standard with a perceived relative advantage is likely to get adopted in the face of switching costs and positive network externalities that favor incumbent standards. In this case, respondents saw the relative advantage of the new standard primarily in terms of cost, as Lintel servers are viewed as a low-cost alternative to proprietary Unix servers. On the other hand, the incumbent Unix and Windows platforms had substantial advantages in complementary assets both in terms of software libraries and user skills, which presented a major hurdle for the Lintel platform.

In our interviews, we found that organizations were, in fact, weighing relative advantage, switching costs, and network externalities in their adoption decisions, as theory predicts. Of greater interest, we found that these factors are moderated by the scope of technology use and the organizational decision process. The interaction of the primary and moderating factors thus influence (1) which organizations are more likely to adopt the new standard and (2) when and how the standard will be adopted within these organizations.

The set of moderators is at the level of the information system (in this case, the server) and its intended use. They include (1) whether it is a special-purpose or general-purpose use; (2) whether the application impacts only the MIS department or whether it involves the core business of the broader organization; (3) whether the server is employed for a new use or whether it involves switching over an existing application from another platform; and (4) the timing of deployment, particularly when existing hardware is becoming obsolete or needs to be replaced. Each of these moderators explains intraorganizational differences in where and when a new standard is adopted. Adoption is more likely when the scope of deployment is more limited, as in the case of uses that affect only the IS department or in the case of special purpose computers. It also is more likely for new uses than for switching existing applications or when existing hardware is due for replacement.

The second set of moderator factors involves the decision-making process within the organization. These include (1) the skills, preferences, and distribution of power among decision makers; and (2) whether there is a preference or official policy favoring internal standardization. Adoption of the new standard is more likely when it is compatible with existing skills and preferences of more powerful decision makers but less likely when organizations have a policy of standardizing on a single server platform.

Factors Influencing Standards Adoption

As predicted by the foregoing theoretical discussion, we found that the main factors influencing adoption of the Lintel standard were its relative advantage and its compatibility with existing complementary assets. In addition, we found that support from third-party vendors was an important factor in encouraging adoption of a non-proprietary standard.

Relative Advantage

The relative advantage of Lintel platforms compared to those based on proprietary operating systems was perceived by MIS departments in terms of cost, performance, and fit to specific tasks.

Cost. The most often-mentioned advantage of the Lintel platform was cost. The commodity PC hardware used by Lintel systems was much cheaper than proprietary RISC-based Unix systems although not cheaper than Wintel servers using similar hardware. Seven of the 11 companies interviewed mentioned hardware cost as an important relative advantage of Linux.

The second advantage is software cost. Linux and its updates can be downloaded for free, making it cheaper than either a proprietary Unix OS or Windows. However, only three of the 10 companies stated that the cost of software was a significant factor in their decision whether to adopt Linux, perhaps because most organizations valued guaranteed support levels that required service contracts priced similarly to proprietary license fees.[5]

Performance and Reliability. Often-cited factors in the decision process were performance and reliability, but interviewees had mixed views. Lintel platforms generally were perceived as more reliable than Wintel but less reliable than proprietary Unix platforms.

The concern with performance was not absolute but relative to the requirements of specific applications. Several organizations were unwilling to switch mission-critical applications without proof that the Lintel platform reliability matched that of proprietary Unix systems. On the other hand, most organizations were willing to utilize Lintel servers for uses such as print and file servers, Web servers, and for applications that provide their own error recovery. We do have some evidence that this perception is changing. For instance, in 2002, Semico was only using Lintel servers for limited tasks, but in a follow-up discussion in 2005, the CIO reported that adoption was much more widespread in the organization.

Fit to Specialized Tasks. It was clear that Linux fit some tasks especially well, given its Unix roots. This includes Internet applications in particular, which helps to explain why the rapid diffusion of Linux paralleled the Internet boom of the late 1990s. ISP's CIO stated:

[The original partners] all pretty much agreed that Unix was the way to go—it's one of the core infrastructures for the Internet, and so they just realized that that's where all the Internet services and products were most mature, and so they wanted to continue with that. (Interview, 4/9/2003)

We offer the following proposition regarding the perceived relative advantage of a new standard.

- **Proposition 1.** A new standard is evaluated on relative advantage for a specific task and is more likely to be adopted if it has a clear advantage in terms of cost, performance, or fit for that task.

Compatibility

The decision to adopt open source platforms appears to be influenced greatly by the compatibility of the new technology with current technologies, skills, and tasks. Greater compatibility lowers switching costs, as it is easier to redeploy existing assets to the new standard.

Applications. Every firm agreed that compatibility with current applications was a major concern in the adoption decision. For most, the issue was the availability of third-party applications; a few sites were concerned with compatibility with internally developed applications, where the organization would have to pay any conversion cost.

The importance of application compatibility supports research on the role of complementary assets (Farrell & Saloner, 1985; Katz & Shapiro, 1986). The impact does not depend as much on the overall pool of complementary assets as on whether the new platform is compatible with applications that the site is now using, considering, or developing. Consistent with West (2005), for server standards we found that most users satisfice (require specific applications) rather than prefer the platform with the largest variety of applications.

Skills. Compatibility with internal IT skills and with skills available in the labor market was another key issue; because Linux is a Unix clone, adopters were concerned with the internal and external supply of skilled Unix administrators. Our respondents were split between organizations that primarily used Unix-based servers (so-called Unix shops) and those that were primarily Windows-based (Microsoft shops). In Tushman and Nadler's (1986) terms, the transition to Linux is incremental for Unix shops where skills are easily transferable, but discontinuous for Microsoft shops that lack such skills.

For smaller organizations, in particular, compatibility with current skills was a major concern. SouthU's CIO said development systems were not chosen for their acquisition costs but for "the x-ty thousand dollars a year plus training to have someone write in it" (Interview, 6/25/2003). On the other hand, those companies that were already heavy Unix users (i.e., Semico, Biotech, and NewMedia) stated that this greatly eased the shift to Linux. A fourth (ISP) selected Linux at the time of inception, largely due to the Unix background of the CIO (our informant).

The CIO of NewMedia felt that having his engineers make a transition from Solaris to Linux was not difficult but that going to Windows would be harder. By contrast, FastFood has a mix of mainframe, Unix, and Windows servers but is predominantly a Microsoft shop with Windows skills; the interviewee predicted that this would be an obstacle to widespread adoption of Linux. Both FastFood and SouthU felt that it would be more difficult to find system administrators with the necessary skills to handle the more complex requirements of a Linux environment. From these findings, we offer the following proposition

- **Proposition 2.** MIS managers will choose standards that are compatible with existing applications and skills. A new standard is more likely to be adopted the more existing applications and skills can be redeployed to that standard.

Vendor Support

The Linux operating system is not owned by any one organization but is developed by an open source community. A potential concern about adopting an open source platform is the lack of support services from a single reputable vendor, as is available on proprietary platforms. The CIO of SouthU stated:

I'm nervous about open source. I'm not paying anybody to support it, and thus, I'm depending largely on goodwill and luck and skill of my own people and soliciting solutions from other people for free. That explanation looks amateurish when you offer it to a dean or to a faculty when a production system is down in my opinion. (Interview, 6/25/2003)

Most of our respondents were reassured by the availability of Linux support from major systems vendors or Linux distributors. As FastFood stated, "Support from major vendors like IBM and HP would be important to us. It gives a little bit of a safety net" (Interview, 3/6/2003).

Three larger companies—FastFood, Biotech, and Semico—cited vendor support as being important in considering open source.[6]

Major vendor support also prompted several respondents to increase their belief in the long-term viability of Linux. In the words of Semico's CIO, "If the world is moving to a Linux standard, even over a very long time frame, you don't want to be on the wrong path following a proprietary standard." This concern over standards viability is consistent with the standards literature, and we find that vendor support not only has value as a complementary asset but is a signal of viability. Based on this finding, we frame the next proposition.

- **Proposition 3.** Organizations are more likely to adopt a new technology standard if it is supported by reputable vendors who can provide necessary technical support and whose sponsorship increases the perceived future viability of the standard.

SYSTEM-LEVEL MODERATORS

We find that the scope of deployment of a new system is a moderator of the impacts of switching costs and network externalities, as limited scope deployments do not impact the broader organization and do not require compatibility with the full range of an organization's applications. The timing of deployment likewise can moderate switching costs. When deployed for a new use, or when an existing technology is reaching obsolescence, switching costs can be much lower. In both cases, the likelihood of adopting a new standard is greater.

Scope of Deployment

Depending on the nature of the system, IT adoption decisions can be made entirely within the information systems (IS) department or may be influenced by others inside or outside the organization. Swanson (1994) defines three types of information technology innovations. Type I innovations are restricted to the IS task and mainly affect the IS unit; Type II innovations support administration of the business and affect the broader organization; Type III innovations involve the core technology of the business and affect both the organization and its customers and other external partners. In this typology, adoption of a server platform would be a Type I innovation whose use has little impact beyond the IS department. As Biotech's CIO said, end users "don't know, don't care" what platform is used on the server side. As such, switching costs associated with adopting a new standard are lower than in the case of Type II or III technologies, in which users within and beyond the organization are affected.

Respondents also suggested that adopting a new platform was easier for single-purpose systems than for general purpose ones. We found many examples of single-purpose servers, whether running an SAP module (Semico) or a Web or print server (several sites). For example, in order to save money, E-store shifted its Web servers from Unix to Lintel; whether they could run any other application was irrelevant. In the case of a single-purpose system, network externalities associated with a large library of applications do not favor incumbent standards the way they do with general-purpose systems such as desktop PCs or mainframe computers.

These characteristics of the scope of deployment lead to two related propositions.

- **Proposition 4.** The cost to an organization of adopting a new technology standard will be lower for Type I technologies whose impact is limited to the MIS organization.

- **Proposition 5.** The cost to an organization of adopting a new technology standard will be lower for special-purpose information systems than general-purpose ones, and the impact of network externalities associated with a large software library is minimized.

Timing of Deployment

While not reviewed every time a new server is purchased, our organizations faced server platform decisions fairly frequently, suggesting that firms are not locked in by previous decisions for a long period of time. Standards decisions mostly came up under two circumstances: New uses, and hardware retirement or obsolescence.

New Uses. Most adoptions of Lintel were for new applications, most commonly when Internet infrastructure was being created or expanded, cited by eight of the 11 organizations.

Biotech's associate director said:

Q: *Are most of your Linux applications ones that you switched from other platforms or are they new applications?*
A: *I would say that since I've been here it's mostly been new applications. We don't have a lot of time to go back through our existing systems and say, "Hey, can we do this better on Linux? What are the cost benefits here?" If we did have that time, we would do it.* (Interview, 9/5/2003)

Hardware Retirement or Obsolescence

For an existing use, the current hardware may be phased out, or the cost of keeping it running becomes prohibitive. For instance, both Semico and Biotech considered Lintel on systems where the platform hardware or operating system was reaching end of life. Others considered changes when newer hardware offered superior price/performance.

When a server is being employed for a new use or when current hardware is being phased out, the cost of adoption or switching is minimized. Therefore, we predict Proposition 6.

- **Proposition 6a.** When a technology is deployed for a new use, switching costs are reduced or eliminated. Thus, organizations are more likely to adopt a new standard for new uses than for existing applications.
- **Proposition 6b.** Switching costs are lowest when existing technologies are becoming obsolete or at the end of a replacement cycle. Thus, organizations are more likely to adopt a new standard when the incumbent systems are due to be replaced for other reasons.

ORGANIZATIONAL EFFECTS

Individual-Level Factors

While we studied the decisions of organizations, the actual decisions were made by individuals. Not surprisingly, the interpretation of various attributes and relative advantages and disadvantages are subjectively interpreted through the biases of the existing MIS decision makers. Respondents identified similar goals, including making life easier for MIS staff, reducing costs, ensuring compatibility with key applications, and having staff with skills necessary to support the platform. However, we found differences between organizations in their approaches toward meeting those goals with a crucial split between MIS departments with IT skills predominantly in Unix or Windows.

The Unix shops preferred staying with a Unix variant than switching to Windows, arguing that Unix IT programmers and administrators easily could learn Linux, thus minimizing switching costs. An alternate (but unstated) explanation might be that staying with a Unix clone provided job security for the existing staff and reinforced their power within the organization.

Likewise, small Windows shops such as SouthU or BeachCo. would not consider a standard that could not be supported by current staff. The skills of Windows administrators and programmers could not be redeployed easily to work on the Lintel platform, so switching to Lintel would require hiring new staff. In one case, the CIO had strong knowledge of the Wintel platform, so a switch to Lintel would dilute the value of his expertise.

In our interviews, we saw how the delegation of power influenced the evaluation of the various standards attributes and, thus, the ultimate adoption outcome. For example, the CIO of SouthU decided to standardize on Wintel servers across the organization but deferred to our other informant when deciding to switch to the Apache Web server for a particular application, stating his trust in the informant's expertise in this area. At Semico, the CIO stated that the person driving the server decision for the SAP module was a systems administrator, because the decision was not considered strategic. When Biotech hired a Linux expert and enthusiast as Associate Director of IT Infrastructure, that enthusiast stated:

My internal policy is for every product we deploy, I always ask, "Can we do this on Linux?" Sometimes the answer is no, but I'll always ask the question "What's the best OS to do this on?" And implicit in that statement is Linux is one of the choices. (Interview, 9/5/2003)

Internal Standardization

Studies of consumer standards decisions assume that individuals will select one standard for each product category. Organizational technologies such as computer servers are different since organizations often support multiple standards adopted at different times for different functions. Organizations must decide whether to standardize their choices internally or to support more than one standard.

In our sample, some had standardized completely or mostly on a single platform, while others supported mixed server platforms. Both those who had standardized—and even some who had not—recognized the tangible benefits of internal standardization, such as ease of hardware and software administration and maintenance. The CIO of E-store stated, "We run the same configuration everywhere. We try very hard to keep it standard across the board. Why create your own problems?" (Interview, 12/12/2003). Another benefit was reducing the scope of skills needed. SouthU specifically avoided non-Windows platforms, because it did not have and did not want to hire skills to support other platforms. Even at Semico, which supported multiple platforms, the CIO stated that it would be advantageous to have one ubiquitous server platform; one reason that he was considering adopting Lintel was that if he did so, he eventually would need only one staffer with a deep knowledge of Linux rather than maintaining deep expertise on a number of platforms.

Respondents also identified disadvantages of complete standardization, such as the risk of becoming too dependent on one vendor. Biotech's CIO said that he specifically avoided standardizing on Windows in order to limit Microsoft's leverage. Internal standardization also limits an organization's ability to select technologies that are best suited for a particular task. For instance, Semico moved one module of SAP to a Lintel server while keeping the more critical database engine on an HP-UX server, thus taking advantage of the low cost of Lintel for one task while enjoying the greater perceived reliability of HP-UX and PA-RISC hardware for a more demanding task. E-store used a similar staged Lintel deployment.

We found that internal standardization reduced the likelihood of adopting a new standard. In our sample, both Beach Co. and SouthU standardized on Wintel servers and, thus, were not generally considering the Lintel standard. ISP was unique since it had standardized on Linux at its inception (a startup in the late 1990s). All the remaining sites were mixed shops either adopted or actively evaluating Linux for some uses.

The effect of internal standardization was to increase the scope of deployment of a standard, as organizations would have to adopt the new standard across the board. This increases switching costs, as all applications eventually must be moved to the new standard. It also increases the importance of network externalities associated with the size of the library of complementary software, as all of the organization's applications must support the new standard.

- **Proposition 7.** A policy of internal standardization increases the scope of deployment of a new standard, hence increasing switching costs and the importance of complementary assets. Organizations that value internal standardization thus are less likely to adopt a new standard.

DISCUSSION

This study advances our understanding of how organizations consider and select IT standards, combing empirical data with previously disjointed research in economics of standards and MIS technology adoption to produce a moderated model of standards selection. This moderated model subsumes competing predictions about the reasons for standards (non-) adoption.

Linking Organizational to Aggregate Adoption Patterns

Long-established models of de facto standards competition predict aggregate patterns of standards adoption but say little about who the earliest adopters will be. As with other new technologies, there are conditions under which standards do not become widely adopted, as the diffusion of innovation literature suggests (Rogers, 1983; Moore, 1991). Both theories offer little insight into the motivations or patterns of organizational adoption.

From our field study, we offer a model (Figure 1) that incorporates familiar constructs (Table 3) from de facto standards literature (network effects, switching costs) as well as from the diffusion of innovations (relative advantage, compatibility). But in addition to these (long-accepted) main effects, we identify two important moderators that influence when, how, and why organizations adopt new standards.

New platform standards start out with a twofold disadvantage: the switching costs associated with moving from an old standard to a new standard and the large library of complements (fueling indirect network effects) that have accrued to the incumbent standard. We found that organizations are most able and willing to adopt new standards when the scope of deployment is limited; such scope increases the more that the decision about an individual system standard is linked (either economically or organizationally) to other decisions and commitments both inside and outside the organization. Thus, a limited scope of deployment (such as a back-office server) reduces the switching costs incurred by the firm. In parallel to this, more limited uses such as specialized systems that serve a single purpose allow adoption of standards that have a few key complements, even if they lack the larger library enjoyed by the incumbent standard.[7]

One key implication of this finding is the impact of intraorganizational standardization. The use of organizationwide standardized architectures (Feld & Stoddard, 2004) has been claimed to offer important efficiency and control benefits for IT managers. But such standardization inherently broadens the scope of deployment for any adoption decision, making it more difficult for such organizations to gain the information and skills necessary to evaluate and use new standardized technologies.

The timing of potential deployment also provides organizations with key windows of opportunity for considering new platform standards. One, of course, is when faced with new IT uses.

Figure 1. Moderated model for organizational adoption of a new standard

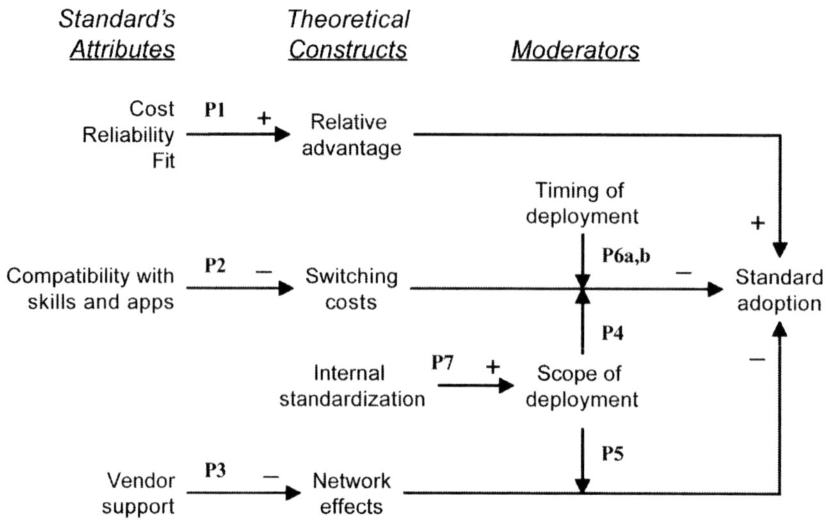

Table 3. Construct definitions for causal model

Construct	Definition
Relative advantage	The degree that the new standard has an advantage relative to the incumbent standard(s) in use, as perceived by users. These include cost, performance, reliability and fit to particular tasks. The relative advantage is measured in terms of performing a particular task or set of tasks, and may be perceived quite differently by different individuals or organizations.
Switching costs	The costs that would have to be borne to switch to a new standard. These include new training for users and for IT staff, buying new software licenses for packaged applications, porting custom applications to a new platform, and changing work processes to accommodate the requirements of a new standard.
Internal standardization	The extent to which the organization attempts to use the same platform standard for all uses throughout an organization.
Network effects	The relative advantage of the incumbent standard over the new standard, as results from the number of others using each standard, and the availability of external complements for each standard. This value is based users' perceptions of both current conditions and expectations of future network size and availability of complements.
Decision maker preferences	The extent to which the preferences, biases and interests of the decision make
Timing of deployment	The degree to which the timing of deploying a new system reduces switching costs, either because there is no existing system or because the existing system no longer has value.
Scope of deployment	The extent to which a technology deployment impacts the organization and interacts with complements. This includes four variables: Technology type (I, II or III); general versus special purpose use; new use versus switching; and degree of internal standardization.
Standard adoption	The likelihood that the new standard will be adopted for a given task

For example, when the World Wide Web became popular in the mid-1990s, firms not only considered a new de facto standard for Web server applications[8] but also installed new computer hardware and operating systems to run these Web servers; here, the adoption of the Apache Web server encouraged firms to adopt the Linux platform standard.

To a lesser degree, firms also have a greater likelihood to consider new standards when the existing systems are due to be replaced anyway, such as when hardware becomes obsolete or software license terms expire, since this reduces the switching costs for using new technologies. In our study, decision makers recognized this window of opportunity and seemed inclined to widen their evaluation processes to consider new standards rather than (as in other times) be constrained implicitly by the prospect of switching costs.

Finally, the attributes of a new standardized product, whether its relative advantage, switching costs, or network effects are subjective evaluations, are made by one or more decision makers, not an externally verifiable objective measure. When comparing different organizations making similar standards decisions, we found (as did West, 1999) that the decision makers differed in their interpretations of the relative advantage and switching costs relevant to their organizations. Even allowing for differences in requirements, organizations with very similar needs arrived at differing conclusions (e.g., NorthU vs. SouthU). Even for the most objective measure (the size of the external library of complements), decision makers differed in their interpretations of the importance of the differences in software availability; in this case, when comparing Linux to the more established Unix and Windows platforms.

LIMITATIONS AND FUTURE RESEARCH

In addition to our findings being particularistic to a given standards decision, our research design also limits our conclusions beyond the specific sample. While multiple qualitative case studies can build theory in emergent areas that is grounded in empirical data, such theory always runs the risk of being idiosyncratic and not generalizable to the entire population and, thus, should be tested using other methods (Eisenhardt, 1989). There is also the risk of attempting to generalize from one particular standards contest to another (whether VCRs or server platforms), as the dynamics and history of standards contests clearly differ from case to case (Grindley, 1995).

The moderating effects of deployment timing and scope as well as the corresponding implications for theory and practice call for further research to establish their generalizability. For example, in looking at 30 years of computer industry platform competition, Bresnahan and Greenstein (1999) identified the importance of new uses for reducing the barriers protecting the existing platform. But they also concluded that each new platform eventually achieved indirect entry, as its domain gradually overlapped that of the incumbent.

Their study specifically looked at general-purpose computer systems. We contend that special-purpose computer systems are different and, thus, always may retain a niche in the intraorganizational standardization decision. At the intraorganizational level, strong internal standardization would reduce the likelihood of considering special-purpose systems, while at an interorganizational level, general-purpose computer systems such as those studied by Bresnhanan and Greenstein (1999) have benefited strongly from economies of scale and scope. Thus, it remains an empirical question as to how frequently such limited scope opportunities occur, how long the opportunity remains, and whether they inevitably lead to competition with established, general-purpose technologies.

There is also the unanswered question of generalizing to other layers of a standards architecture (in the sense of West and Dedrick, 2000). Our findings suggest that strong internal platform standardization rules out platform experimentation and, thus, is a risk to the standardization prescriptions of Feld and Stoddard (2004). Their normative advice focuses on the benefits of experimentation above the platform layer, while assuming that the need to change platforms arises rarely, at best. So, how often do significant new platform opportunities arise? While the platform definition here (Bresnahan & Greenstein, 1999) is defined narrowly as a processor and an operating system, does the process of limited scope deployment also apply to deeper platforms such as a computer system plus middleware? Does it apply to other non-platform standards decisions such as peer-to-peer networking protocols or application file formats? These remain open empirical questions.

ACKNOWLEDGMENTS

Earlier papers from this study were presented at the HBS-MIT Sloan Free/Open Source Software Conference (June 2003), the Workshop on Standard Making: A Critical Research Frontier for Information Systems (December 2003), and HICSS-37 (January 2004). We thank the participants for their useful feedback and also acknowledge the helpful guidance provided by editor Kai Jakobs and four anonymous JITSR reviewers.

REFERENCES

Arthur, W.B. (1996). Increasing returns and the new world of business. *Harvard Business Review, 74*(4), 100-109.

Benbasat, I., Goldstein, D.K., & Mead, M. (1987). The case research strategy in studies of information systems. *MIS Quarterly, 11*(3), 369-386.

Bresnahan, T.F., & Greenstein, S. (1999). Technological competition and the structure of the computer industry. *Journal of Industrial Economics, 47*(1), 1-40.

Brynjolfsson, E., & Kemerer, C.F. (1996). Network externalities in microcomputer software: An econometric analysis of the spreadsheet market. *Management Science, 42*(12), 1627-1647.

Chau, P.Y.K, & Tam, K.Y. (1997). Factors affecting the adoption of open systems: An exploratory study. *MIS Quarterly, 21*(1), 1-24.

David, P.A. (1987). Some new standards for the economics of standardization in the information age. In P. Dasgupta, & P. Stoneman (Eds.), *Economic policy and technological performance* (pp. 206-239). Cambridge: Cambridge University Press.

Davis, C.K. (2000). Managing expert power in information technology. *Journal of Database Management, 11*(2), 34-37.

Dutton, W.H. (1981). The rejection of an innovation: The political environment of a computer-based model. *Systems, Objectives, Solutions, 1*(4), 179-201.

Eisenhardt, K.M. (1989). Building theories from case study research. *Academy of Management Review, 14*(4), 532-550.

Eveland, J.D., & Tornatzky, L.G. (1990). The deployment of technology. In L.G. Tornatzky, & M. Fleischer (Eds.), *The processes of technological innovation* (pp. 117-147). Lexington, MA: Lexington Books.

Farrell, J., & Saloner, G. (1985). Standardization, compatibility & innovation. *Rand Journal of Economics, 16*(1), 70-83.

Feld, C.S., & Stoddard, D.B. (2004). Getting IT right. *Harvard Business Review, 82*(2), 72-79.

Gabel, H.L. (1987). Open standards in computers: The Case of X/OPEN. In H.L. Gabel (Ed.), *Product standardization and competitive strategy* (pp. 91-123). Amsterdam: North Holland.

Gallagher, S., & Park, S.H. (2002). Innovation and competition in standard-based industries: A historical analysis of the U.S. home video game market. *IEEE Transactions on Engineering Management, 49*(1), 67-82.

Gallaugher, J.M., & Wang, Y.-M. (2002). Understanding network effects in software markets: Evidence from Web server pricing. *MIS Quarterly, 26*(4), 303-27.

Glaser, B.G., & Strauss, A. (1967). *The discovery of grounded theory: Strategies of qualitative research*. London: Wiedenfeld and Nicholson.

Greenstein, S.M. (1993). Did installed based give an incumbent any (measurable) advantages in federal computer procurement? *Rand Journal of Economics, 24*(1), 19-39.

Greenstein, S.M. (1997). Lock-in and the costs of switching mainframe computer vendors: What do buyers see? *Industrial and Corporate Change, 6*(2), 247-274.

Grindley, P. (1995). *Standards, strategy, and policy: Cases and stories*. Oxford: Oxford University Press.

Hanseth, O., & Braa, K. (2000). Who's in control: Designers, managers—Or technology? Infrastructures at norsk hydro. In C.U. Ciborra (Ed.), *From control to drift—The dynamics of corporate information infrastructures* (pp. 125-147). Oxford: Oxford University Press.

Isaak, J. (2006). The role of individuals and social capital in POSIX standardization. *International Journal of IT Standards & Standardization Research, 4*(1), 1-23.

Katz, M.L., & Shapiro, C. (1985). Network externalities, competition & compatibility. *American Economic Review, 75*(3), 424-440.

Katz, M.L., & Shapiro, C. (1986). Technology adoption in the presence of network externalities. *The Journal of Political Economy, 94*(4), 822-841.

Klemperer, P. (1987). The competitiveness of markets with switching costs. *Rand Journal of Economics, 18*(1), 138-150.

Krechmer, K. (2006). The meaning of open standards. *International Journal of IT Standards & Standardisation Research, 4*(1), 43-61.

Liebowitz, S.J., & Margolis, S.E. (1994). Network externality—An uncommon tragedy. *Journal of Economic Perspectives, 8*(2), 133-150.

Liebowitz, S.J., & Margolis, S.E. (1997). *Winners, losers & Microsoft: Competition and antitrust in high technology*. Oakland, CA: Independent Institute.

Markus, M.L. (1983). Power, politics and MIS implementation. *Communications of the ACM, 26*(6), 430-444.

Moore, G.A. (1991). *Crossing the chasm*. New York: HarperBusiness.

Morris, C.R., & Ferguson, C.H. (1993). How architecture wins technology wars. *Harvard Business Review, 71*(2), 86-96.

Orlikowski, W.J. (1993). CASE tools as organizational change: Investigating incremental and radical changes in systems development. *MIS Quarterly, 17*(3), 309-340.

Pinsonneault, A., & Kraemer, K.L. (1993). The impact of information technology on middle managers. *MIS Quarterly, 17*(3), 271-292.

Pinsonneault, A., & Kraemer, K.L. (1997). Middle management downsizing: An empirical investigation of the impact of information technology. *Management Science, 43*(5), 659-679.

Rogers, E.M. (1962). *Diffusion of innovations*. New York: Free Press.

Rogers, E.M. (1983). *Diffusion of innovations* (3rd ed.). New York: Free Press.

Shapiro, C., & Varian, H.R. (1999). *Information rules: A strategic guide to the network economy*. Boston: Harvard Business School Press.

Sheremata, W.A. (2004). Competing through innovation in network markets: Strategies for challengers. *Academy of Management Review, 29*(3), 359-377.

Swanson, E.B. (1994). Information systems innovation among organizations. *Management Science, 40*(9), 1069-1092.

Tam, K., Hui, K.L. A choice model for the selection of computer vendors and its empirical estimation. *Journal of Management Information Systems, 17*(4), 97-124.

Teece, D. (1986). Profiting from technological innovation: Implications for integration, collaboration, licensing and public policy. *Research Policy, 15*(6), 285-305.

Tornatzky, L.G. & Klein, K.J. (1982). Innovation characteristics and innovation adoption-implementation: A meta-analysis of findings. *IEEE Transactions on Engineering Management, 29*(1), 18-45.

Tushman, M.L., & Nadler, D. (1986). Organizing for innovation. *California Management Review, 28*(3), 74-92.

von Weizsäcker, C.C. (1984). The costs of substitution. *Econometrica, 52*(5), 1085-1116.

West, J. (1999). Organizational decisions for I.T. standards adoption: Antecedents and consequences. In K. Jakobs, & R. Williams (Eds.), *Proceedings of the 1st IEEE Conference on Standardisation and Innovation in Information Technology* (pp. 13-18). Aachen, Germany.

West, J. (2003). How open is open enough? Melding proprietary and open source platform strategies. *Research Policy, 32*(7), 1259-1285.

West, J. (2005). The fall of a Silicon Valley icon: Was Apple really Betamax redux? In R. Bettis (Ed.), *Strategy in transition* (pp. 274-301). Oxford: Blackwell.

West, J. (2006). The economic realities of open standards: Black, white and many shades of gray. In S. Greenstein, & V. Stango (Eds.), *Standards and public policy*. Cambridge: Cambridge University Press.

West, J., & Dedrick, J. (2000). Innovation and control in standards architectures: The rise and fall of Japan's PC-98. *Information Systems Research, 11*(2), 197-216.

West, J., & Dedrick, J. (2001). Proprietary vs. open standards in the network era: An examination of the linux phenomenon. In *Proceedings of the 34th Annual Hawai'i International Conference on System Sciences*, 5011.

Wu, M.-W., & Lin, Y.-D. (2001). Open source software development: An overview. *IEEE Computer, 34*(6), 33-38.

Yin, R.K. (1994). *Case study research, design and methods* (2nd ed.). Newbury Park, CA: Sage Publications.

ENDNOTES

1. Such effects were originally referred to as *network externalities*. However, after the analysis of Liebowitz and Margolis (1994) suggested that actual externalities are rare, subsequent researchers have used the term *network effect* (Shapiro & Varian, 1999).
2. This is one reason some researchers and standardization professionals are loathe to refer to any de facto proprietary software as a standard, even if it performs the same technical role of modularity and enabling complementary products as does a multi-vendor standard. Recent research has suggested that previous bifurcation between the extremes of open and proprietary standards is oversimplified, and stakeholders disagree in the importance they ascribe to the various dimensions of "open"-ness (Krechmer, 2006; West, 2006).
3. While there effectively may be only one supplier of Windows implementations, there are multiple implementations of the Windows platform, in that competing PC vendors provide their own hardware implementations that share a common OS implementation (West & Dedrick, 2000).
4. Some advocates (particularly of free software) have claimed that copyleft licenses can be used to compel the return of individual changes and, thus, prevent forking. In practice, Linux forking continues to be a problem, not because the changes are unavailable but because the various forks are addressing heterogeneous needs and have weak incentives to adopt modifications irrelevant to their own needs.
5. The role of price could be seen in NorthU and SouthU, who said that Linux did not have cost advantage because Microsoft's education discounts meant that its server products cost almost the same as Red Hat's products.
6. In the abstract, support by software developers for the Linux platform conceivably could include any implementation that conforms to the Linux API standards. As a practical matter, support service commitments by application or systems vendors are limited to a specific subset of implementations (e.g., Red Hat or SuSE). Of course, Linux distributors provide support services only for their own particular compilation of Linux and related software, where the act of compilation defines a specific implementation.
7. This is an organizational-level manifestation of the aggregate observation made by Bresnahan and Greenstein (1999) that new platforms can become established if they serve new niches. Bresnahan and Greenstein (1999) consider new-to-the-world uses, while our study looks at new-to-the-organization uses.
8. Some could argue that a choice among the Apache, NCSA, Netscape, or Microsoft Web servers is merely a choice of specific implementations of the open HTML and HTTP standards. However, during key periods of competition, each of these servers had slightly different interpretations of the standard and/or proprietary extensions. More seriously, each of these servers, in turn, comprised a middleware platform on which other technologies could be layered through plug-ins, modules, and other internally or externally sourced complements, as when IBM decided to base its WebSphere e-commerce system on the Apache Web server (West, 2003).

This work was previously published in The Journal of IT Standards & Standardization Research, 4(2), edited by K. Jakobs, pp. 1-23, copyright 2006 by IGI Publishing, formerly known as Idea Group Publishing (an imprint of IGI Global).

Section IV
IS Perspectives

Chapter IX
Standards for Business Component Markets:
An Analysis from Three Theoretical Perspectives

Heiko Hahn
University of the Federal Armed Forces, Germany

Klaus Turowski
University of Augsburg, Germany

ABSTRACT

The idea of component-based software systems and markets for the exchange of components dates back to the late 1960s. However, so far no large-scale components markets can be found. The purpose of this chapter is to present a more in-depth analysis of the conditions that have to be met for the successful realization of this idea. Three perspectives are presented: first, a system-theoretic perspective; second, an economic perspective; and third, a knowledge-codification perspective of standardization. As we want to argue, the problem should be considered as an empirical question that depends to a large extent on the future technological development and its outcome, like specification techniques, software verification standards, and the performance and maturity of existing systems.

INTRODUCTION

Products often consist of different parts; that is, they are rather product systems consisting of different components[1] than they are monolithic products. The parts are to a varying degree independent and interchangeable and might be produced by one integrated firm or by different firms. In the latter case, the components can be purchased prefabricated on the market or from a supplier with whom the customer has an intensive business relationship and who produces the parts according to the special requirements of the customer.

In the case of enterprise resource planning (ERP) software like SAP R/3, the dominant design is monolithic; that is, all customers purchase and install basically the same software package, which is than adapted by the use of parametric configuration to the different needs of the firms. As we will discuss in the next section, the choice between different architectonic principles (i.e., modular or monolithic) can be motivated by certain factors, and given these factors, some architectonic designs might be more appropriate than another for a given situation.

The advantages of a modular design are well known, and the idea to use markets for the exchange of software components has been discussed since the late 1960s (McIlroy, 1976). However, despite the advantage of component-based software systems and the success of component-based systems in other industries (for example, platform strategies in the automotive industry), large-scale markets for software components can hardly be found (Dietzsch & Esswein, 2001).

In this chapter we will address the problem of establishing markets for business components from three distinct perspectives. The first is a system-theoretic perspective discussing the principle of modular design. The second perspective is an economic analysis discussing the preconditions for market-based coordination. The last perspective focuses on standardization, especially as a process of knowledge codification. It takes up certain aspects of standardization that are also discussed in the two other sections to discuss the underlying problems more in depth.

SYSTEM-THEORETIC PERSPECTIVE

Modular Design and the Principle of Information Hiding

The concept of modular design is central to system theory. Simon (1962, p. 476) characterizes the term *nearly decomposable system*, that is, a modular system, as follows:

(1) in a nearly decomposable system the short run behaviour of each of the component subsystems is approximately independent of the short-run behaviour of other components; (2) in the long run the behaviour of any one of the components depends in only an aggregate way on the behaviour of the other components.

For technical systems, the influence that different components have on each other is limited and mediated by interfaces that allow abstracting from the lower level implementation. The interface is the device that mediates the interaction between the different components and prevents a direct communication. Parnas (1972) introduced the principle of information hiding, which emphasizes the importance of indirect communication and a black-box approach. According to the principle of information hiding, the design decision should be led by the goal to make the different parts of a software program as independently interchangeable as possible so that the system can accommodate changes easily. Information hiding provides a criterion for the decoupling of the different systems: The interface level that connects the different parts of which the system consists should be invariant against changes at the implementation level. To reach this goal, the system has to be separated into parts that are stable (i.e., interface level) and are likely to change (i.e., implementation level). The implementation of the modules should be as hidden from the other parts of the system as possible; that is, it should be a black box to other components to prevent a designer of a module from creating interdependencies by referring to specific aspects of the implementation of other modules that are likely to change.

The value of this principle has been elaborated and tested based on simulations by Baldwin and Clark (2000), who introduced so-called modular operators that can be deduced from a modular architecture. Inter alia, they analyzed the value of the possibility to split up a system into modules, to substitute different modules for another, or to augment the functionality.

Besides the decoupling of the different subsystems, there is also a second important property in the theory of nearly decomposable systems: The structure is hierarchical which means that the components are part of other components that are themselves part of larger components (Simon, 1962). For software systems, components will be part of frameworks that can be considered as high-level components (Szyperski, Dominik, & Murer, 2002; Turowski, 2000). A module is therefore not only interchangeable and substitutable for a module that fulfills the same interface specification, but it is also part of a modular architecture that is created by the horizontal (same level of abstraction) and vertical composition of components (i.e., composition inside the black box consisting of lower level components).

According to Fellner and Turowski (2000), a component can be defined as follows. A component consists of different (software) artifacts. It is reusable, is self-contained, is marketable, provides services through well-defined interfaces, hides its implementation, and can be deployed in configurations unknown at the time of development. A business component is a special component that implements a certain set of services out of a given business domain.

The relationship between the concept of modularity and the concept of component-based software systems is obvious: Modularity is an important prerequisite condition for the architecture of component-based systems. For example, to fulfill the condition that a component can be used in combinations unknown at the time of development, modularity is necessary because without independent systems, it is very unlikely that a recombination is possible. Reusability would also be restricted because the market has to be narrowly defined by very interdependent combinations. The criterion of being self-contained is also best realized by splitting up the systems into subsystems with only little external interdependencies but containing all their interdependent elements inside.

Preconditions for Modular Product Architectures

Given the advantages of a modular design, the question arises: Why are all products not modular in the first place? In this section, we want to take a closer look at factors that influence the practical side of the design decision about the appropriate product architecture.

In general, product modularity is realized by a modular product architecture instead of a monolithic or integrated product architecture. Ulrich (1995, p. 420) defines the product architecture as "(1) the arrangement *of functional elements*; (2) the mapping from functional elements to *physical components*; (3) the specification of the *interfaces*." The functional elements represent the set of functionality that is offered by a system to elements inside or outside the component. Depending on the level of abstraction, the description of the functional elements will be more detailed at lower levels and rather abstract at higher levels (Ulrich). Ulrich's definition of a modular architecture is based on the relationship between elements and components. He calls a system *modular* if a one-to-one mapping of the functional elements to the components exists and *integrated* if all functional elements are mapped to just one component. The interface definition, finally, specifies the interaction between the components. It should guarantee that all those components that comply with the specification are able to interoperate with each other. The aspect of interdependence is not considered directly in this definition.

However, given that the leading principle of modularity is that the system should be split up into subsystems with elements of high internal but low external interdependencies, the interdependency between the elements becomes a determining factor. For practical design decisions, interdependency emerges because the system has to meet functional requirements and nonfunctional requirements. Ulrich (1995) discusses the relationship between modular design, interdependency, and performance. According to Ulrich (p. 432), product performance can be defined "as how well the product implements its functional elements"; that is, performance is used as an umbrella term for nonfunctional requirements. Nonfunctional requirements can be more important than functional requirements. The term is applied to a rather wide range of aspects like security, usability, reliability, or even ethical considerations (c.f. Sommerville, 2004, for an overview). Not all nonfunctional requirements are influenced by the design decision about the appropriate level of system modularity with which the discussion in this chapter is concerned. Furthermore, the relevance of a nonfunctional requirement depends on the intended use of the component; for example, embedded systems might have other requirements than business application systems. As Sommerville stresses, nonfunctional requirements might depend on rather emergent systemic properties. Such systemic properties call for a joint rather than independent coordination of the elements.

Schilling (2000) also analyzed the problem of systemic properties (here called *synergistic specificity*) but makes no distinction between functional and nonfunctional requirements: "The degree to which a system achieves greater functionality by its components being specific to one another can be termed its *synergistic specificity*; the combination of components achieves synergy through the specificity of individual components to a particular combination" (p. 316). Similarly, Ulrich and Ellison (1999) introduce the concept of holistic requirements. In general, the degree to which a requirement has to be considered holistic or synergistic specific is influenced by the number of other components that reciprocally influence the nonfunctional requirements the component or the whole system can meet in certain combinations.

Based on a formal approach, Milgrom and Roberts (1995) discuss the concept of complementarities as a systemic property. Complementarities have to be considered if the return from a joint configuration of the elements yields a higher return than the sum of independent changes. Coordination partly compensates for lower performance of the subsystems: "when complementarities are present, 'fit' is more important, that is, even mistaken variations from a plan are less costly when they are coordinated than when they are made independently" (Milgrom & Roberts, p. 186). In the case of information systems, for example, it might be more important to guarantee the coordination of the different components, using a best-of-breed approach within certain limits defined by the relationship between performance differences and gains from coordination, than using perfect components favoring a centrally coordinated approach.

From a theoretical perspective, Fleming and Sorenson (2001) have argued that a modular architecture limits the search for new high-performance combinations. Based on Kauffman's N-k model (Kauffman, 1993), they stress that in the case of high interdependencies between the performance of the components, even minor changes in one component can have a tremendous impact on the performance of the system. In the case of noninterdependent components, the performance is independent from any other components. Fleming and Sorenson consider as the best situation one of intermediate interdependence because if the interdependence is too high, searches for high-performance combinations will become too erratic, which makes a rational design decision almost impossible.

However, a modular design approach does not imply architectures of independent modules in an absolute sense (e.g., independent of architectures with well-defined interfaces) but rather a design of moderated interdependencies. The problem is not that interdependencies exist, but how difficult it is to predict and to manage interdependent systems under the condition of evolutionary changes and to coordinate the independent and specialized supplier of the subcomponents that the system of freely interchangeable (compatible) components consists of.

In certain situations, however, integrated systems might be the only feasible solution. The system might not be easily split into interchangeable components, or performance considerations might favor an integrated design because it can be easier optimized for a single purpose taking the different interdependencies individually into consideration. An integrated design can be designed to be leaner. As defined above, components can be deployed in configurations unknown at the time of development. Hence, the reuse of components would be very restricted if not every component had a built-in slack that allows some variance in the requirements they can meet. Furthermore, the use of an architecture that mediates the communication and introduces several levels of abstraction to hide the complexity of the system might also reduce the performance of the overall design. However, integrated systems are not necessarily leaner in comparison to modular systems, especially if they are used in a dynamic environment. Monolithic systems are best suited for static optimization situations; but often the situation is not static but changing, and the systems have to serve diverse purposes and not a single purpose. Current monolithic ERP software shows this. If any new set of functionality is added to ERP software, the now larger system will not be redeveloped from scratch. Instead, the functionality is added on top of the existing functionality and directly integrated into the system, which makes the system increasingly complex. Optimization becomes ever more difficult as the complexity of the system grows.

The PC industry can be considered an example of a modular design successful in a dynamic environment. The technological progress is mainly realized at the component level, and the technological development since the early 1980s has had only a limited influence on overall architecture. This does not mean that no interdependencies exist. For example, it does not make sense to buy a high-performance graphics accelerator and a low-performance or any other outdated component. There might also be evolutionary changes that make the combination of modules from very different product generations practically impossible.

In fact, most consumers prefer to buy all-new packages. A large and, for most consumers, satisfying performance increase is possible by the combination of newer components still incorporating the basically unchanged architecture. So far, no architectonic bottleneck has rendered the technological progress at the component level futile—at least not for the time being. In other words, there has been no need to invent a completely new basic architecture to benefit from the radical increase in the performance of the different components since the early 1980s. However, nothing is said about the technological progress inside the components, and it should be noted that a modular design is not the logical endpoint to the development of some technology as the design might switch back and become more integrated again with new technological approaches (Schilling, 2000).

In general, it is likely that modular and integrated designs might compete against each other as both can be considered so-called dominant designs. According to Tushmann and Murmann (1998, p. 232), dominant designs "end eras of ferment and initiate eras of incremental technological change." A dominant design can be considered as a next step in the mostly evolutionary rather than revolutionary development of an architectonic design approach to a special functional solution. A dominant design does not need to be modular. Both modular as well as integrated designs can function as dominant designs. For example, the dominant design of ERP software since the 1990s clearly had an integrated design. The PC architecture instead has successfully relied on a modular design. However, as it is discussed in Abernathy and Utterback (1978), the advent of a new dominant design often changes the nature of competition. Before a dominant design is established, competition is based on product innovations and custom-made production processes. After the emergence of a dominant design, products and production processes become more standardized. However, standardization and price competition should rather favor modular designs that allow specialization of component suppliers to realize economies of scale and scope. The difference between the two alternative designs is the level of competition. In the case of monolithic systems, competition takes place between systems. In the case of modular systems, architectures and components compete against each other.

Christensen, Verlinden, and Westerman (2002) argue that some kind of typical technology life cycle exits. Often at some point in the history of the application of a technology, the performance of the products starts to overshoot the needs of average customers or of those consumers that are at the lower end of the market. In parallel, the knowledge about the technology and the technologies applied increase in this evolutionary process. As the technological knowledge matures, the interconnections between the functional elements become evident and increasingly better understood so that the decoupling of the different subsystems becomes possible.

Two main implications can be drawn from that perspective. First, it is important to control for the performance demands of the customers relative to the performance offered by standardized systems. If there is still a demand for new and better designs across all groups including those at the lower end of the market, a more integrated solution might still be favorable. Second, and more fundamental, the feasibility of a modular design is an empirical question. At a certain point in time, the knowledge may have matured enough and appropriate technological approaches (e.g., programming and specification techniques) have been found that allow separating the different elements and considering the different requirements the system has to meet.

Critical Aspects of Standards and Modular Design

Standards for the specification of the architecture and of the functional and nonfunctional requirements can be considered as the main coordination devices for the different companies that make up the value chain for the production of business application systems. The fulfillment of certain quality criteria are an additional necessary condition. To profit from the division of labor and specialized producers of components and architectures, a mean for the coordination of the firms must be found. Case-based coordination has to be replaced by a formal coordination based on clearly specified interfaces. Idealistically, the standards contain all information that has to be specified. Therefore, standards allow abstracting from the complexity of the system and defining and limiting the scope of information that needs to be exchanged between the partners of a transaction.

Before we can start a more in-depth discussion, a very important decision has to be made with respect to what kind of standard is meant and what purposes the standard should serve. Though not disjoint classes, a basic distinction is made between specification standards and (software) quality standards.[2] The latter includes standards for evaluating the quality of the software itself or the software development process. Typical standards are the ISO 9000 (International Organization for Standardization) family or standards especially created for evaluating software development processes like the capability maturity model (CMM; Paulk, Charles, & Curtis, 1995). CMM classifies the software development processes of firms according to five levels with increasing maturity. These standards have a positive effect on the evolution of component-based software systems because the reuse of software will be enhanced by the development of more trustworthy software. Component-based software development is a form of software reuse (see Sommerville, 2004, for an overview of software reuse) that practices the reuse of flexible architectures (frameworks) and interchangeable components and not only prefabricated and preintegrated monolithic systems. The independent development of prefabricated products presumes a highly trustable (industrialized) software development process because of the higher requirements for externally sourced third-party software. For example, with many suppliers, the responsibility for any malfunction that is not easily attributable to any single component has to be settled. From the perspective of system theory, specification standards are of a higher concern. Given that the system is split up, standards specify the requirements that the components have to fulfill to become part of the system and verify that the system and all other components still work as expected. Given the large number of requirements that have to be considered simultaneously, the decision might become even so (time) complex that the test for complete compatibility becomes undecidable (Schaefer, 1999). For most of the criteria, robustness must be given so that it is sufficient to control for a limited but critical number of criteria.

It is furthermore important to realize that the term *standard* is multidimensional or multilayered. For example, IT standards often include only the technical but not the semantic level. However, for systems that offer more than a generic functionality that is invariant to the situation of single firms, the underlying semantic concepts have to be documented and specified to avoid the risk of a semantic mismatch (Dioan, Madhavan, Domingos, & Halevy, 2004). Without the specification of all relevant dimensions of compatibility, the requirement that the component can be deployed in configurations unknown at the time of development will hardly be met. Semantic aspects are often only rudimentarily documented by corporate data models and sometimes are not consistent across different divisions. In general, the semantic dimension of software products is not easily determinable because no standards for an exact semantic description exist or they have only limited relevance in practice.

Another important aspect is the maturity of the standards, that is, how often standards change and how strong the impact of such changes is on existing components and architectures. This aspect deals with the interdependency of the technological and architectonical development. Changes of the standards can pose a serious risk on investments in existing components if they make these investments obsolete.

Summary of the System-Theoretic Perspective

We have discussed modularity as the central property of component-based systems. Standards are the practical devices that allow the free combination and composition of components because they specify and limit the interdependencies that have to be considered. As a summary, we want to present the following three propositions as the main result of the discussion of the system-theoretic perspective.

- **Proposition 1:** The more dynamic and the less stable the technological development is, that is, the more frequently changes occur, the less likely the establishment of a modular architecture is.

Figure 1. Factors of product modularity

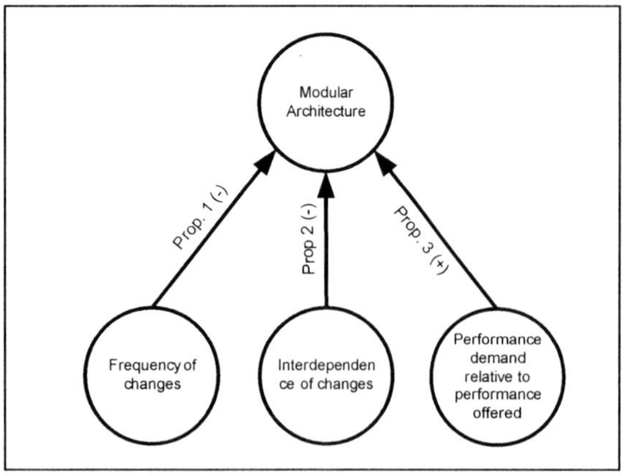

- **Proposition 2:** The higher the interdependence between the technological development and the architecture of the system is, the less likely the establishment of a modular architecture is.
- **Proposition 3:** The more customers are satisfied by the performance of the systems already offered, the more likely the establishment of a modular design is.

ECONOMIC PERSPECTIVE

The economic perspective that is presented in this section provides an analysis of the preconditions of markets as coordination mechanisms for the demand and supply of software functionality. Originally, the idea of component markets was an integral part of the vision of component-based software development (McIlroy, 1976). For a deeper understanding of the obstacles that have to be overcome, we will present the concept of a market and then critically discuss the preconditions for a pure market-based coordination. The role of standards will be discussed again at the end of the section.

Neoclassical Markets

The discussion begins with an outline of the neoclassical definition of markets and with the

Figure 2. Characteristics of market-based coordination

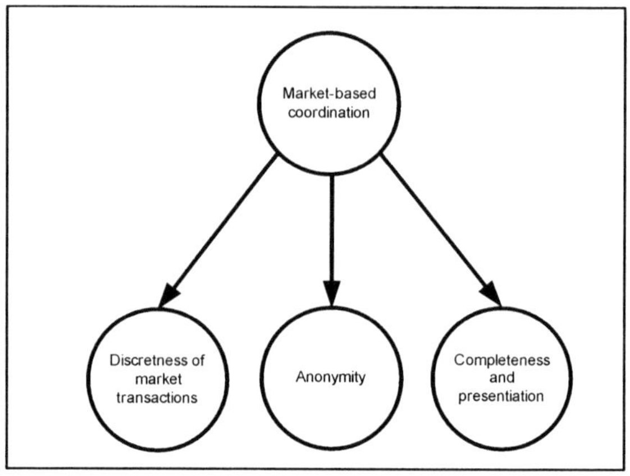

explicit and implicit assumptions underlying the neoclassical modeling of markets.[3]

In the neoclassical terminology, the market is defined as an abstract place where commodities are frictionlessly exchanged (Walras, 1954). The bidding process determines the value of the goods. Complete information is assumed to exist and the costs for contractual activities like negotiation and enforcement are assumed to be zero. The contract is complete; that is, all relevant future contingencies are already considered (E. Furubotn & Richter, 1991; MacNeil, 1978). In other words, the contract can be realized at present; hence, the contracting parties negotiate only ex ante. All conceivable and verifiable circumstances are covered by the contract. Business relationships or more general, and the identities of the parties in a transaction are irrelevant because everything is covered by the contract. Even if the initial agreement should fail to materialize, remedies are easily found and enforced. Contract relationships can be considered as discrete without any binding side effect steaming from the current transaction relationship in respect to future transactions. This situation was summarized by MacNeil (p. 738) as "sharp in by clear agreement, sharp out by clear performance."

In summary, pure market-based coordination should manifest itself in the three criteria of discrete and complete transaction that allow the realization of the contract at present (MacNeil, 1978) and acting anonymously; that is, the identity of the transaction partner and any underlying long-term relationship to secure the transaction are irrelevant.

By analogy, we can define a (neoclassical) business component market as an abstract place where components are frictionless exchanged and their value is determined by the bidding process between the supplier and the user of the functionality offered by the business components. The transaction object is perfectly specified by the underlying legally binding contracts.

Transaction Costs and Markets

To define markets in the neoclassical sense outlined above, it is necessary to abstract from the institutional arrangements in which the transactions are embedded. This abstraction is helpful for deriving analytic conclusions about perfect markets. However, because of the level of abstraction, neoclassical economic theory is neutral to the choice of governance structure used for the coordination of economic activities (E. G. Furubotn & Richter, 2001). Because transactions are assumed to be frictionless, neoclassical economic theory supposes that transaction costs are nonexistent. Hence, the neoclassical analysis is of only limited validity for our analysis because it neglects the defining role transaction costs play for the choice between different institutional arrangements (markets, internal coordination, hybrid or cooperative modes of coordination, etc.). Markets are not pregiven, but their use as coordination devices is dependent on certain criteria, of which transaction costs are the most important and the main focus of our analysis in this section.

Transactions are embedded in an institutional framework. By the institutional framework, we mean "the humanly devised constraints that structure human interaction. They are made up of formal constraints (e.g., rules, laws, constitutions), informal constraints (e.g., norms of behavior, conventions, self-imposed codes of conduct), and their enforcement characteristics" (North, 1994, p. 350). The primary economic function of institutions is the reduction of uncertainty by providing a structure that allows, for example, stable expectations in respect to the behavior of the transaction partners and to the sanctions mechanisms that come into effect if the contract should fail to materialize. As Coase (1988) points out, coordination situations that come closest to the ideal of perfect markets, for example, commodity and stock exchanges, are in fact highly regulated by the underlying institutional framework in which the transactions are embedded.

For the discussion of the economic perspective in this section, the limits of neoclassical economic theory that have been addressed by (new) institutional economics[4] are of special importance. Institutional economics is not a closed theory but consists of several strands like, for example, property-rights theory, principal-agent theory, and transaction cost economics (TCE). Another related field is information economics. Institutional economics and neoclassical analysis are not distinct approaches but complement each other.

The main insight from transaction cost analysis is that the choice of the coordination mechanisms itself is governed by economic considerations and "an exercise in comparative institutional analysis" (Williamson, 1996, p. 17). Markets are only one possible institutional arrangement to coordinate transactions. The foundations of TCE were laid by Coase (1937), but the most influential framework for a transaction-cost-based analysis was provided by Williamson (1975, 1985). In the framework presented by Williamson and given the basic behavioral assumptions of opportunistic behavior and bounded rationality underlying his analysis, transaction costs are mainly influenced by the so-called critical dimensions of transactions, that is, frequency, uncertainty, and, most important, asset specificity.

Transactions that differ in respect to these three critical dimensions are expected to be governed by different coordination mechanisms; for example, if highly specific investments have to be made, some mechanism needs to be provided for safeguarding the risks steaming from the possibility of a holdup situation.

For our analysis, however, we take a more fundamental interpretation of transaction costs as a starting point that is more concerned with the problem of first defining and second securing property rights that are exchanged in software markets. Property-rights theory considers an exchange first of all not as an exchange of (physical or nonphysical) goods, but of bundles of rights that separate allowed actions from prohibited actions in respect to certain (physical or nonphysical) goods or resources (Demsetz, 1967). The value of a contract depends on the bundles of property rights that are transferred. Transaction costs arise because of the costs that are associated with the exchange of property rights. According to Barzel (1989, p. 2), transaction costs are "the costs associated with the transfer, capture, and protection of rights." Property-rights theory assumes that a more comprehensive specification and strict enforcement of property rights enhances the efficient use of scarce resources (E. G. Furubotn & Pejovich, 1972). For achieving predictive validity, a different property-rights regime should imply differences in the efficiency of the use of resources.

Selling components via component markets based on advanced auction mechanisms would give the buyer access to functionality. From an abstract point of view, the right to invoke some functionality means the transaction object and the component seller would get the price for the provision of the functionality as determined in the bidding process. However, the situation faced by a buyer of business application software is much more complicated. First, not some singular functionality but a larger set of functionality is normally bought. Second, there is no tabula rasa situation, but the successful use of the functionality is dependent on the fulfillment of certain criteria like compatibility or a set of nonfunctional requirements. Third, despite the progress in the field of software engineering, the problem of software quality remains.

The problem therefore is first the specification of the contract itself, which has to be done ex ante and autonomously for market-based transactions. Practically, comprehensive specification techniques need to be used that allow the specification of all relevant information. Problems related to the development of such all-embracing specification schemes for business components will be discussed in depth in the next section. Besides the specification of the transaction object, the assessment of the quality of the product is a second equally important precondition from an economic point of view. The buyer must be in the position to verify that the software really fulfils all its promised requirements. In summary, the customer must be able to give a valid specification of the company's needs and to verify that the software fulfils all requirements specified; that is, comprehensive standards have to cover both aspects: specification and validation, and quality and verification (c.f. Balci, 1997, for a discussion of validation and verification for simulation models).

Information economics has developed three categories of the assessment qualities of products.

- **Search qualities:** Search qualities of goods can be assessed prior to the purchase of the good (Nelson, 1970).
- **Experience qualities:** Experience qualities of goods can be judged only after the use or the consumption of the product. Most appropriate is a trial-and-error method to find out the most suitable product (Nelson, 1970).
- **Credence qualities:** Credence qualities cannot be evaluated in normal use or it is too

costly to evaluate these quality properties. Furthermore, there can be a long time lag between use and the knowledge of the quality of the product, which blurs the distinction between experience and credence quality (Darby & Karni, 1973).

In general, all of the three categories are involved in assessing a product. The relevance of the different characteristics depends on the subjective judgment of the buyer. If a component could be completely evaluated by its formal specification, only search quality would be involved. This would make market-based transactions more feasible because all relevant information could be obtained and verified before the purchase, favoring efficient market coordination. However, such comprehensive specification schemes still need to be developed, if ever possible. The institutional framework offers only a limited remedy. For example, given the inherent complexity of today's software, comprehensive quality guarantees that are normal for traditional hardware goods are absent for software. Taking the Uniform Computer Information Transaction Act (UCITA) (Chow, 2001) as a case in point[5], the comment to section 403 states that the:

... question for merchantability is not whether errors exist but whether, the program still comes within the middle belt of quality in the applicable trade or industry, i.e., whether it is reasonably fit for the ordinary purposes for which such programs are used in accordance with average levels of quality and reasonable standards of program capability. (p. 136)

Though the UCITA is only legally binding in two states of the USA and failed its goal to become a uniform act in all states, from the point of view of property-rights theory, the vagueness of the document is still remarkable. It seems that it is left to interpretation what the common expectation or what reasonable fitness or average levels of quality are. Irrespective of any deeper technical, economic, or cost reasons that may render a more comprehensive liability for software infeasible at the moment, this vagueness in respect to liability leads from a purely economic perspective to a situation of less well-defined property rights in comparison to normal hardware goods. For the buyer, it is not clear of what quality the software service will be and the right to invoke the functionality and the sanction mechanism are therefore only partly specified. Because the institutional framework does not provide an appropriate level of uncertainty reduction, any additional measure has to be negotiated individually by the partners to a transaction.

A major problem arises if the quality of products cannot be communicated as reliable in the market or at reasonable costs, or if the buyer cannot differentiate between products of bad (so called "lemons") and of good quality (Akerlof, 1970). Akerlof showed that such situations can provide incentives to deliver products of low quality if the price is paid in relation to average quality, and that all sellers rather than an individual seller profit from the returns for good-quality products. In other words, while the cost savings achieved by offering low-quality products accrues to one special provider of a good, the cost of reputation damage has to been borne at least partly by the whole industry and leads to lower average quality and therefore lower prices paid. If it is impossible to discriminate between different offerings, lemons will be traded for the same price as products of high quality, eventually driving out the good quality as the seller of a high-quality product cannot recover its investment in high quality.

In the case of component-based software systems, the problem of quality evaluations is exacerbated if the system is composed from components that are offered by different vendors. In respect to our analysis, the basic difference between buying an integrated software system and buying a component-based system is that the scope of any responsibility, however specified (e.g., formal guarantee or informal reputation damage), is rather limited to the product bought. Therefore, in the case of the failure of a component-based system, the component that caused the malfunction has to be singled out, and it has to be verified that this special component indeed caused the problem. Even more problematic would be a failure caused by an ill-defined component standard. A guarantee provided for components but not for the system as a whole is incomplete for the buyer. From the point of view of property-right theory, the responsibility for failures of the system is possibly not perfectly specified. All suppliers that offer components for a certain market defined by some common set of standards have to take the cost of the negative reputation effect of low-quality products. The costs are the highest for that party that has most invested in the reputation of that technology.

An obvious solution from the perspective of property-rights theory would be to restrict access to the market to only those components that meet certain quality criteria or certification schemes. The certification scheme has to prove to be effective in discriminating between components of different levels of quality. The right to sanction and to prescribe the certification schemes that have to be met would be given preferably to a single party. This could be the party that is most affected by a defection of the quality standards, for example, the owner of the business relationship with the final customer.

Specific Investments and Markets

Similar to the problem of assessing the transaction object, the need to undertake specific investments contradicts the idea of market transactions, especially those of discrete buying decisions. Specific investments play a crucial role in traditional transaction costs analysis based on Williamson's "operationalization," where it is considered as the main source of transaction costs (Riordan & Williamson, 1985). If a company realizes that a software system based on an alternative standard is better suited for its needs than its current system, most of the investments in the existing system will be lost in the cases of a switch to a system that is based on a different design. The investments that are specific to a special standard are sunk costs. Specific investment leads to a locked-in situation because the value of the investment cannot be regained through the market. In the case of high specific investments, any changes would be accompanied by high switching costs. Given the assumption of self-interested or even opportunistic agents, there is the need to take appropriate precautionary steps to guard the specific investments from appropriation by the other party. Hence, without a mutual locked-in situation, there is a high risk of holdup. In the case of proprietary standards, for example, the owner of these standards could raise the prices for additional components or releases. Without contractual arrangements, the investor depends solely on the other party's benevolence. Williamson (1985) calls the transition from precontractual to postcontractual circumstances the "fundamental transformation" because even a situation of ex ante perfect competition can lead to a situation of ex post monopoly.

Investments in component-based software systems are largely specific. First, the decision about a special component model normally involves specific investments. Once a component model has been selected, successive components cannot be chosen independently. Integration costs and investment in human capital like training are also specific. Information costs like the effort to screen the market and to evaluate the different offerings are sunk costs. The migration to component-based software systems is probably a sequential process. Therefore, there will be more than one occasion on which a company has to screen the market for the best offer. Because of the costs involved, component buyers could prefer companies that offer a full range of products, reducing these transaction costs. This effect depends on how different the offerings in the marketplace are (c.f. Klemperer & Padilla, 1997, for a discussion of the relationship between specific investments and product-line range from a largely neoclassical perspective). Finally, the effort invested in mutual trust and reputation is a specific investment as well. In conclusion, specific investment, especially in proprietary standards, can make a customer dependent on the firm that owns the standard. Contrary to the idea of discrete transactions typical for real market-based coordination, customers may be temporarily locked in to a business relationship with one special firm that is shielded from market competition.

Critical Aspects of Standards and Market-Based Coordination

There are two aspects critical for the discussion here. First, standard-based coordination is an alternative mode to relationship-based coordination. The more comprehensively the standards cover all coordination-relevant information, the more favored a market-based coordination is. If the information contained in the standard is not sufficient, alternative coordination modes need to be applied. An obvious alternative is a more case-based, service-intensive coordination that is based on a closer business relationship. The result might be systems that are highly idiosyncratic to the special situation the firm faces, making an interchange of components and business partners more difficult.

Standards also have an influence on the possibilities of a stricter division of labor because they allow a better control of the different stages of the value chain. If the quality of the outcome of one stage of a component supplier is not measurable, for example, and different independent suppliers exist so that the responsibility for any malfunction is not easily detectable, then there is a risk of opportunistic behavior and the provision of low-quality products because, as already discussed from a property-rights perspective, the main costs have to be borne by the firms that have the business relationship. However, as Barzel (2004, 2005) argues, standards for quality assessment allow the transfer of private knowledge about quality into public knowledge, reducing measurement costs. The result in the decrease of the measurement costs makes a higher degree of vertical specialization more feasible because of the reduction of transaction costs.

The second aspect is the problem of network effects or, more in general, of supply-side and demand-side economies of scale. Components, once standardized to serve the needs of a wider market, tend to have large fixed costs (first copy costs) and only little or zero marginal costs. These supply-side economies of scale may lead to a natural monopoly (Varian, 2001). Demand-side economics provide the foundation for so-called network effects.[6] A short summary will be given here (c.f. Economides, 1996). Strong direct and indirect network effects are characteristic for markets of network goods. Hence, the value of a network good not only depends on its inherent properties, but also on its network value. Direct network effects depend on the number of people that can directly connect to the network. Indirect network effects depend on the number of complementary products and services that are available for the products that incorporates a special standard (Katz & Shapiro, 1985). The more frequently the good is purchased, the more people know about the product, the more additional services are available, and the like. For a company, it is therefore important to reach a critical mass and not to get stranded with an obsolete standard. Before the standard has reached a critical mass, the risk is high for both sides. After the standard has reached its critical mass, the seller risks that the standard owner abuses its market power as discussed above. Firms try to attract consumers with predatory pricing, hoping that they can exploit this consumer base when the customers are locked in. Firms are therefore reluctant to commit themselves to one standard and try not to become too much dependent on one standard. Open standards with ex post competition between different vendors are therefore a credible commitment because they reduce the risk of ex post holdup situations (c.f. Farrell & Gallini, 1988, for a discussion). If there are only short technology circles, making a standard obsolete fast, firms may even be reluctant to commit to any standards. This could make it difficult for firms to recover their high investments if they need large economies of scale. In network industry, the expectations of the agents play an important role. A commitment of the whole industry to one open standard could overcome the problem of specific investments and reluctant buyers allowing free market competition. However, given the disruption steaming from the technological development, this dependence might be a short-term rather than a long-term phenomenon as systems become more and more similar in the market process and the pressure increases to offer open systems and the specialization of component suppliers, which makes the legitimization of proprietary systems increasingly difficult.

Summary of the Economic Perspective

In this section, we have discussed market-based coordination. Standards play a pivotal role for making real market-based transaction possible if they are comprehensive for providing a valid base for specification and verification. From a theoretical perspective, negatively related to standards-based coordination is the degree to which personal relationships are used as coordination mechanisms. Finally, highly proprietary standards lead to nonmarket relationships.

As a summary, we want to present the following three propositions as the main result of the discussion of the economic perspective.

- **Proposition 4:** The more comprehensive the standards for specification and verification are, the more market-based the coordination will be.
- **Proposition 5:** The more personal relationships are involved for the coordination, the less market-based the coordination will be.
- **Proposition 6:** The more open the standards are, the more market-based the coordination will be.

STANDARDIZATION FROM A KNOWLEDGE-CODIFICATION PERSPECTIVE

In this section are discussed the preconditions for a purely formal, specification-based coordination of the different firms including the final buyers of the systems. They altogether represent the value chain for the production and use of business application systems. Christensen et al. (2002) list three conditions that have to be met at the interface level before value-added activities can be decoupled: First, there must be the ability to specify attributes and parameters; second, metrics must exist to measure the attributes unambiguously; and third, firms must understand the interdependencies between them. As already

Figure 3. Factors of market-based coordination

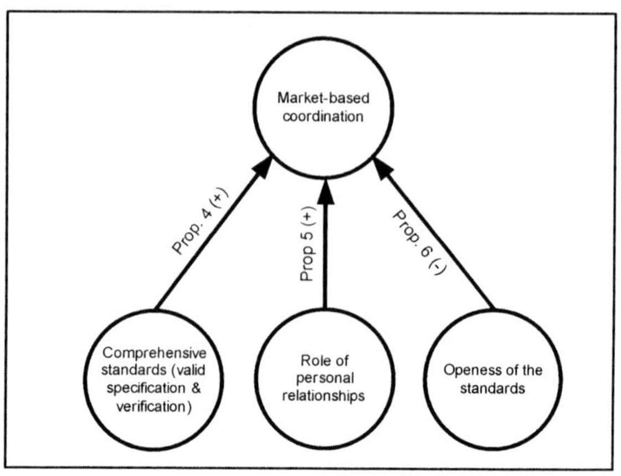

discussed in the section about the system-theoretic perspective, for systems that offer more than generic functionality, the underlying semantic concepts have to be specified and any semantic mismatch has to be avoided. Hence, domain-specific knowledge has to be specified. With respect to the standardization processes for the derivation of new standards, the process may differ in terms of (a) whose participation in the standardization process is required and (b) to what extent a user needs to acquire additional knowledge before he or she can use a (specification) standard. For example, in the case of the DVD or CD-ROM, the different firms that participated in the standardization process were probably able to agree on a specification without concrete knowledge of the specific needs of a single user. Similarly, a buyer of a CD-ROM or a DVD does not need to worry about whether some DVDs will serve his or her special needs since they are probably rather generic and similar to those of other users. A specification for DVD has to meet rather generic requirements to be valid, and it seems that almost an instantaneous consumption of this standard is possible. The knowledge required is rather unspecific in respect to the needs of individual users, and the buying decision can be based on generic evaluation criteria, for example, like those that can be found in special-interest magazines. Uncertainty is rather caused by the question whether enough other participants in the market will use the same standard, that is, if a possible network effect will have a positive or negative influence on the success and value of a product that incorporates this standard. In fact, because the demand is rather homogenous, network effects might play an important role. However, uncertainty seems to be less caused by questions about the content and validity of the standard.

Looking at component-based software systems, a strong requirement would be the need to actually involve many or all customers from different industries (instead of some representative average or lead customer; Hippel, 1986) to derive valid semantic concepts of basic and higher order terms like the customer, order, and so forth. The actual implementation of these concepts might depend on criteria like the industry, company size, production process, and so on. However, even if most users do not need to get involved in the standardization process, firms need to acquire some special knowledge before they can use the standard. A firm that wants to buy a component needs to know (a) what the specification provided with the component means and (b) what kind of specification requirements the components have to meet given the specific demands of the firm. Hence, the customer needs to translate his or her needs into a specification language and compare it with the information that is provided along with the components. This information has to be codified in the same specification language or at least in one the buyer is able to understand. Furthermore, the specification language must cover all the information that is important for that customer. Otherwise, an alternative medium for transmitting this information is necessary, which raises information and transaction costs for assessing qualities other than those covered by the standard. The more independent and autonomous the producers of the different components are, the more likely a semantic mismatch is. Also, before firms can agree on shared semantic concepts, they must be able to make their knowledge explicit and translate and codify it into a specification language that allows them to communicate all relevant aspects with their business partners.

Cowan, David, and Foray (2000) and Cowan and Foray (1997) have defined the core of knowledge codification as a process that includes three critical tasks: the development of a shared language, and the development of models, and the exchange of messages using the language and models.

- **Development of a language:** The language represents the basic infrastructure for the model development and message generation. The relevant community must share the language competence.

- **Development of models:** Codification should lead to the definition of basic concepts specified in models based on the shared language. The derivation of models is, similar to model building in general, a process of the abstraction of real-world or artificial sociotechnical systems. The model building is a creative process and a precondition for low-cost communication of knowledge via messages.
- **Creation of messages:** According to Cowan et al. (2000), codification is a process of knowledge generation that modifies and converts already existing knowledge into information that is easily transferable via messages. Codification structures knowledge via a language and models and allows the exchange of knowledge between different members of the same epistemic community via messages.

The challenge for knowledge codification arises because of the scope of information that has to be covered for a comprehensive specification. Traditional software production is still to a certain degree craftsmen-like, and the quality of the outcome depends on the skills of individual programmers. For an engineering-based production mode, a systemic change is necessary that relies less on the implicit knowledge of expert programmers but more on the use of specification techniques. Knowledge codification may be a misleading term[7] because it is not the (implicit) knowledge of the current production mode that has to be converted into codified knowledge, but a codified mode of coordination has to be found that allows the production of a functionally equivalent solution to the craftsmen-like production mode, eventually replacing it with specialist producers of components and architectures. Both modes have to compete against each other. Firms might either prefer to rely on the special skills of gifted

Figure 4. Factors for codified coordination

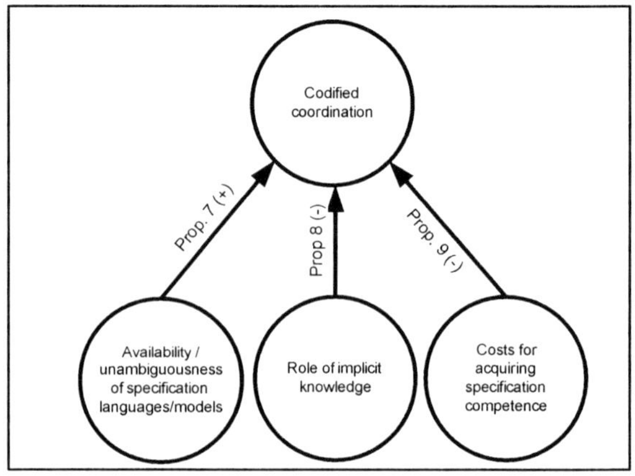

programmers and system analysts who produce nonstandardized outcomes or rely on software engineers who are trained in using standards for the specifications of the production of goods that might be done by a third party, which either reuses existing solutions or builds a new one according to the specification. The teaching of the software engineers and the development of the knowledge infrastructure (language and models) are investments that need to be recovered by the multiple uses of the outcomes.

The tasks involved for knowledge codification are the creation of a specification language and the development of models that specify the underlying semantic concepts. A contract is basically the creation of a message, that is, out of the universe of possible models that can be specified based on the language. The messages specify which concept the company uses. The language and the models must be flexible enough to represent the special situation of the company. The use of codified knowledge can therefore be considered as dependent on the factors that are summarized in the following three propositions (Figure 4).

- **Proposition 7:** The more available unambiguous and comprehensive specification languages and models are, the more codified the coordination is.

- **Proposition 8:** The more implicit the knowledge for the coordination is, the less codified the coordination is.
- **Proposition 9:** The higher the costs for the acquisition of the relevant competence for using the language and the models are, the less codified the coordination is.

CONCLUSION

In this chapter, we have discussed the concept of component-based software development and the use of markets for the coordination of the different component producers and customers from three different perspectives. The first was a system-theoretic perspective discussing the preconditions for modularity as the basic concept underlying component-based systems. The second was an economic perspective discussing the preconditions for a purely market-based coordination. The last perspective analyzed the standardization process from a knowledge-codification perspective. All three perspectives show that the establishment of component-based systems and markets for the exchange of business components are dependent on certain factors that still need to be fulfilled. The success of component-based software for business application systems and markets for the exchange of business components is therefore an empirical question that depends to a large extent on future technological development and its outcomes, like specification techniques, software verification standards, and the performance and maturity of existing systems. Component-based systems have to compete against monolithic systems and show their superiority in terms of performance, complexity, and costs.

REFERENCES

Abernathy, W. J., & Utterback, J. M. (1978). Patterns of industrial innovation. *Technology Review, 80*(7), 40-47.

Akerlof, G. A. (1970). The market for "lemons": Quality uncertainty and the market mechanism. *The Quarterly Journal of Economics, 84*(3), 488-500.

Balci, O. (1997). *Verification, validation and accreditation of simulation models.* Paper presented at the Winter Simulation Conference.

Baldwin, C. Y., & Clark, K. (2000). *Design rules: The power of modularity.* Cambridge, MA: The MIT Press.

Barzel, Y. (1989). *Economic analysis of property rights.* Cambridge, MA: Cambridge University Press.

Barzel, Y. (2004). Standards and the form of agreement. *Economic Inquiry, 42*(1), 1-13.

Barzel, Y. (2005). Organizational form and measurement costs. *Journal of Institutional and Theoretical Economics, 161*(3), 357-373.

Chow, S. Y. (2001). The effects of UCITA on software component development and marketing. In G. T. Heineman & W. T. Councill (Eds.), *Component based software engineering: Putting the pieces together* (pp. 719-730). Boston: Addison-Wesley.

Christensen, C. M., Verlinden, M., & Westerman, G. (2002). Disruption, disintegration and the dissipation of differentiability. *Industrial and Corporate Change, 11*(5), 955-993.

Coase, R. H. (1937). The nature of the firm. *Economica, 4*, 386-405.

Coase, R. H. (1984). The new institutional economics. *Zeitschrift für die Gesamte Staatswissenschaft (Journal of Institutional and Theoretical Economics), 140*(1), 229-231.

Coase, R. H. (1988). The firm, the market, and the law. In R. H. Coase (Ed.), *The firm, the market, and the law* (pp. 1-31). Chicago: The University of Chicago Press.

Cowan, R., David, P. A., & Foray, D. (2000). The explicit economics of knowledge codification and tacitness. *Industrial and Corporate Change, 9*(2), 211-253.

Cowan, R., & Foray, D. (1997). The economics of codification and the diffusion of knowledge. *Industrial and Corporate Change, 6*(5), 595-622.

Darby, M. R., & Karni, E. (1973). Free competition and the optimal amount of fraud. *The Journal of Law and Economics, 16*, 68-88.

David, P. A. (2000). *Path dependence, its critics and the quest for "historical economics"* [Working paper]. Stanford University, Department of Economics.

Demsetz, H. (1967). Toward a theory of property rights. *The American Economic Review, 57*(2), 347-359.

Dietzsch, A., & Esswein, W. (2001). Gibt es eine "softwarekomponenten industrie"? Ergebnisse einer empirischen untersuchung. In H. U. Buhl, A. Huther, & B. Reitwiesner (Eds.), *Information Age Economy: 5. Internationale Fachtagung Wirtschaftsinformatik 2001* (pp. 697-710). Heidelberg, Germany: Physika-Verlag.

Dioan, A., Madhavan, J., Domingos, P., & Halvevy, A. (2004). Ontology matching: A machine learning approach. In S. Staab & R. Studer (Eds.), *Handbook on ontologies* (pp. 385-403). Berlin, Germany: Springer.

Economides, N. (1996). The economics of networks. *International Journal of Industrial Organization, 14*(6), 673-699.

Farrell, J., & Gallini, N. (1988). Second-sourcing as a commitment: Monopoly incentives to attract competition. *Quarterly Journal of Economics, 103*(4), 673-694.

Fellner, K. J., & Turowski, K. (2000). Classification framework for business components. *Proceedings of the 33rd Annual Hawaii International Conference on System Sciences.*

Fleming, L., & Sorenson, O. (2001). Technology as a complex adaptive system: Evidence from patent data. *Research Policy, 30*(7), 1019-1039.

Furubotn, E., & Richter, R. (Eds.). (1991). *The new institutional economics: A collection of articles from the Journal of Institutional and Theoretical Economics.* Tübingen, Germany: Mohr.

Furubotn, E. G., & Pejovich, S. (1972). Property rights and economic theory: A survey of recent literature. *Journal of Economic Literature, 10*(4), 1137-1162.

Furubotn, E. G., & Richter, R. (2001). *Institutions and economic theory: The contribution of the new institutional economics.* Ann Arbor: University of Michigan Press.

Hippel, E. v. (1986). Lead users: A source of novel product concepts. *Management Science, 32*(7), 791-805.

Hodgson, G. (1988). *Economics and institutions.* Cambridge: Polity Press.

Katz, M. L., & Shapiro, C. (1985). Network externalities, competition, and compatibility. *American Economic Review, 75*(3), 424-440.

Kauffman, S. (1993). *The origins of order.* New York: Oxford University Press.

Klemperer, P., & Padilla, J. (1997). Do firms' products lines include too many varieties? *RAND Journal of Economics, 28*(3), 472-488.

Liebowitz, S. J., & Margolis, S. E. (1995a). Are network externalities a new source of market failure? *Research in Law and Economics, 17*(1), 1-22.

Liebowitz, S. J., & Margolis, S. E. (1995b). Path dependence, lock-in, and history. *The Journal of Law, Economics and Organization, 11*(1), 205-226.

MacNeil, I. R. (1978). Contracts: Adjustment of long-term economic relations under classical, neoclassical, and relational contract law. *Northwestern University Law Review, 72*(6), 854-905.

McIlroy, M. D. (1976). Mass produced software components. In P. Naur & B. Randell (Eds.), *Software engineering* (pp. 88-95). New York: Petrocelli Charter.

Milgrom, P., & Roberts, J. (1995). Complementarities and fit: Strategy, structure, and organizational change in manufacturing. *Journal of Accounting and Economics, 19*, 179-208.

Nelson, P. (1970). Information and consumer behavior. *The Journal of Political Economy, 78*(2), 311-329.

Nightingale, P. (2003). If Nelson and Winter are only half right about tacit knowledge, which half? A Searlean critique of "codification." *Industrial and Corporate Change, 12*(2), 149-183.

North, D. C. (1994). Economic performance through time. *The American Economic Review, 86*(3), 359-368.

Parnas, D. L. (1972). On the criteria to be used in decomposing systems into modules. *Communications of the ACM, 15*(12), 1053-1058.

Paulk, M. C. W., Charles V., & Curtis, B. (1995). *The capability maturity model: Guidelines for improving the software process.* Reading: Addison-Wesley.

Riordan, M. H., & Williamson, O. E. (1985). Asset specificity and economic organization. *International Journal of Industrial Organization, 3*(4), 365-378.

Schaefer, S. (1999). Product design partitions with complementary components. *Journal of Economic Behavior and Organization, 38*, 311-330.

Schilling, M. A. (2000). Toward a general modular systems theory and its application to interfirm product modularity. *Academy of Management Review, 25*(2), 312-334.

Simon, H. A. (1962). The architecture of complexity. *Proceedings of the American Philosophical Society, 106*, 467-482.

Söllner, F. (2001). *Die geschichte ökonomischen denkens.* Berlin, Germany: Springer.

Sommerville, I. (2004). *Software engineering* (7th ed.). Harlow, United Kingdom: Addison Wesley.

Szyperski, C., Dominik, G., & Murer, S. (2002). *Component software: Beyond object-oriented programming* (2nd ed.). New York: ACM Press.

Turowski, K. (2000). Establishing standards for business components. In K. Jakobs (Ed.), *Information technology standards and standardisation: A global perspective* (pp. 131-151). Hershey, PA: Idea Group Publishing.

Tushmann, M. L., & Murmann, J. P. (1998). Dominant designs, technology cycles, and organizational outcomes. *Research in Organizational Behavior, 20*, 231-266.

Ulrich, K. T. (1995). The role of product architecture in the manufacturing firm. *Research Policy, 24*(3), 419-440.

Ulrich, K. T., & Ellison, D. J. (1999). Holistic customer requirements and the design-select decision. *Management Science, 45*(5), 641-658.

Varian, H. R. (2001). *Economics of information technology.* Berkeley, CA: University of California.

Walras, L. (1954). *Elements of pure economics or the theory of social wealth.* London: Allen & Unwin.

Williamson, O. E. (1975). *Markets and hierarchies: Analysis and antitrust implications. A study in the economics of internal organization.* New York: Free Press.

Williamson, O. E. (1985). *The economic institutions of capitalism.* New York: The Free Press.

Williamson, O. E. (1996). Efficiency, power, authority and economic organization. In J. Groenewegen (Ed.), *Transaction cost economics and beyond* (pp. 11-42). Boston: Kluwer Academic Publishers.

ENDNOTES

1 The terms part, component, and module are often used synonymously in the literature if not in the context of a programming language. We follow the literature in the first part but define the term business component later. The term 'business component' has therefore a concrete meaning for this paper.

2 They are not disjoint because a specification standard can specify quality requirements and a quality standard may require a certain kind of specification standard used.

3 We present the neoclassical definition as starting point because neoclassical economic theory represents the mainstream of economic theory (Söllner, 2001).

4 There exists an older tradition of (American) Institutional Economics. Both consider transactions as the basic unit of analysis. But New Institutional Economics is closer to the neoclassical mainstream and both are also rather critical of each other. See (Hodgson, 1988) for a scholar in the tradition of (American) Institutional Economics critical of New Institutional Economics and the critique of (Coase, 1984, p. 230): "The American Institutionalists were not theoretical but anti-theoretical particularly where classical economic theory was concerned. Without a theory they had nothing to pass on except a mass of descriptive material waiting for a theory, or a fire".

5 This version can be downlaoded from http://www.law.upenn.edu/bll/archives/ulc/ucita/2002final.pdf

6 There is a vivid debate about the practical relevance of network effects (or network externalities), c.f. (Liebowitz & Margolis, 1995a, 1995b) and (David, 2000) for different views. Much of the outcomes of the models that are used for (neoclassical) network effect analysis depend on the assumptions underlying the models. It is therefore always important to avoid any overgeneralization.

7 (Cowan et al., 2000) from which some of the arguments of this section are borrowed, make intensive use of a 'codebook' as a mean for codification or knowledge documentation to avoid the „semantic and taxonomic confusion" (p.216) about the concept of implicit knowledge. This codebook is mainly considered as dictionary for an epistemic community. However, there is a high risk of using rather metaphysical concepts. It is very unlikely finding someone actually carrying such a codebook, i.e. the concepts is not very usefully for any empirical analysis, cf. (Nightingale, 2003) for an author who is very critical of mainstream concepts of codification.

Chapter X
Should Buyers Try to Shape IT Markets Through Nonmarket (Collective) Action?
Antecedents of a Transaction Cost Theory of Network Effects

Kai Reimers
RWTH Aachen University, Germany

Mingzhi Li
Tsinghua University, China

ABSTRACT

This chapter develops a transaction cost theoretic model of network effects and applies it to assessing the chances of users to influence, through collective action, the range of technological choices available to them on IT markets. The theoretical basis of the model is formulated through a number of empirically refutable propositions that overcome some conceptual and empirical difficulties encountered by the traditional interpretation of network effects as (positive) network externalities. The main difference between our model and modeling network effects as network externalities is that network effects are seen as caused by the costs of purchasing and marketing new technology, that is, transaction costs, rather than by the benefits of using a new technology. A first application of the model suggests that users can significantly improve the chances of replacing an established technology with a new, potentially superior one if they set up an organizational structure that serves as a conduit of information exchange and knowledge sharing. This, however, would call for a rather different type of collective user action than exists today in the form of user groups.

INTRODUCTION

As information technology increasingly permeates all kinds of business operations, firms' dependence on IT increases too. This would be of no concern if IT markets were perfectly elastic, not only in terms of quantity supplied but also in terms of responsiveness to new requirements as they arise. However, throughout its entire history, the IT industry has been characterized by the emergence of de facto standards that proved rather long lived and difficult to replace. Bresnahan and Greenstein (1999) have described this phenomenon as the dominance and persistence of technological platforms and shown how the history of the IT industry can be described as a history of long periods of stable platforms punctuated by rare events of platform replacement. While Bresnahan and Greenstein have refrained from evaluating this phenomenon from a normative point of view, corporate IT users should be worried by it in light of much of the rich economic literature on standards. Specifically, that literature revolves around the concept of positive network externalities (PNE) according to which network products become more valuable to buyers as the network grows in size (Katz & Shapiro, 1985). As the PNE concept has also been applied to standardization processes in the realm of information technology, where standards ensure that different components can easily be assembled into larger systems (Katz & Shapiro), it seems that users quickly are locked in to a particular platform even when other technologies may have emerged that could much better serve their needs were it not for the accumulated benefit of the existing technology's large installed base that prevents that new technology from being successful on the market. Thus, apart from hoping for favorable idiosyncratic historical events, the only way to have users dislodge an existing platform and replace it with another one based on a better technology would consist of massive collective action in which users jointly specify a new platform and collectively commit themselves to these specifications. This form of collective user action would be an instance of user-driven ex ante standardization. Although such collective action has been attempted in the past (Dankbaar & van Tulder, 1992), to our knowledge there is no successful instance of it so far. Also, from a theoretical point of view, it has been argued that users do not have the ability to engage in such action, both individually and collectively (Jakobs, 2000).

This reasoning would apply if the emergence and dominance of technological platforms in the IT industry were indeed driven by positive network externalities. Bresnahan and Greenstein (1999) have convincingly shown how the structure and evolution of the IT industry can be explained by the phenomenon of sunk costs, that is, costs related to the production of system components that are irreversible such as investments in programming, training, and research. In this chapter we want to extend that analysis from a transaction cost perspective, which is largely excluded from Bresnahan and Greenstein's work but which we deem a necessary complement to their work. We will also link that perspective to the discussion of network effects by providing a transaction cost interpretation of network effects as an alternative to the standard interpretation of network effects as positive network externalities. We will show that the interpretation of network effects as positive network externalities leads to a number of conceptual problems that are avoidable through a transaction-cost-based interpretation. We also show how that new interpretation allows for a different form of collective user action, one aimed at stimulating the offering of alternative technological platforms and thus increasing the range of choices corporate users have on IT markets.

In the next section we will briefly recapitulate the literature on network externalities as it is relevant to the focus of this chapter. In the third section, several problems resulting from interpreting network effects as network externalities will be identified. In the fourth section we will discuss antecedents of a transaction cost theoretic model of network effects and provide some empirically refutable propositions. In the fifth section, the model is applied to analyzing the way users of technological products can influence the range of choices available to them on IT markets. The concluding section summarizes the results and discusses some possible normative implications of our model regarding the organization of collective user action.

NETWORK EFFECTS AS POSITIVE NETWORK EXTERNALITIES

The classical formulation of the PNE concept has been provided by Katz and Shapiro (1985, 1986).[1] They also introduced the distinction between direct and indirect network externalities. Their major findings are that (a) network products will be supplied in a smaller quantity than is socially optimal (because consumers ignore the positive externality they exert on other consumers) and (b) there is a strong tipping tendency if two networks are competing, leading to consumers being locked in by the winning network. This latter effect, however, may be reduced by consumer heterogeneity and product differentiation. Reflecting on the emerging literature, Katz and Shapiro (1994) later conclude that there is no general theoretical support for an "excess inertia problem" when two systems are competing, meaning that the emergence of a new technology need not be prevented by an existing network and its accompanying positive network externalities if that would be socially optimal. Indeed, they agree with Farrell and Saloner (1986) that there can be "excess momentum" as well, meaning that consumers may be too eager to adopt a new technology, thus creating a bandwagon that leaves users of the old technology stranded.

Farrell and Saloner (1985, 1986) use a game theoretic setting in order to explore the possible lock-in effect in more detail. Apart from finding evidence of possible excess inertia, they identify the possibility of the opposite problem, excess momentum, as mentioned above. They also ask whether communication among suppliers might eliminate these problems, which, they find, is not the case although it might reduce them. Extending the early work of Katz and Shapiro, Farrell and Saloner explicitly distinguish between two types of network externalities that are responsible for excess inertia and excess momentum respectively, namely, those users of the existing technology exert on users of the new technology and those users of the new technology exert on users of the existing technology. Besen and Farrell (1994), also reflecting on the emerging literature, emphasize the strong tipping tendency in markets characterized by network externalities. Moreover, they stress that it is the expected network size rather than the existing network size that matters. This point is made by Economides (1996) as well. Krugman (1991) demonstrates under which conditions the expected network size rather than the existing network size determines the competitive outcome. He finds that if any of the following conditions are fulfilled, it is rather the existing network size that determines competitive outcomes: if future flows of costs and benefits are strongly discounted; switching to the new technology takes a long time; or external economies of scale are low.[2]

Dranove and Gandal (1999) and Gandal (1994, 1995) have conducted a number of studies that seek to provide an empirical basis for the notion of positive network externalities. As put forward in these works, the argument mainly rests on the empirically validated observation that users are willing to pay a premium for products that are compatible with a dominant product. This is interpreted as evidence of the presence of indirect network externalities since compatibility with a dominant product increases the potential market for suppliers of complementary products, which, in turn, will increase the variety of complementary products actually offered. This, it is argued, is the effect that is valued (and paid for) by consumers. Moreover, a two-way positive feedback effect between the sales of complementary products has been found (such as between CDs and CD players; c.f. Gandal, Kende, & Rob, 1997). Closely related to these results are Schilling's (2002) strong statistical support for her hypothesis that the relative size of the installed base of a certain technology as well as the availability of complementary products are associated with that technology's success in the marketplace, which is seen as evidence for positive network externalities.

However, a number of models demonstrate that similar effects can be constructed without the assumption of positive network externalities, namely, in so-called "mix-and-match" markets, that is, in markets where complementary products must be assembled by users to form systems. The reasons put forward differ. Economides (1989) argues that in a regime of compatibility, profits (and prices) are higher than under incompatibility because price elasticities are greater in the latter case (a similar line of argument can be found in Matutes & Regibeau, 1988). As a consequence, vendors have strong incentives to provide compatible products, which, following a similar logic as above, increases the variety of complements on offer and thus creates an equivalent to indirect network externalities. Desruelle, Gaudet, and Richelle (1996) base their model on fixed costs that lead to economies of scale in systems markets, which they interpret as a type of network effect. When the production of some components of a system's good is characterized by economies of scale due to the existence of fixed costs, the number of components supplied will increase with the number of users, implying a higher variety of components; that is, more components are available as the network increases in size.

Arthur (1996) subsumes network effects under the broader concept of increasing returns, which represent an instance of positive feedback mechanisms. Apart from network effects, Arthur identifies two other mechanisms that account for increasing returns, namely, fixed costs and "customer groove-in." In contrast to the PNE concept, Arthur sees network effects as supply-side phenomena in that they are created through positive feedback among compatible products (similar to the mix-and-match models discusses above). The positive feedback loop created through fixed costs is the same as in natural monopolies. Customer groove-in refers to the lock-in effect that familiarity with a complex technological product brings about. Thus, Arthur's discussion also shows that economic phenomena cited as evidence for the existence of demand-side externalities (i.e., PNE) can be explained by alternative mechanisms as well.

The most prominent critique of the PNE concept has been put forward by Liebowitz and Margolis (1990, 1994, 1995, 1996, 1998). Their criticism rests basically on three arguments. Their first argument is that the distinction between direct and indirect network externalities is a crucial one and should not be blurred. Specifically, they point out that indirect network externalities resemble pecuniary externalities that are not socially harmful because they represent a transfer of wealth between producers and consumers (c.f. Church, Gandal, & Krause, 2002, for a response to that critique). This leads to their second main criticism, which holds that most effects described by using the PNE concept can be derived with traditional models of natural monopoly as well. Third and finally, they claim that another crucial distinction is missing in most PNE-based models, namely, that between remediable network effects and those that are not. Only the former can be called true network externalities. They demonstrate that the circumstances under which true network externalities (as opposed to network effects, which are not remediable) emerge are extremely rare and that all cases that are commonly used to demonstrate the existence of positive network externalities do not fall into this category.

WEAKNESSES OF THE PNE CONCEPT

Apart from the criticism voiced by Liebowitz and Margolis, there are two other weaknesses in the interpretation of network effects as positive network externalities. The first concerns the actual mechanism through which positive network externalities are supposed to increase the benefit of buyers. This lack of clarity also contributes to some confusion surrounding the concepts of lock-in and self-fulfilling prophecy. If positive network externalities are assumed to arise due to the requirements of interacting with existing components of a system, that is, through compatibility features, it is the existing size of a network that would be decisive. Then, it might be said that buyers are locked in because they cannot change to a newer (possibly better) technology if this is incompatible with existing components. In this case, buyers are not concerned with the future size of the network and, by implication, with the variety of components offered in the future.

However, if the latter is their true concern, any expectation of network size can be self-fulfilling if the future is not discounted too strongly and the adoption of the network does not take too long, as Krugman (1991) has demonstrated. In this case, the concept of lock-in (or path dependency) cannot be reasonably applied. In fact, we would expect to see the opposite: frequent adoption and rapid diffusion of new technologies.

Second, indirect or virtual network effects seem to result from the combined effects of decisions concerning the degree of modularity and the degree of compatibility. These concepts, however, need to be clearly separated.

Modularity designates an approach that decomposes a system into modules so that interactions or interdependencies within modules are hidden from other modules (Langlois, 2002). This, of course, requires that modules have clearly defined interfaces; that is, they need to be vertically compatible.[3] Increased modularization of a system (implying increased levels of vertical compatibility) allows for increased levels of specialization and thus increased levels of scale economies (Langlois).[4]

It might be argued that an increased degree of modularity leads to an increased variety of complements because economies of scale imply falling prices as the network grows (if components are supplied competitively). Thus, indirect network effects may emerge as the market for complements grows as well, implying an increased number of different complements offered, that is, an increased variety of complements (c.f. Desruelle et al., 1996, for a similar argument). It is probably because of this implied virtual network effect that Economides (Economides, 1996, p. 16) claims that although the "mix-and-match literature" does not assume a priori the existence of positive network externalities, "it is clear that demand in mix-and-match models exhibits network externalities" (see also the discussion in the previous section).

A similar effect, however, may occur as a result of decisions about horizontal compatibility (see Endnote 3). If two firms offering substitute components agree on horizontal compatibility, they effectively increase the network size from the perspective of suppliers of complementary products, thus creating a similar indirect network effect (as would be the case if Microsoft and Apple agreed to making their operating systems compatible). This link between horizontal compatibility and indirect network effects has led scholars to analyze the incentives of firms to offer their products under a regime of compatibility rather than as a proprietary system (which is accordingly called a regime of incompatibility; c.f. Economides, 1989; Matutes & Regibeau, 1988).

Thus, although decisions about the degree of modularity (vertical compatibility) and about the degree of horizontal compatibility can have the same effect (increased product variety), the mechanisms accounting for these effects are quite different. More importantly, these two mechanisms have different direct effects. Increased modularity (vertical compatibility) reduces component prices due to economies of scale while increased horizontal compatibility allows for greater freedom and ease in combining products of different vendors, thus reducing implementation and integration costs for users.

To summarize, the PNE concept suffers from a lack of clarity in distinguishing between two sets of phenomena. First, it must be decided in which way buyers are said to benefit from increases in network size, which will determine if the current network size (the installed base) or rather the expected network size is decisive for their buying decisions. Second, indirect network effects, as increased product variety, may be a result of either increased horizontal compatibility or increased modularization of systems; these two mechanisms, however, have different direct effects. These two phenomena should be clearly separated for either a positive theory of network effects and a normative theory on regulation or, as in this chapter, of buyer behavior.

A TRANSACTION COST THEORETIC INTERPRETATION OF NETWORK EFFECTS

In order to tackle the problems mentioned in the previous section, it is necessary to adopt a more substantial approach toward new technology that goes beyond stylized examples such as the telephone network or the introduction of video recorders. First of all, it must be defined what is meant by new technology. If Microsoft brings a new version of its operating system Windows to market, can we call this an introduction of new technology? Presumably, the introduction of a new version of a computer operating system differs fundamentally from the construction of a large-scale system such as the cellular telephone system.

Next, the distinctions that have been identified above need to be made and operationalized in order to incorporate them in a model of network effects.

Finally, it must be clearly stated how costs and benefits depend upon various factors for both suppliers and buyers. These factors should include the phenomenon of network size in an appropriate specification. This statement should be such as to facilitate empirical validation or refutation.

The following discussion will address each of these issues in turn. The results of this discussion will then be combined in a simple model.

THE MEANING OF NEW TECHNOLOGY

The analytical problem of clarifying the meaning of the term *new technology* rests on the observation that what might look like a radical innovation or technological revolution from a macroscopic point of view appears to be an evolutionary process of incremental innovation from a microscopic point of view (Geels, 2002; Schumpeter, 1939).[5] In order to avoid that difficulty, the term *new technology* will be defined with respect to interaction properties between vendors and buyers of new technology rather than with respect to some properties of the technology itself.

Any technological artifact (machine), that is, any physical artificial product incorporating technological knowledge, must embody a number of trade-off decisions concerning various performance and cost characteristics (Dosi, 1982). This is clearly true for design quality vs. cost considerations. However, different performance characteristics must also be traded off against one another. For example, when IBM introduced its System/360 mainframe computer together with software automating systems operations (the operating system OS/360) in 1964, it assumed that users valued the comfort of having an operating system more than accompanying reductions in processing speed (Fisher, McKie, & Mancke, 1983).

Thus, in order to appropriately assess the value of a certain machine, users must be informed about the kind of trade-off decisions embodied in the new technology. This knowledge is both costly to acquire (for buyers) and to communicate (for vendors), and thus constitutes part of the overall transaction costs. We call the trade-off decisions embodied in a technological artifact its *trade-off positions*.

The term *new technology* can then be defined as follows. Whenever the existing knowledge of users is not appropriate to evaluate a machine's trade-off positions but must be newly acquired, the machine represents an instance of new technology. In contrast, if the performance characteristics of a machine are improved without changing its trade-off positions, users can rely upon the existing evaluation knowledge to assess its price and performance characteristics.[6]

TWO TYPES OF TRANSACTION COSTS

From this definition of new technology it follows that any firm offering new technology on the market must communicate knowledge about how to assess its new product properly or rely on potential buyers acquiring that knowledge by themselves, that is, without the help of vendors. These costs are clearly an instance of transaction costs, which we call vertical transaction costs because they relate to communication requirements between buyers and sellers rather than another type of transaction costs, which will be introduced below. Vertical transaction costs are not just a logical derivative of our definition of new technology but can be identified empirically as well. One of the few instances of the replacement of an established technological platform that punctuate the history of the IT industry was the creation of the PC platform by IBM in the early 1980s. According to Bresnahan and Greenstein (1999, pp. 24-25), "IBM's marketing capability and reputation helped overcome the standards persistence built around the established 8-bit platform....This was the first competitive replacement of an established computing platform by another." While Bresnahan and Greenstein attribute that feat to a historical coincidence of several factors—the strong position of IBM in other segments, a low risk of cannibalization, technological transition (from 8-bit to 16-bit processors), the choice of an open architecture by IBM so that other firms could easily develop complementary products—the superior marketing ability of IBM as being conducive in that platform succession stands out as a highly plausible reason.[7] Furthermore, a striking empirical regularity found by Goldenberg, Libai, and Muller (2002) is also supportive of the existence of vertical transaction costs. That regularity refers to the phenomenon of a "saddle" in sales figures of new technological products, meaning that an initial peak in sales is often followed by an extended and clearly discernible decline before sales pick up again to start a sustained period of growth. Goldenberg et al. convincingly explain that phenomenon by a communication gap between early and main market adopters. In simulations using a cellular automata model, the authors show that the saddle phenomenon (which occurs in one third to one half of cases depending upon saddle definition) is highly likely when there is a very low level of communication between early and main market adopters relative to communication levels within these groups. This communication gap is important because under such conditions main market adopters cannot rely upon early adopters for guidance in their adoption decisions. Together, these two observations not only make the existence of vertical transaction costs in the marketing of new technologies highly plausible, but also show that they can be incurred by both sellers and buyers (as marketing expense and communication activity respectively).[8]

In addition to these vertical transaction costs, there is a second type of transaction cost that needs to be considered when analyzing the special case of IT markets. IT markets are increasingly characterized by systems competition, meaning that the products offered by IT vendors are but components of a larger system that has to be assembled by the buyer or a buyer's agent (c.f. Church et al., 2002; Desruelle et al., 1996; Economides, 1989; Matutes & Regibeau, 1988; see also the discussion in the second section). A firm offering new technology on a market characterized by systems competition (systems markets) has to not only communicate new evaluation knowledge to buyers, but also persuade other firms to offer complementary products. This requirement constitutes a second type of transaction costs for vendors offering information technology, which we term horizontal transaction costs since they relate to interactions between laterally connected firms.[9]

Distinguishing between these two types of transaction costs provides for the possibility of identifying two types of network effects that are linked to the size of the existing network and the size of the expected network respectively.

Knowledge necessary for evaluating new technology properly can travel along the network of existing users by direct exchange (e.g., word of mouth at conferences). Thus, transaction costs resulting from the need to communicate evaluation knowledge to potential buyers, that is, vertical transaction costs, will vary with the size of the existing network (i.e., the number of past buyers) because acquiring or communicating evaluation knowledge becomes easier as the network increases in size.

In contrast, the difficulty of persuading potential developers of complementary products to actually offer such products on the market, that is, the extent of horizontal transaction costs, will correspond to the perceived future size of the relevant market. The theoretical basis for this proposition is that the costs of developing new technology are largely fixed, implying that economies of scale dominate the calculus of potential suppliers of complements. Thus, the second type of transaction costs will vary with the expected network size.

The roles of the existing and the expected network size in the decisions of users and vendors of new technology can be summarized by the following two propositions:

- **Proposition 1:** The transaction costs (for vendors and/or buyers) resulting from communicating or acquiring the knowledge necessary for potential buyers to accurately value products incorporating new technology, that is, vertical transaction costs, vary with the number of past buyers of these products.
- **Proposition 2:** The transaction costs resulting from the need to persuade potential suppliers of complementary products to actually develop them, that is, horizontal transaction costs, vary with the expected total number of buyers of products incorporating that technology over its whole life cycle.

DEGREE OF STANDARDIZATION AND MODULARIZATION

As shown above (see "Weaknesses of the PNE Concept"), both the degree of standardization (i.e., horizontal compatibility) and the degree of modularization (i.e., vertical compatibility) can be related to a kind of network effect. However, this approach involves several causal steps, implying an increased degree of indirectness that makes empirical evaluation quite problematic.

The direct effects of decisions concerning the degree of standardization and the degree of modularization respectively are quite different.

As pointed out above, increased degrees of modularization imply higher levels of economies of scale, which translate into lower prices if components are supplied competitively. In contrast, increased standardization reduces implementation costs for buyers since the likelihood of being able to combine these components with existing or future complementary products without additional cost increases (Bresnahan & Chopra, 1990).[10] Both, however, might have similar indirect effects, namely, an increase in the variety of complementary products offered, as shown in the section "Weaknesses of the PNE Concept."

Note that by this conceptualization, compatibility with (dominant) standards enters the calculus of users on the cost side and not on the benefit side as is usually assumed in the PNE literature (which assumes that users prefer compatibility because of the then larger range of complementary products available; see the section "Network Effects as Positive Network Externalities"). However, these two approaches are not mutually exclusive as the one adopted here models variety as the ease of integration with existing or future complementary products. This seems plausible because ex post compatibility can always be achieved by technical means (adapters, converters), which, however, come at a cost (which can be saved when standards are available). Thus, the direct effects of increasing the degrees of horizontal compatibility are modeled as cost savings rather then benefit increments.

Also note that our concept of implementation costs is closely related to the concept of switching costs (Greenstein, 1993). On the one hand, a similar relationship as that between implementation costs and the degree of standardization exists for switching costs, namely, the higher the degree of standardization, the lower the switching costs. On the other hand, switching between different brands or systems with standardized interfaces is but one element of implementation costs. Extending existing systems through adding new components is another aspect that is not captured by the concept of switching costs. Thus, the notion of implementation costs is broader than the concepts of product variety and of switching costs taken individually while being flexible enough to include both.

The roles of standardization and modularization in the decisions of vendors and users of new technology can be summarized by the following two propositions.

- **Proposition 3:** As the degree of modularization (vertical compatibility) increases, the unit costs of supplying (developing, manufacturing, and distributing) mutually compatible components decrease, which translates into lower prices if components are competitively supplied, and product variety increases.
- **Proposition 4:** As the degree of standardization (horizontal compatibility) increases, implementation costs for buyers decrease and product variety increases.

OUTLINES OF THE MODEL

By conceptualizing network effects as cost savings, the question arises on how to model the benefit of adopting new technology. For vendors, the benefit of offering new technology on the market is measured by revenues collected. Adopting a Schumpeterian perspective on competition, the main advantage of offering new technology (as opposed to existing technology) consists of creating a temporary monopoly that allows for setting prices close to reservation prices, that is, the maximum willingness of buyers to pay for a product. Reservation prices, in turn, are determined by net benefits of users (benefit minus costs of using new technology). As substitute products are offered on the market, prices will be driven down toward cost levels. Thus, the only remaining variable in the calculus of buyers and vendors to be determined is the benefit of using new technology.

In the PNE literature, a typical way to model the decision calculus of buyers is to distinguish between an intrinsic value of a product and the value added through the network effect. The intrinsic value is generally not specially considered and assumed to be distributed according to the requirements of the models used.

Since we have suggested a shift of the network effect from the benefit to the cost side of the calculus, the intrinsic value of the product remains the only factor on the benefit side of buyers' decision calculus. Rather than proposing hypotheses concerning functional forms for this intrinsic value, we will treat it as an unconstrained variable in the model.

From the previous discussion, it is now possible to specify the foundation of a transaction cost theoretic model of network effects in the following way.

The calculus of buyer *j* of new technology can be represented by the following equation:

$$r_j = b_j - c_j(s) - tc_j(x^i), \qquad (1)$$

with

r_j being the reservation price,
b_j the benefit of using new technology,
c_j the costs of implementing new technology,
s the degree of standardization of the new technology,
tc_j the transaction costs of acquiring knowledge about trade-off positions embodied in the new technology (vertical transaction costs), and
x^i the size of the existing network (installed base).

The calculus of vendor *i* of the new technology can be formulated accordingly:

$$\pi_i = \sum r_j - PC_i(m) - TC^c_i(x^i) - TC^s_i(x^e), \qquad (2)$$

with

π_i being the profit of supplying new technology,
PC_i the production and distribution costs of supplying new technology,
m the degree of modularization,
TC^c_i the transaction costs of communicating knowledge about trade-off positions embodied in the new technology (vertical transaction costs),
TC^s_i the transaction costs of persuading other firms to supply complementary products for the new technology (horizontal transaction costs), and
x^e the expected network size.

Recognizing that it does not matter whether vertical transaction costs are carried by buyers or vendors, we redefine r and define TC^{cm} as follows:

$$\check{r} \equiv b - c(s)$$
$$TC^{cm} \equiv TC^c + tc,$$

which yields

$$\pi_i = \sum \check{r}_j - PC_i(m) - TC^{cm}_i(x^i) - TC^s_i(x^e). \qquad (3)$$

As a firm will offer new technology only if it expects profits to be higher than those of offering existing technology, Equation 3 must be compared with the profit of supplying existing technology. Since prices for existing technology will be determined by average cost and the intensity of competition rather than reservation prices, the profit of supplying existing technology is:

$$\pi^o_i = x^{oe} * \{PC^o_i/x^{oi} + z_i(t)\} - PC^o_i(m) - TC^{ocm}_i(x^{oi}) - TC^{os}_i(x^{oe}), \qquad (4)$$

with

π^o_i being the profit of offering existing technology,
x^{oe} the expected size of the network of existing technology,

PC^o_i the production and distribution costs of supplying existing technology,
x^{oi} the size of the installed base of existing technology,
z_i the profit margin of existing technology,[11]
TC^{ocm}_i the sum of vertical transaction costs, and
TC^{os}_i the horizontal transaction costs.

New technology will only be offered if there is at least one firm for which the following holds:

$$\pi > \pi^o. \quad (5)$$

HOW BUYERS CAN INFLUENCE NEW-PRODUCT DEVELOPMENT DECISIONS OF VENDORS

There are two principal ways in which buyers can try to influence the range of choices of technological alternatives available to them on the market.

First, they can try to take on an active role by specifying their requirements concerning technological products and then communicating these to potential vendors. This approach, however, is likely to meet with serious coordination problems as the mixed results of user groups' attempts at exerting direct influence over product specifications indicate (Burrows, 1999). This is because new technology, by its very nature, is still open to change so that buyers likely disagree on the future course of desirable technological development since any one buyer would like to see their special requirements implemented. Thus, reaching consensus on the course of future technological development seems unlikely (Jakobs, 2000).

Moreover, buyers need to commit to these specifications in a credible way, implying that they would have to shun those vendors whose products do not comply with the required technical specifications. This type of collective commitment is even more difficult to achieve due to the free-rider problem (Olson, 1965). Consider the example of the up to now biggest effort in forcing vendors of new technology to adopt buyer-defined specifications, the GM-initiated and -sponsored MAP[12] process. This effort failed due to an insufficient degree of coordination among large buyers to pose a credible threat to vendors who were hesitant to adopt these specifications (Bresnahan & Chopra, 1990; Dankbaar & van Tulder, 1992).

Note that this possible course of action is the only way buyers can influence the range of technological options offered on the market if one resorts to the concept of positive network externalities for both of its possible mechanisms, increasing benefits due to an increase in the size of the existing as well as of the expected network.

The interpretation of network effects proposed in this chapter, however, suggests a second possibility for influencing the course of technological development through collective action. If the effect of the existing network size is to facilitate the acquisition of knowledge about trade-off positions embodied in new technology, users could try to artificially create such an effect even if the existing network of users of the new technology is relatively small. By promoting knowledge about trade-off positions embodied in new technological products, they can significantly reduce the transaction costs of supplying or purchasing new technology. If users succeed in establishing this type of communication platform, the effect may be sufficiently large to trigger the development of new technology that otherwise would not be developed.[13]

In order to explore this idea formally, consider the formulation of a vendor's calculus as developed above for a preliminary analysis. In order to isolate the effects a communication platform among users might have on the transaction costs of supplying or acquiring new technology, assume the following: (a) Production costs and horizontal transaction costs are the same for both the new and the existing technology, (b) similarly, the ultimate expected network size is identical for both technologies, that is, $x^e = x^{oe}$, and (c) reservation prices are the same for all buyers. Then, inequation 5 is reduced to:

$$x^e * \check{r} - x^e * \{PC^o_i/x^{oi} + z_i(t)\} > TC^{cm}_i(x^i) - TC^{ocm}_i(x^{oi}). \quad (6)$$

One way to analyze the decision situation of users who are considering collective action consists of expressing the extra benefit that a user derives from deploying new technology as compared to the benefit of deploying existing technology as a function of the existing network size disadvantage. As this extra benefit provides a source of extra profit for the vendor due to the temporary monopoly situation it enables, the extra profit of a potential provider of new technology could then be related to the existing network disadvantage too. If there would be a way of knowing how artificially, that is, by the collective action of users, reducing the network size disadvantage impacts this extra profit of the potential vendors of new technology, users could assess the likelihood of the success of such efforts.

For this purpose, define:

$PC^o/x^{oi} + z(t) \equiv p$
$\check{r} \equiv p + tp$
$x^{oi} \equiv x^i + nd$,

with

- p being the market price of existing technology,
- tp the temporary profit of supplying new technology corresponding to the extra benefit of deploying new technology rather than existing technology, and
- nd the existing network disadvantage of new technology.

Then, inequation 6 becomes:

$$tp > 1/x^e * \{TC^{cm}_i(x^i) - TC^{ocm}_i(x^i + nd)\}. \quad (7)$$

In order to further explore this relationship, it seems practical to determine a functional form for TC^{cm} and TC^{ocm}.

Since the effect of the existing network size on potential buyers consists of providing a communication channel between existing and potential buyers of technological products, it is possible to use diffusion theory for determining a proper functional form. Diffusion theory claims that innovations spread through a population of potential users as a result of a type of "infection" mechanism. As potential users learn about new technology from current users, they become infected and start to use the technology themselves (c.f. Rogers, 1983). Accordingly, the diffusion of innovations follows an *S*-shaped or logistic pattern known from the spread of diseases because, initially, when the fraction of current users is still small, the chances of infecting new users through direct contact are still high. This is because the likelihood of a current user meeting a potential (uninfected) user is higher than that of meeting another current user. Each newly infected user will, in turn, start to infect other potential users, thus creating the accelerating section of the *S*-shaped diffusion pattern. However, as the share of current users increases, the probability of a current user meeting uninfected members of the population begins to decrease because more often than not one current user will meet other current users, causing a slowing down in the speed of diffusion (and thus accounting for the decelerating section of the *S*-shaped diffusion curve).

Whereas diffusion theory measures the spread of knowledge in a given population of potential users as a function of time, the underlying mechanism consists of communication between current users and potential users. The more direct (and for our purposes the more useful) way of modeling that dependency consists of expressing the spread of evaluation knowledge as a function of the number of current users (who, via their communication activity, trigger the spread of knowledge), which then can also be represented by a logistic functional form. In that specification, current users inform potential users about the new technology who, in turn, inform further potential users and so on. However, as the knowledge thus spread tends to be diluted as it is handed from one potential user to the next, it will quickly become useless, and the diffusion process would trickle away were it not for old and new current users that keep communicating their knowledge to new potential users, thus driving the whole diffusion process. However, as the mechanism accounting for the accelerating part of the original logistical diffusion curve is much weaker in this specification, this part becomes negligible and the overall

Figure 1. Deriving TC^{cm} and TC^{ocm} (indicating required levels of spending for each value of x^i)

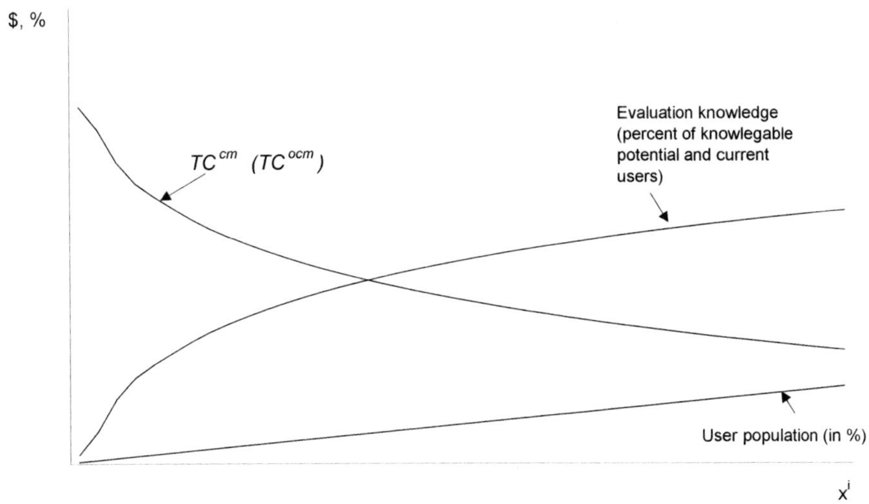

Figure 2. The shape of F

shape of the diffusion curve is dominated by the decelerating part. An example of such a diffusion curve is depicted in Figure 1.[14]

Since an existing network of users facilitates communication of the evaluation knowledge necessary for making purchasing decisions, the costs of communicating that knowledge, that is, vertical transaction costs, decrease as the network and thus evaluation knowledge grows when seen from the vendors' perspective since an ever larger share of the total transaction cost burden will be taken over by the network. Thus, as evaluation knowledge about trade-off positions embodied in new technology travels through the population of current and potential users, transaction costs of communicating and acquiring that knowledge decrease accordingly. This relationship is also depicted in Figure 1.

According to inequation 7, the extra benefit tp of new technology must be bigger than the difference between $TC^{cm}(x^i)$ and $TC^{ocm}(x^i + nd)$ times the reciprocal of the expected network size if the new technology is to be offered on the market. Assuming that TC^{cm} and TC^{ocm} have the same functional form and identical parameters, a function representing inequation 7, denominated by F, can be derived from the shape of TC^{cm}/TC^{ocm}. This function is depicted in Figure 2.

As the logistical diffusion curve is dominated by decelerating growth rates, F is dominated by decelerating growth rates, too, with a negligible section displaying accelerating growth rates. This implies that collective user action, which mimics the communication effect of an installed base, would have the biggest effect for its initial efforts; more importantly, it suggests that there is no critical threshold that needs to be overcome for such collective efforts to show an effect. Thus, users have a means of actively affecting the range of choices offered to them on technology supply markets by reducing the degree of a new technology's superiority required to induce vendors to actually develop and supply that technology.[15] From the point of view of positive theory, in turn, the implications of this model can then be summarized by the following proposition.

- **Proposition 5:** Other things being equal, the frequency with which a new technology seen as superior by buyers replaces entrenched technologies increases with the degree of collective user action aimed at improving the communication flow among current and potential users.[16]

CONCLUSION

Based on a transaction cost theoretic interpretation of network effects, the analysis presented in this chapter suggests how users can actively influence the range of technological choices available to them on markets for technology-intensive products through collective action other than by collectively specifying the desired technology and committing to only buy products incorporating that particular new technology, that is, user-driven ex ante standardization. This latter alternative would be the only way for users to collectively affect the course of technological development if one relies on models that interpret network effects as positive network externalities. Thus, the transaction cost theoretic interpretation of network effects presented in this chapter provides a new perspective for analyzing the collective behavior of buyers that transcends the analytical boundaries of the interpretation of network effects as network externalities. For example, based on an analysis of the so-called Wintel (Windows/Intel) monopoly from a PNE perspective, Takahashi and Namiki (2003) have argued that government intervention is necessary to break this monopoly because the intrinsic technological superiority of potential challenger technologies (such as Java or Linux) is not sufficient to overcome the installed-base advantage of the Windows network. The possibility of reducing the extent of technological superiority necessary for dislodging an established technology through collective action among users is not considered.

In addition, these theoretical considerations also provide for a more positive view on the possibility of rationalizing IT purchasing decisions. For example, Tingling and Parent (2002) show in an experiment that purchasing managers are willing to sacrifice a slight superiority of a particular technology if competitors choose the inferior technology, which they interpret as an indication of irrational behavior. However, as the technology underlying their experiment (a plug-in for a Web browser) has strong external compatibility requirements, these results can also be seen to suggest that in the face of a high degree of technological uncertainty, herd behavior might indeed be the best course of action unless collective action among potential buyers to reduce that uncertainty is allowed for (which was not the case in that experiment).

According to the theoretical considerations presented in this chapter, the type of collective user action required for widening the range of technological choices on markets for new technology consists of providing a communication platform for the exchange of technical and economic information about new technology. An initial simple model incorporating these theoretical considerations suggests that such collective user action would have the strongest effect in its initial phase, implying that no threshold value has to be reached before collective user action could be effective. The model is based on propositions that are formulated in a way allowing for relatively easy empirical refutation. Moreover, the model predicts that the frequency of instances in which a superior technology (i.e., one regarded as superior by users) replaces an entrenched technology is related to the extent of information exchange among current and potential users (other things being equal), a proposition that lends itself to empirical testing as well.

The model presented in this chapter also has a rather practical implication. Collective action in the form of user groups tends to be organized around existing products of specific vendors. This is, in no small part, due to their need for obtaining sufficient financial resources for their own operation and administration, which is frequently provided by vendors.[17] However, such practices rule out the communication effect of collective user action, which has been described in this chapter as facilitating the supply of a greater variety of technologies on IT markets.[18] Indeed, forming user groups around existing products and brands may increase the degree of path dependency and lock-in due to the ease of acquiring product-specific knowledge, an effect coined customer groove-in by Arthur (1996).[19]

Achieving the communication effect described in this chapter requires that collective user action comprise users of different technologies so that the sharing of knowledge regarding their respective economic and technical characteristics becomes possible. Vendors of existing products will normally not have sufficient incentives to incorporate new technologies into their products lest they cannibalize their own revenues. Thus, entrenched technology is most likely replaced by newcomers (Martin & Mitchell, 1998; Tushman & Anderson, 1986). However, any supplier of a competing product incorporating new technologies would face the daunting task of overcoming the huge communication disadvantage created by the size of the user network of the existing technology (thus increasing the supplier's transaction costs relative to those of suppliers of the existing technology). Collective user action may help to reduce the magnitude of this disadvantage only if it mixes users of different technologies and, by implication, brands. This, at least, is what the model we have presented in this chapter would suggest. Thus, increasing the variety of technologies offered on technology supply markets, which would be one way of increasing the responsiveness of technology supply markets to changing user requirements, would require a different type of collective user action than is represented by today's user groups, one actively managed (and financed) by users formed around specific types of user requirements rather than products and brands.

There are several organizational ways in which users could try to bring about this communication effect, which we want to briefly discuss in order to stimulate further research on this topic.

- Conferences
- Professional organizations
- Trade fairs
- Industry associations

Among these, we think that industry associations represent the most suitable organizational option with regard to pursuing the goals discussed in this chapter. Conferences, professional organizations, and trade fairs bring together users with very different backgrounds in terms of technical requirements and experiences and would thus serve the purpose well, especially when compared with traditional user groups that focus on one particular product. However, unless conferences are set up to be organized around related topics in a regular and recurring fashion, they may not be sufficient to reliably circulate knowledge about upcoming technologies among potential users (and therefore fail to ensure the intended communication effect since potential vendors of a new technology need to rely on this ability if the new technology's inherent degree of superiority is not sufficient to overcome the advantages of an entrenched technology's installed-base communication effect). Professional organizations tend to promote the interests of individuals rather than corporate interests and thus may be reluctant to spend extra resources on the purpose considered here (which promotes the interests of firms rather than of individuals). Trade fairs, by their very nature, are dominated by vendors that would make collective user action relatively more difficult. For example, it would be difficult for buyers to set up discussion platforms on which competing products incorporating different technological approaches could be discussed without transforming these discussion platforms into race courses for competing vendors' products, which would be counterproductive in terms of discussing the trade-off positions of the underlying technologies (which, in turn, would be an essential aspect of the communication effect described in this chapter). Industry associations avoid all these problems while also bringing together different technical requirements and expertise, albeit at the cost of a much narrower focus. Still, this drawback seems tolerable as establishing technological alternatives in one or a few industries would already serve the purpose described in this chapter; that is, the goal is not to dislodge established technologies globally, but to offer alternatives in certain areas of applications. If such niche applications accumulate, however, they may well start a chain reaction that could result in the rise to dominance of a new technology (see Geels, 2002, for an illustration of this process using the example of the rise of the steamship).

However, the discussion of the possibility of using the communication effect of collective user action for facilitating greater responsiveness of technology supply markets and the tentative suggestions regarding the alternative forms of organization for such collective user action are derived from a theoretical model of technology supply markets that needs to be empirically validated before it can be used as a basis for recommendations to practitioners. If, indeed, the propositions on which this model is built can be shown to hold empirically, it would seem necessary to reconsider parts of current standardization literature.

ACKNOWLEDGMENT

We would like to thank Mary Brabston and the other participants of the International Federation of Information Processing Working Group 8.6 (IFIP WG 8.6) working conference The Diffusion and Adoption of Networked Information Technologies, October 6 to 8, 2003, Helsingor, Denmark, for their helpful comments and feedback on an earlier version of the chapter. In addition, we have gratefully received extensive and very helpful comments from four anonymous reviewers. The second author acknowledges financial support from the National Science Foundation of China (Project Numbers 70231010 and 70321001).

REFERENCES

Abernathy, W. J., & Clark, K. B. (1985). Innovation: Mapping the winds of creative destruction. *Research Policy, 14*, 3-22.

Andersen, E. S. (1991). Techno-economic paradigms as typical interfaces between producers and users. *Journal of Evolutionary Economics, 1*(2), 119-144.

Arthur, W. B. (1996, July-August). Increasing returns and the new world of business. *Harvard Business Review*, 100-109.

Besen, S. M., & Farrell, J. (1994). Choosing how to compete: Strategies and tactics in standardization. *The Journal of Economic Perspectives, 8*(2), 117-131.

Bresnahan, T. F., & Chopra, A. (1990). The development of the local area network market as determined by user needs. *Economics of Innovation and New Technology, 1*, 97-110.

Bresnahan, T. F., & Greenstein, S. (1999). Technological competition and the structure of the computer industry. *Journal of Industrial Economics, 47*(1), 1-40.

Burrows, J. H. (1999). Information technology standards in a changing world: The role of users. *Computer Standards & Interfaces, 20*, 323-331.

Church, J., Gandal, N., & Krause, D. (2002). *Indirect network effects and adoption externalities.* University of Calgary, Department of Economics.

Clark, K. B. (1995). *Notes on modularity in design and innovation in advanced ceramics and engineering plastics* (Working Paper No. 95-073). Boston: Harvard Business School.

Dankbaar, B., & van Tulder, R. (1992). The influence of users in standardization: The case of MAP. In M. Dierkes & U. Hoffmann (Eds.), *New technology at the outset: Social forces in the shaping of technological innovations* (pp. 327-350). Frankfurt; New York: Campus.

Desruelle, D., Gaudet, G., & Richelle, Y. (1996). Complementarity, coordination and compatibility: The role of fixed costs in the economics of systems. *International Journal of Industrial Organization, 14*(6), 747-768.

Dosi, G. (1982). Technological paradigms and technological trajectories. *Research Policy, 11*, 147-162.

Dranove, D., & Gandal, N. (1999). *The DVD vs. DIVX standard war: Network effects and empirical evidence of vaporware.* Tel Aviv, Israel: Tel Aviv University, Foerder Institute of Economic Research, Faculty of Social Sciences.

Economides, N. (1989). Desirability of compatibility in the absence of network externalities. *American Economic Review, 79*(5), 1165-1181.

Economides, N. (1991). Compatibility and the creation of shared networks. In M. E. Guerin-Calvert & S. S. Wildman (Eds.), *Electronic services networks: A business and public policy challenge* (pp. 39-55). New York: Praeger.

Economides, N. (1996). The economics of networks. *International Journal of Industrial Organization, 14*(6), 673-699.

Farrell, J., & Saloner, G. (1985). Standardization, compatibility, and innovation. *Rand Journal of Economics, 16*(1), 70-83.

Farrell, J., & Saloner, G. (1986). Installed base and compatibility: Innovation, product preannouncements, and predation. *The American Economic Review, 76*(5), 940-955.

Fisher, F. M., McKie, J. W., & Mancke, R. B. (1983). *IBM and the U.S. data processing industry: An economic history.* New York: Praeger.

Gandal, N. (1994). Hedonic price indexes for spreadsheets and an empirical test for network externalities. *RAND Journal of Economics, 25*, 160-170.

Gandal, N. (1995). Competing compatibility standards and network externalities in the PC software market. *Review of Economics and Statistics, 77*, 599-608.

Gandal, N., Kende, M., & Rob, R. (1997). *The dynamics of technological adoption in hardware/software systems: The case of compact disc players* (Working Paper No. 21-97). Sackler Institute of Economics Studies.

Geels, F. W. (2002). Technological transitions as evolutionary reconfiguration processes: A multi-level perspective and a case-study. *Research Policy, 31*(8-9), 1257-1274.

Goldenberg, J., Libai, B., & Muller, E. (2002). Riding the saddle: How cross-market communication can create a major slump in sales. *Journal of Marketing, 66*, 1-16.

Greenstein, S. M. (1993). Did installed base give an incumbent any (measurable) advantages in federal computer procurement? *RAND Journal of Economics, 24*(1), 19-39.

Jakobs, K. (2000). *Standardisation processes in IT: Impact, problems and benefits of user participation.* Braunschweig, Germany: Vieweg.

Katz, M. L., & Shapiro, C. (1985). Network externalities, competition, and compatibility. *The American Economic Review, 75*(3), 424-440.

Katz, M. L., & Shapiro, C. (1986). Technology adoption in the presence of network externalities. *Journal of Political Economy, 94*(4), 822-841.

Katz, M. L., & Shapiro, C. (1994). Systems competition and network effects. *Journal of Economic Perspectives, 8*(2), 93-115.

Krugman, P. (1991). History versus expectations. *Quarterly Journal of Economics, 106*, 651-667.

Langlois, R. N. (1992). Transaction-cost economics in real time. *Industrial and Corporate Change, 1*(1), 99-127.

Langlois, R. N. (2002). Modularity and technology and organization. *Journal of Economic Behavior and Organization, 49*(1), 19-37.

Liebowitz, S. J., & Margolis, S. E. (1990). The fable of the keys. *Journal of Law and Economics, 33*, 1-25.

Liebowitz, S. J., & Margolis, S. E. (1994). Network externality: An uncommon tragedy. *The Journal of Economic Perspectives, 8*(2), 133-150.

Liebowitz, S. J., & Margolis, S. E. (1995). Path dependence, lock-in, and history. *Journal of Law, Economics and Organization, 11*, 205-226.

Liebowitz, S. J., & Margolis, S. E. (1996). Should technology choice be a concern of antitrust policy? *Harvard Journal of Law and Technology, 9*(2), 283-318.

Liebowitz, S. J., & Margolis, S. E. (1998). Network externalities (effects). *The New Palgrave's Dictionary of Economics and Law.* MacMillan.

Martin, X., & Mitchell, W. (1998). The influence of local search and performance heuristics on new design introduction in a new product market. *Research Policy, 26*(7/8), 753-771.

Matutes, C., & Regibeau, P. (1988). "Mix and match": Product compatibility without network externalities. *RAND Journal of Economics, 19*(2), 221-234.

Olson, M., Jr. (1965). *The logic of collective action: Public goods and the theory of groups.* Cambridge, MA: Harvard University Press.

Rogers, E. M. (1983). *Diffusion of innovations* (3rd ed.). New York: The Free Press.

Schilling, M. A. (2002). Technology success and failure in winner-take-all markets: The impact of learning orientation, timing, and network externalities. *Academy of Management Journal, 45*(2), 387-398.

Schumpeter, J. A. (1939). *Business cycles: A theoretical, historical, and statistical analysis of the capitalist process.* New York: McGraw Hill.

Takahashi, T., & Namiki, F. (2003). Three attempts at "de-Wintelization": Japan's TRON project, the US government's suits against Wintel, and the entry of Java and Linux. *Research Policy, 32*(9), 1589-1606.

Tingling, P., & Parent, M. (2002). Mimetic isomorphism and technology evaluation: Does imitation transcend judgment? *Journal of the AIS, 3*(5), 113-143.

Tushman, M. L., & Anderson, P. (1986). Technological discontinuities and organizational environments. *Administrative Science Quarterly, 31*, 439-465.

von Burg, U., & Kenney, M. (2003). Sponsors, communities, and standards: Ethernet vs. token ring in the local area networking business. *Industry and Innovation, 10*(4), 351-375.

Witt, U. (1997). "Lock-in" vs. "critical masses": Industrial change under network externalities. *International Journal of Industrial Organization, 15*(6), 753-773.

ENDNOTES

1 *There are many products for which the utility that a user derives from consumption of the good increases with the number of other agents consuming the good.* (Katz and Shapiro, 1985, p. 424).

2 Krugman employs a dichotomous specification of network externalities implying that either all or no actor switch. Thus, for a self-fulfilling prophecy to take place perfect coordination of expectations would be required, which is unlikely. Allowing for continuous externalities might change the results.

3 Vertical compatibility means that two complementary products can be combined without additional cost; horizontal compatibility means that component interfaces are standardized so that two substitute components can be combined with one and the same complementary product without additional cost (cf. Economides, 1991).

4 Note that in the literature the concept of modularity is often seen as also comprising horizontal compatibility (i.e. in addition to vertical compatibility) which allows for recombining modules and thus easy adaptation to varying requirements (mix-and-match capability; cf. for example Clark, 1995). Langlois (2002) uses the terms internal and external modularity to allow for this extension. In this paper, we use the concept of modularity in the narrow sense of implying only vertical but not horizontal compatibility.

5 *... there is as little contradiction between them [the macroscopic and the microscopic point of view] as there is between calling the contour of a forest discontinuous for some and smooth for other purposes.* (Schumpeter, 1939, p. 227).

6 This definition of new technology bears some resemblance with the notion of "Techno-Economic Paradigms" (TEP) which has been proposed by Andersen (1991). The TEP concept is itself an extension of Dosi's (1982) Technological Paradigms which has been modified to describe the interface between developers and users of new technology across a market interface. Improvements

of performance characteristics without changing embodied trade-off positions correspond to Dosi's/Anderson's concept of "normal technological progress." Our definition of new technology can also be related to the widely used distinction between competence-enhancing and competence-destroying innovations introduced by Abernathy and Clark (1985). While it might be argued that new technology as defined by us corresponds to evaluation knowledge being destroyed, the distinction between competence-enhancing and -destroying innovations has generally been applied to the analysis of using rather than evaluating new technology (ibid.; Tushman and Anderson, 1986).

7 Bresnahan and Greenstein, while frequently mentioning IBM's superior marketing capability and its role in that platform succession, do not consider this factor in their explanatory framework, probably because they explicitly choose to exclude coordination costs related to standards adoption decisions (1999, p. 11).

8 While the example of the PC platform refers to corporate users, the saddle phenomenon identified by Goldenberg et al. refers to consumer goods. It might be argued that vertical transaction costs are less important in business markets because corporate users are more resourceful and thus in a better position to evaluate new technology. However, we think that the phenomenon of technology-hype which is characteristic of IT markets clearly indicates that corporate users often rely on their peers with regard to technology acquisition decisions implying a significant level of uncertainty regarding the value of new technological products in that market among corporate users and thus the importance of vertical transaction costs.

9 For an illustration of this need see von Burg and Kenney (2003) who describe how the creation of a "community" of IT firms facilitated the emergence and, finally, dominance of the Ethernet technology vis-à-vis the IBM-sponsored Token Ring technology for Local Area Networks. Our concept of horizontal transaction costs corresponds to Langlois' (1992) notion of dynamic transaction costs: "Ultimately, the costs that lead to vertical integration are the (dynamic) transaction costs of persuading, negotiating with, coordinating among, and teaching outside suppliers in the face of economic change or innovation." (p. 116).

10 The model used by Bresnahan and Greenstein (1999) to explain the history of the IT industry is based on the notion of sunk costs which comprise expenses for developing components that can be combined into a platform, i.e. system, including both vendor and buyer related expenses (such as those for training). In this paper we are not concerned with explaining the structure of the IT industry therefore ignoring the possibility of costs being irreversible (i.e. sunk). On the other hand, we extend Bresnahan and Greenstein's model by not only including production costs (which we differentiate between vendor and buyer-related costs according to the principles of modularity and standardization) but also transaction costs as described above.

11 We assume that z decreases with time since the longer the product is on the market, the more any proprietary advantage the technology might originally have had will disappear; defining z(0) as the profit margin at the time the new technology is introduced on the market and z(t') = 0, the value of t' provides a dynamic measure for the degree of competitive intensity in a given industry.

12 MAP: Manufacturing Automation Protocol.

13 Note the analogy of this argument with the explanation of the saddle phenomenon by

Goldenberg et al. (2002). Goldenberg et al. argue that a strongly pronounced temporary decline in sales, i.e. a "saddle," may lead vendors to prematurely withdraw a new product from the market which effect would be avoided if there were a stronger communication link between early and main market adopters.

14 This curve has been created using the following assumptions: (1) every current user can inform four potential users per unit of time; (2) every potential user "infected" by a current user can inform another two potential users per unit of time; (3) every one of these latter potential users can only inform one other potential user per unit of time who cannot communicate the knowledge any further; (4) the likelihood of meeting a current user (who cannot be informed any more) rather than a potential user increases in proportion to the share of current users.

15 Note the difference of this approach to that suggested by Witt (1997) who demonstrated how in a model based on the notion of positive network externalities the likelihood of an entrenched technology to be replaced by a new (possibly superior) one can be expressed as a function of the new technology's superiority: "In principle, there is always a chance of overcoming what appeared ... to be an inescapable "lock-in" situation produced by increasing returns to adoption—if the new variant is sufficiently superior to the established one." (ibid., p. 768). Thus, the cumulated positive externalities of the old technology have either to be compensated by the new technology's superiority and/or by some subsidizing scheme to reach a critical network size in order to be dislodged; in any case, buyers would have no means of affecting the course of technological development other than by developing the new technology themselves or collectively committing to buy (only) products incorporating the new technology. A similar line of argument can be found in Arthur (1996) who suggests that "a new product often has to be two or three times better in some dimension—price, speed, convenience—to dislodge a locked-in rival" (p. 105).

16 Although we concur with the criticism of Liebowitz and Margolis regarding the concept of positive network externalities as reported above (see section 'Network Effects as Positive Network Externalities'), we are not as optimistic about the possibility of existing technology being replaced by new technology as they are (cf., for example, Liebowitz and Margolis, 1994). This is because we think that transaction costs play a larger role in technology markets than these authors are willing to concede. Thus, we adopt an intermediate position between theories of lock-in which predict market failure in the presence of positive network externalities and Liebowitz and Margolis' position who assume that there are no significant barriers to the adoption of new technology in the face of an entrenched technology. As a result, frequency of technology replacement becomes a contingent variable in our model.

17 However, not all user groups are vendor-financed and there is a history of grass-roots user groups in the field of corporate computing which, however, are still built around specific vendors' systems (we are indebted to one anonymous reviewer for pointing out this possibility apart from making many more very helpful comments).

18 Note the difference between the terms of technological variety and product variety. Product variety means the variety of complementary products and tends to increase lock-in effects while technological variety implies a greater choice of different technologies incorporated in products on offer at any given moment in time.

[19] This type of product-specific use-knowledge represents an instance of asset specificity in terms of transaction cost economics; thus, lock-in may also be interpreted as a form of hold-up strategy for which no adequate governance structures exist when considering the bilateral relationships between IT customer and supplier.

Chapter XI
Comparing the Standards Lens with Other Perspectives on IS Innovations:
The Case of CPFR

M. Lynne Markus
Bentley College, USA

Ulric J. Gelinas, Jr.
Bentley College, USA

ABSTRACT

Conceptual labels influence researchers' observations and analytic insights. This article aims to clarify the contributions of standards label by contrasting it with other ways of viewing the same entity and applying it to the IT-enabled supply chain innovation of collaborative planning, forecasting, and replenishment (CPFR). Proponents have labeled CPFR not only as a standard but also, at different decreasing levels of abstraction, as a business philosophy, methodology, and set of technologies. By comparing the analytic leverage offered by the different labels, we conclude that there is value in combining the standards perspective with other conceptual lenses. The specific case of CPFR also raises an interesting question for future research: Can information systems innovations justifiably be considered standardized in practice, if they are not standardized at all relevant levels of abstraction?

INTRODUCTION

This article was motivated by our investigation of a particular information system (IS) innovation known as CPFR (collaborative planning, forecasting, and replenishment) used by interdependent organizations to improve supply chain performance. Proponents repeatedly referred to CPFR as a standard or as standards-based, but we could not see exactly how the standards label applied or how it added value relative to other ways of looking at the innovation. (As explained later, CPFR also can be analyzed as a business philosophy, methodology, and set of technologies.) Knowing that conceptual labels can affect researchers' observations and analytic insights, we decided to compare several partially overlapping conceptual perspectives and apply them to this innovation.

The four conceptual lenses applied in this article are (1) philosophy (frame, organizing vision); (2) methodology (procedure, process); (3) technology (tool, technical infrastructure); and (4) standard (standardization). The first three concepts represent an innovation's logical and temporal progression from abstract idea to concrete implementation (Iivari, Hirschheim, & Klein, 2001). The fourth concept can be thought of as the end point of a process about which actors implicitly or explicitly reach agreement and widely adopt solutions to matching problems (Brunsson, Jacobsson & Associates, 2000; Cargill, 1989; de Vries, 1999). Thus, the four labels overlap, as shown in Figure 1.

Figure 1. Overlapping concepts

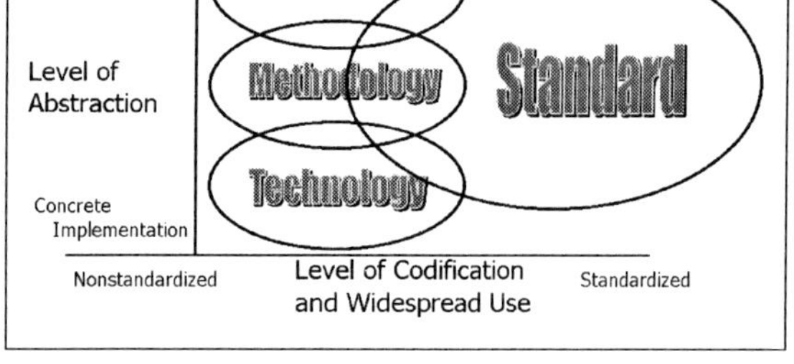

Through our analysis of CPFR, we found that, despite their overlaps, each perspective provides unique insights. We also found that the CPFR innovation cannot be considered standardized yet at any level of abstraction. This observation raises an intriguing question for future standards and standardization research: Unless it has achieved standardization at all levels of abstraction, can an IS innovation truly be considered to be standardized?

THEORETICAL BACKGROUND

The label standard can be applied to such entities as products, processes, services, materials, equipment, systems, interfaces, protocols, functions, methods, and activities (de Vries, 1999). Regarding IS and information technology (IT), the term standard can be applied to technology specifications or products such as GSM (Iversen, 2000) or the Windows operating system, to methodologies such as ISO 9000 (Brunsson et al., 2000) or the capability maturity model of software development, to business processes such as those addressed by the RosettaNet Consortium, and so forth. Calling these entities standards implies that they differ in essential ways from nonstandardized specifications, products, methodologies, processes, and so forth. This observation raises questions about the overlaps and unique contributions of different conceptual labels applied to the same phenomenon.

There is as much debate about the definitions of core concepts in the IS field (Alter, 2005; Orlikowski & Iacono, 2001) as about the definitions of standards and standardization (Brunsson et al., 2000; de Vries, 1999; Soderstrom, 2002). Nevertheless, IS innovations can be analyzed at multiple levels of abstraction with concepts such as philosophy, paradigm, and organizing vision, at the most abstract; concepts such as tools,

Table 1. Four conceptual categories related to IS innovations compared

Conceptual Category	Definitions	Relevant Theory About Origins, Evolution	Similarities With Standards Theory	Contrasts With Standards Theory
Philosophy, frame, organizing vision	Socially constructed set of goals, causal attributions, and justifications for a proposed course of action; broadly shared idea about the application of an IS innovation	Emerges over time through discourse and negotiation among actors, often with conflicting interests; may exhibit a life cycle or "career" from first proposal through popularity (widespread adoption) to decline	Emergent process of sharing an organizing vision is similar to the emergent norms version of standards theory, but not to the version of standards theory that emphasizes deliberate design/selection	Visions are emergent and often implicit, in contrast with many standards, which are deliberately designed or selected and explicitly recorded; visions have careers in which they can change in response to experience; visions need not deal with "matching problems", whereas standards do
Methodology, procedure, process	Organized collection of IT-related concepts, methods, techniques, beliefs, etc.; codified set of goal-oriented procedures intended to guide the work and cooperation of various parties	Specifically designed or selected to achieve uniformity and/or compatibility; often believed to be necessary and sufficient for success; however, IT methodologies are rarely faithfully followed in practice, but rather are heavily customized to local conditions, if followed at all	Methodology theory and standards theory have similar notions of deliberate development or selection, explicit recording, common purpose of solving matching problems, and voluntary (versus mandated) implementation	Methodologies are often viewed as the best or only way to accomplish a goal, whereas standards are often viewed as somewhat arbitrary, in the sense that another solution could be equally good; methodology theory emphasizes the likelihood of customization and the rarity of widespread adoption in contrast to the expectation that standards will be widely adopted as published

continued on next page

techniques, and technical infrastructures at the most concrete; and concepts such as methodology, procedure, and process at an intermediate level of abstraction (Iivari et al., 2001). When a philosophy or idea about how an IS innovation should work in practice is extremely well understood, software tools and technologies can be developed to embody those ideas. When the idea or philosophy is less understood, it is sometimes possible to specify procedures and methodologies without necessarily being able to concretize these processes in software tools.

Despite the multiplicity of labels and definitions, considerable theory and empirical research have emerged around each of the four conceptual perspectives examined in this article and depicted in Figure 1: (1) philosophy (frame, organizing vision); (2) methodology (procedure, process); (3) tools (techniques, technical infrastructure); and (4) standard (standardization). Although space does not permit an exhaustive review of prior theory and research, we try to capture the spirit of each literature. Table 1 summarizes the discussion, providing a definition of each conceptual category, a sketch of relevant theory and research, and an analysis of the similarities and differences among the conceptual lenses.

Philosophy (Frame, Organizing Vision)

Swanson and Ramiller (1997) defined *organizing vision* as an idea or philosophy, held by members of an organizational community, about how adopters should apply an IS innovation. Similar notions include technology *frames* (Iacono & Kling, 2001;

Table 1. continued

Conceptual Category	Definitions	Relevant Theory About Origins, Evolution	Similarities With Standards Theory	Contrasts With Standards Theory
Technology, tool, technical infrastructure	Technologies include tools and techniques used by participants in a work system; technical infrastructures include computer networks, programming languages, and other technologies shared by several work systems; distinction between technologies and technical infrastructures not always clear	Deliberately developed or selected to improve work system outcomes; early in the technology life cycle, much design variation exists; eventually a dominant design emerges, which spurs widespread adoption; technology use is most successful/beneficial when technology is well integrated with other aspects of the work system (such as business processes) and with the technical infrastructure	Similar notions of deliberate development or selection, emergence of de facto standards (dominant designs) and network effects leading to widespread adoption	Technology perspective distinguishes between tool and technical infrastructure; tools can be standardized, while implementation vis-à-vis a technical infrastructure remains unstandardized
Standard, standardization	Deliberate acceptance by group of people with common interests or backgrounds of a quantifiable metric that influences their behavior and activities by permitting a common interchange); limited set of solutions to actual or potential matching problems directed at benefits for the party or parties involved, balancing their needs, and intending and expecting that these solutions will be repeatedly or continuously used during a certain period by a substantial number of the parties for whom they are intended	Standards can emerge over time through repeated interactions among interdependent organizations or through market forces; they can be dictated by a government body; or they can be designed or selected by a formal standards body, a consortium of organizations, an industry sector, or a specific company; emergent standards are often implicit, rather than explicit; deliberately designed or selected standards are usually explicit and voluntary; not all standards exhibit network effects, but some certainly do		Standards theory covers both emergent and designed/selected standards; can be applied to IS innovations at multiple levels of abstraction; emphasizes the actors involved in creating designed/selected standards; generally focuses either on the development of standards or on standards implementation, with limited efforts to bridge the two

Orlikowski & Gash, 1994) and *paradigms* (Iivari et al., 2001). Examples include the notion that organizations should conduct clean-sheet business process redesign (i.e., business process reengineering), manage their customer relationships strategically (i.e., customer relationship management), and integrate core enterprise systems (i.e., enterprise resource planning). Organizing visions that center on the business problems innovations purport to solve and often predate workable solutions that companies can implement. Constructed through discourse (Heracleous & Barrett, 2001), organizing visions serve to mobilize resources (e.g., encouraging vendors to develop enabling software) and to legitimize innovation, promoting widespread adoption (Greenwood, Hinings, & Suddaby, 2002). Organizing visions evolve over time, as influential parties (e.g., vendors or industry associations) promote their views, and practical experience accumulates. Organizing visions have careers in which they rise or fall in popularity, eventually being replaced (Ramiller & Swanson, 2003; Swanson, 2002).

Methodology (Procedure, Process)

Methodologies are defined as "codified set[s] of goal-oriented 'procedures' ... intended to guide the work and cooperation of ... various parties" (Iivari et al., 2001, p. 186). Examples include software development and business process reengineering methodologies. Methodologies are more concrete than philosophies (Iivari et al., 2001) and, thus, can be viewed as a later stage in the evolution of IS innovations, characterized by more certain knowledge of cause-effect relationships. In contrast, with organizing visions that emerge through interaction, methodologies are specifically designed (or selected) to achieve uniformity and/or compatibility in practice (Brunsson et al., 2000). Methodology developers often believe that following prescribed procedures is necessary and sufficient for success, and that not following methodologies properly leads to failure (Harrington, 1998; Wynekoop & Russo, 1993). However, some researchers describe methodologies as convenient fictions that present an image of control and as too mechanistic to guide practitioners' actions successfully (Nandhakumar & Avison, 1999). Thus, methodologies may be beneficial not for detailed, step-by-step procedures but rather for essential principles and practices, which can be (and usually are) customized to local conditions (Brunsson et al., 2000; Iivari et al., 2001). (A similar theme is found in the literature on EDI standards [cf. Damsgaard & Truex, 2000]). Some research shows that although using a methodology produces superior results to not using one, competing methodologies do not differ greatly in performance (Howard et al., 1999).

Technology (Tool, Technical Infrastructure)

The IS literature differentiates two technology elements: tools (e.g., software and devices) used by work system participants and technical infrastructures (e.g., computer networks and programming languages) shared by several work systems (Alter, 2005). The distinction between tools and technical infrastructures is fuzzy (Alter, 2005), but tool use has been found more beneficial when tools are well integrated with other work system elements such as business processes (Clark & Stoddard, 1996; Lee, Pak, & Lee, 2003) and with well-designed technical infrastructures (Truman, 2000; Zhu, 2004). Like methodologies, tools and infrastructures are designed to achieve particular goals (Ross, 2003), although they can exhibit evolutionary processes (Anderson & Tushman, 1990; Orlikowski, 2000). Early in their lifecycles, much variation exists in the designs of tools and infrastructures; eventually, dominant designs emerge, spurring widespread adoption (Anderson & Tushman, 1990). Adoption of tools and technical infrastructures also often exhibits network effects (Katz & Shapiro, 1985).

Standard (Standardization)

Standards are defined as "limited set[s] of solutions to actual or potential matching problems directed at benefits for the party or parties involved, balancing their needs and intending and expecting that these solutions will be repeatedly or continuously used during a certain period by a substantial number of the parties for whom they are meant" (de Vries, 1999, p. 13). Standardization often is defined as deliberate design or selection (e.g., the activity of establishing and recording a standard) (de Vries, 1999) or deliberate acceptance by group of people (Cargill, 1989). Consequently, standards usually are viewed as explicit statements of expected behavior, with which compliance is voluntary (Brunsson et al., 2000). However, some authors suggest that products and processes can emerge as implicit norms and come to appear standardized (i.e., similar or compatible) through repeated interaction (Brunsson et al., 2000), imitation, or coercion. In contrast with methodologies that often are assumed the best (or only) way to achieve particular goals, standards (and norms) are viewed as essentially arbitrary in that another solution might work equally well. Not all standards exhibit network effects (de Vries, 1999), but technology standards often do (David & Greenstein, 1990).

Similarities and Differences Among the Perspectives

The philosophy lens emphasizes evolutionary processes by which organizing visions emerge. Consequently, it overlaps only with the view of standards as implicit norms or emergent behaviors. It differs considerably from standards theory, emphasizing explicit design, consensus selection, and formal recording. Whereas the standards lens emphasizes solutions to matching problems, the organizing vision concept is broader and not limited to problems of similarity and compatibility. Furthermore, philosophies are viewed as having careers, in which they are shaped or disconfirmed by accumulating experience.

The methodology lens is similar to the standards lens in that both emphasize deliberate design or selection, explicit recording, solutions to matching problems, and voluntary compliance. A key difference is that methodologies often are viewed as the best or only ways to accomplish goals, whereas standards often are seen as arbitrary solutions. Standards theory emphasizes desirability of compliance and expectations of widespread adoption; by contrast, methodology literature emphasizes frequency of customization and rarity of widespread adoption.

The technology lens is similar to the standards lens in emphasizing deliberate design and selection; the emergent versions of technology theory employ the notions of dominant designs and network effects, also overlapping with some of the standards literature. A key difference is that the technology perspective explicitly differentiates between tools and technical infrastructures, allowing for the possibility of standardized tools coexisting with nonstandardized technical infrastructures.

When both the emergent (norms) and explicit (design/selection) versions of standards theory are considered, the standards lens considerably overlaps all three other levels of abstraction lenses. Even so, the other lenses retain potential analytic leverage. More so than the other lenses, standards theory highlights the role of actors (e.g., standards development organizations, vendor consortia, industry associations, governments) in developing explicit standards. Although the focus on actors is a significant strength, its potential drawback is lack of integration between the development and adoption sides of standardization processes (Fomin & Keil, 1999), because these actors often play a lesser role during adoption.

In the next section, we explore the value of our analysis by applying the four conceptual lenses to the IS innovation of CPFR. CPFR is particularly interesting for this study, because proponents have described it using all four labels: philosophy, methodology, technology, and standard.

THE CASE OF CPFR

Data for our analysis came from published sources and our fieldwork in a supply chain consisting of a large retailer and two suppliers in which CPFR was recently initiated. In CPFR, business partners (e.g., retailers and their suppliers or manufacturers and their suppliers) attempt to improve supply chain performance (e.g., on-time deliveries and lower inventory costs) by collaborating on sales and demand (order) forecasts. Thus, CPFR represents a solution to a particular category of interorganizational matching problems. The industry association, VICS (Voluntary Interindustry Commerce Standards Association), regularly convenes working groups to develop and publish CPFR guidelines and holds educational conferences for industry participants with the goal of establishing CPFR as a standardized industry practice.

The benefits of CPFR can be great (Seifert, 2003). For example, the retailer we call Specialty Superstores claimed that CPFR improved its relationships with suppliers and reduced sales forecast error rates from 40% to 20% on average and to 1% with certain suppliers. Despite publication of CPFR standards documents and numerous reports of benefits, CPFR has not been adopted as widely as VICS expected and, thus, has not yet achieved its proponents' goal of becoming a standardized industry practice.

In the following, we describe the CPFR innovation through each of our four conceptual lenses, trying to keep the perspectives as distinct as possible. Then we discuss the overlaps and differences in the perspectives.

CPFR Described Through the Philosophy Lens

Proponents sometimes describe CPFR as a business philosophy. For example, a leading popularizer defined CPFR as "an initiative among all participants in the supply chain intended to improve the relationship among them through jointly managed planning processes and shared information" (Seifert, 2003, p. 30). According to VICS, "CPFR is simply the latest *embodiment of knowledge and experience* that has been compiled to continually improve a company's internal efficiencies while increasing external effectiveness" (VICS, 2004d, p. 3, emphasis added). By the phrase "simply the latest embodiment," VICS acknowledged that CPFR possibly is not the final result of a long prior career of industry partici-

Table 2. Evolution of the CPFR supply chain innovation

Quick Response (QR) Context: Soft goods industry (1986) Vision: QR was envisioned as four levels of successively more sophisticated technologies and applications: 1) point-of-sale technology and price lookup, 2) automatic inventory replenishment and sales and inventory forecasting, 3) pre- and post-season planning and support for cross-docking, 4) seasonless retailing and the transfer of inventory management functions to suppliers. Practical Experience: Retailers did not implement QR according to those levels (Palmer & Markus, 2000); but of the complexity of the program, they picked and chose among its components. Although the early focus of QR was on the flow of materials (Heard, 1994), the emphasis shifted to the flow of information, as people came to view short lead times as the way to reduce inventory. As a result of experiments by companies like Procter & Gamble, Kmart, and Wal-Mart (Harvard Business School, 1995; Koch, 2002), QR evolved into "the earliest example of continuous replenishment planning" (VICS, 2004d).
Efficient Consumer Response (ECR) Context: Grocery industry (mid 1980s) Vision: "The first robust initiative created to enable integration in the supply chain…." (Barratt & Oliveira, 2001, p. 267), ECR was an umbrella concept for efforts to provide better value for consumers while reducing costs for retailers and suppliers (Harris, Swatman, & Kurnia, 1997; Palmer & Markus, 2000). ECR promised to move the supply chain from a "push" to a "pull" system with replenishment based on store point of sale data. ECR focused on trust-based relationships and improvements in four core business processes: efficient store assortment, efficient promotions, efficient product introductions—together known as "category management" on the demand side—and efficient replenishment—or "supply chain management" on the supply side. Benefits were expected to result from combining the demand and the supply side into a single framework. ECR also involved the use of enabling technologies for capturing and transmitting point of sale data (Universal Product Codes, scanners, and EDI); Activity-Based Costing and cross-docking were also recommended. Practical Experience: Although ECR was a much broader concept, it was often implemented solely as Continuous Replenishment Planning/Vendor Managed Inventory (Mathews, 1994a, 1994b), with the problems noted below.

Table 2. continued

Continuous Replenishment Planning (CR or CRP) Vendor-Managed Inventory (VMI) Context: Consumer product retailing and grocery Vision: In CR/CRP/VMI suppliers automatically replenished retailers' inventory without orders, based on point of sale data. Practical Experience: CR/CRP/VMI did not solve the problems they were intended to solve (Barratt & Oliveira, 2001; Bruce & Ireland, 2002). Warehouse data was often used in lieu of point of sale data, thus reducing the visibility of actual consumer demand (Holmstrom et al., 2002). Vendors often did not push retailer data back into their organizations to improve their own production and replenishment systems. VMI was not viewed as an appropriate approach for "innovative" products (for which demand cannot be predicted from past sales) (Fisher, 1997; Sandoe, Corbitt, & Boykin, 2001) or for products that the retailer planned to promote. "VMI … shifted the ownership and responsibility from retailer to supplier" (Bruce & Ireland, 2002), which some vendors believe is not in their best interests unless the partners can come to an equitable agreement on responsibility for 'shrink' [i.e., theft]" (Frankel et al., 2002). Marketing promotions and new product introductions involve demand uncertainty not addressed by basing replenishment on history data and inventory balances. Wal-Mart and Warner-Lambert began experimenting to address problems with VMI, resulting in Collaborative Forecasting and Replenishment (CFaR).
Collaborative Forecasting and Replenishment (CFaR) Context: Consumer product retailing and grocery (1995) Vision: CFaR added sales and order forecasts to CRP. CFaR is "a formalized way for manufacturers and retailers to collaborate on future demand for products. By posting selected internal data on a shared Web server, supply chain partners could share and jointly develop more accurate forecasts" (Verity, 1997, p. 12). CFaR can also involve the sharing of partners' strategies or the use of complex decision models (Raghunathan, 1999). Practical Experience: Although it provided considerable benefits to retailers and manufacturers (Caldwell, Stein, & McGee, 1996), CFaR still fell short of proponents' visions of supply chain management excellence. They began to promote CPFR.
Collaborative Planning Forecasting and Replenishment (CPFR) Context: Consumer products retailing and supply chain management (1998) Vision: CPFR is "an initiative among all participants in the supply chain intended to improve the relationship among them through jointly managed planning processes and shared information" (Seifert, 2003 p. 30). CPFR went beyond CFaR by "enabl[ing] the forecast calculation to incorporate specific information about how much of an item will actually be available for delivery at some future date." (Verity, 1997, p. 12). Thus, CPFR is seen as the successor of CFaR (Barratt, 2004) and as a further refinement of ECR (Seifert, 2003).

pants trying to solve a common business problem through experimentation and discourse.

VICS was founded in 1986 to implement the Quick Response program in the soft goods industry (called Efficient Consumer Response in the food industry). These earlier visions, described as "a management approach" for effective supply chain coordination (*RIS News*, 2004, n.p.), were extremely broad, encompassing many specific philosophies (such as seasonless retailing), methodologies (cross-docking and forecasting), and technologies (EDI, point of sale technology using Universal Product Code scanning, etc.). The experience of supply chains attempting to realize these visions revealed numerous problems, leading to their progressive refinement and renaming as Continuous Replenishment, Vendor-Managed Inventory, Collaborative Forecasting and Replenishment, and, most recently, Collaborative Planning, Forecasting, and Replenishment. This evolution and some of the practical problems that drove it are summarized in Table 2.

In its current embodiment, the vision of CPFR is described as creating "collaborative relationships between buyers and sellers through co-managed processes and shared information. By integrating demand and supply side processes CPFR® will improve efficiencies, increase sales, reduce fixed assets and working capital, and reduce inventory for the entire supply chain while satisfying customer needs" (www.cpfr.org, accessed June 15, 2004).

CPFR Described Through the Methodology Lens

The CPFR voluntary guidelines, first issued in 1998 (revised in 2002 and 2004), portrayed CPFR (much more concretely than as a management approach or initiative) as a methodology with the core objective of establishing "a common *process* that can be used not only between two trading partners, but across an entire marketplace" (VICS, 2004a, p. 21, emphasis added). The 1998 version of the CPFR methodology consisted of nine required steps:

1. Develop front-end agreement
2. Create joint business plan
3. Create sales forecast
4. Identify exception items for sales forecast
5. Resolve/collaborate on exception items
6. Create order forecast
7. Identify exceptions for order forecast
8. Resolve/collaborate on exception items
9. Order generation

Some analysts argued that following the nine steps faithfully was key to supply chain collaboration success (Barratt & Oliveira, 2001). However, it soon appeared that "quite a few companies are collaborating with critical partners in a manner that is less complete than the full CPFR process" (Seifert, 2003, p. 90). For many CPFR adopters, a major sticking point is step 5—create order forecast. Although some retailers can produce reliable order forecasts at the level required by the nine-step model (by distribution center and store), others cannot. For instance, Specialty Superstores lacks software powerful enough to forecast orders for its 10 million SKU-store combinations. (Over a 104-week planning horizon, order forecasting would require more than 1 trillion data records.) Consequently, some experts argued for new collaboration approaches that did not depend on retailers' abilities to generate order forecasts (Holmstrom et al., 2002).

Even when retailers can produce accurate order forecasts, they might choose not to implement the full nine-step model with all suppliers or for all products. Specialty Superstores do not conduct CPFR processes for products of minor strategic importance. Furthermore, for innovative products (Fisher, 1997) of high strategic value, Specialty Superstores uses ladder plans (manual forecasts in which estimates of future orders gradually are refined as sales data become available) instead of CPFR. As a result, Specialty Superstores conducts the CPFR process with only 2% of its more than 500 vendors on less than 2,000 of its more than 10,000 stock-keeping units (SKUs). In addition, Specialty Superstores conducts two forms of collaboration, dubbed CPFR Heavy and CPFR Lite. CPFR Heavy involves weekly performance of a series of daily activities consistent with the CPFR nine-step model. In CPFR Lite, Specialty Superstore uploads sales forecasts to its CPFR software; sales and inventory data are posted on an extranet for vendors to use as they wish.

From experiences like these, CPFR promoters realized that collaboration did not require all nine steps of the original CPFR model (Seifert, 2003) nor did all companies need to use the same combinations of steps to perform CPFR. In the current version of the CPFR Process Model (VICS, 2004a) (see Figure 2), the nine required steps have been replaced with four activities (strategy and planning, demand and supply management, execution, and analysis) comprising eight tasks (account planning, market planning, market data analysis, demand planning, production and supply planning, logistics/distribution, execution monitoring, and customer scorecard). The new model was designed to account for the fact that "most companies are involved in all of [the activities] at any point in time. There is *no predetermined sequence of steps*. ...collaboration may focus on just a subset of the four activities ... These partial implementations are sometimes called 'CPFR Lite'" (VICS, 2004a, p. 7, emphasis added).

The latest VICS model also describes four CPFR scenarios that further adapt the collaboration approach to particular needs: Retail Event Collaboration, Distribution Center Replenishment Collaboration, Store Replenishment Collaboration, and Collaborative Assortment Planning. Indeed, "trading partners are free to combine scenarios if appropriate" (VICS, 2004a, p. 13).

CPFR Described Through the Technology Lens

According to VICS, "the CPFR process does not fundamentally depend upon *technology*" (VICS, 2004b, p. 20). Indeed, Romanow (2004) contends that early-stage CPFR collaborations are adequately managed with limited technology, such as telephone, FAX, and e-mailed spreadsheets (p. 2). Although experimenting with CPFR does not require advanced information technology, "specialized technology can make the process more scalable" (VICS, 2004b, p. 20); that is, expandable to additional partners and SKUs. Indeed, several vendors (i.e., Manugistics, SAP, i2, and Retek's Syncra) offer CPFR software packages that automate or facilitate collaboration activities, making them more cost effective. Even though Specialty Superstores considers CPFR to be "80% business process and only 20% technology," the company purchased CPFR software to expand its collaborative planning efforts to more vendors and items. In general, the availability of a limited number of CPFR tools from well-established vendors appears to promote widespread adoption of CPFR and its infusion (use with more partners and items) within adopting organizations.

Figure 2.

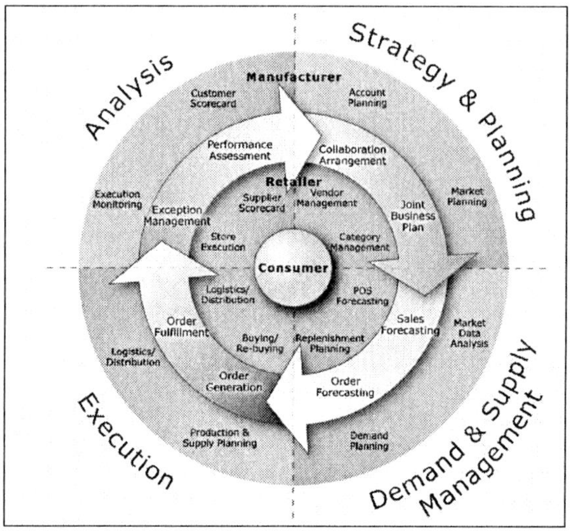

At the level of technical infrastructure, however, potential CPFR adopters face a bewildering variety of non-standardized implementation approaches—a factor likely to retard CPFR adoption and diffusion. For example, "the selection of data transport, security scheme, and middleware is beyond the scope of the CPFR standard, however, and is subject to implementers' agreements" (VICS, 2004b, p. 82). "Trading partners can use EDI messages, XML messages, or both to facilitate CPFR communications ... The EAN.UCC Global Business Standard provides the most comprehensive coverage of the process, with a suite of eleven CPFR-specific XML message types. While there are no EDI mappings for some CPFR messages, some projects use XML to 'fill in' where EDI messages have gaps" (VICS, 2004a, p. 21). Collaborating partners also can deploy a CPFR application in a shared mode in which partners use the same tool via an extranet, an application service provider, or an electronic marketplace, or they can share data across different applications using a peer-to-peer architecture (e.g., company-to-company, company-to-marketplace, marketplace-to-marketplace) (VICS, 2004c). As a consequence of all these options, a "company may need to deploy CPFR *using more than one approach* to collaborate with its full set of trading partners" (VICS, 2004c, p. 4, emphasis added).

In addition, CPFR adopters need to integrate CPFR tools with their other data processing systems (see Seifert [2003] for a more complete discussion.) At Specialty Superstores, CPFR software must work with (1) an enterprise data warehouse fed by sales, inventory, and ordering systems; (2) an extranet used to communicate point of sale data, inventory levels, and order status data to suppliers; and (3) order forecasting software. Specialty Superstores and its partners currently use manual processes to bridge gaps among these components. Although technical integration undoubtedly will increase over time, technology factors currently limit Specialty's ability to expand the CPFR innovation.

CPFR Described Through the Standards Lens

Proponents frequently describe CPFR in terms of standards. For example, CPFR is called a standards-based innovation (Seifert, 2003), because it depends heavily on EDI and XML standards. Indeed, the CPFR Technical Specification describes four areas in which technical standards could apply:

1. Data content and format (EDI and SIL [Standard Interchange Language])
2. Communication vehicle (FTP [transport] and TCP/IP [network protocol])
3. Security measures (e.g., authentication, encryption, non-repudiation, and origin)
4. Application/Middleware (alternatives for location, coordination, and management of the data processing elements [servers, agents]).

More importantly, CPFR is described as a standard because performing CPFR effectively is a matching problem requiring a limited range of solutions (de Vries, 1999). Although many individual supply chains have experimented over time to develop the essential philosophies, procedures, and techniques of CPFR (and its precursors), industry participants long have seen the need for CPFR standards, because without them, full employment of CPFR would involve the inefficient and ineffective use of different methodologies and technologies with different partners. Such a situation occurred with the EDI innovation (Damsgaard & Truex, 2000), and many experts believed that it retarded the diffusion and infusion of EDI. With respect to CPFR, Johnson (2004) noted that without standardization, supply chain collaboration with three partners could mean three interfaces and three processes; whereas, with standards, collaboration with three partners could require only one interface and one process. In addition, as already noted, VICS (2004c) reported that today's CPFR arrangements are not fully standardized, because they require some companies to maintain different processes and technologies with different partners. In addition, because CPFR is performed jointly by pairs of organizations, the innovation involves direct network effects that are consequential for its widespread diffusion (David & Greenstein, 1990).

The VICS organization can clearly be viewed as a developer of sectoral standards (de Vries, forthcoming). Since its founding in 1986, VICS has provided a forum in which industry participants come together to design and/or select among the philosophies, methodologies, and technologies devised by industry participants, consultants, and vendors. VICS' conferences have helped to educate industry participants, promoting the diffusion of CPFR, and its activities have helped to encourage technology providers to develop appropriate enabling technologies. For example, VICS developed mandatory common data formats and specifications for message interchange; commercial CPFR packages have implemented these formats.

Comparing the Lenses on CPFR

The philosophy lens draws attention to CPFR's long evolutionary career, in which the vision changed in response to practical experience. Similar insights might be generated by the view of standards as an emergent process; however, focusing only on the formal standards development activities of the VICS organization might not have revealed the full extent to which practical experiences of adopters disconfirmed earlier supply chain management philosophies.

On the other hand, standards theory sheds light on the specific shape of the CPFR career. The philosophy lens applies to all innovations, whether or not they involve matching problems with a concomitant requirement for a limited solution set that meets the needs of all parties concerned. The standards lens calls attention to matching problems and their solutions and, therefore, helps explain why CPFR's precursors failed in practice. For example, QR and ECR provided a too-broad set of solutions, which overwhelmed potential adopters and retarded diffusion (Frankel, Goldsby, & Whipple, 2002). For another example, VMI "shifted the ownership and responsibility from retailer to supplier" (Bruce & Ireland, 2002, p. 4), which some vendors believe is not in their best interests unless the partners can come to an equitable agreement on responsibility for "shrink" (i.e., theft) (Frankel et al., 2002, p. 64).

The methodology lens highlights the challenges of implementing an innovation that was believed to require several parties to follow an exact, step-by-step procedure jointly. Some supply chains found that the 1998 nine-step VICS model did not fit their collaboration scenarios; some retailers found that they could not comply with the required step of producing an order forecast. In order to increase the adoption of CPFR, VICS eliminated the prescribed sequence of activities and introduced four new scenarios, allowing supply chains greater flexibility to customize the CPFR process. The methodology literature would have predicted the move toward greater flexibility, but standards theory would not have. After all, by increasing the customizability of the CPFR methodology, VICS also was increasing the likelihood that CPFR would never achieve standardization in practice, because even though more companies might adopt a flexible CPFR, many more adopters would find themselves performing the flexible CPFR process differently with different partners.

The challenge of methodology standardization (e.g., IT Service Management [http://www.itil-itsm-world.com/]) has not figured prominently in the standards literature. Brunsson et al. (2000) described a methodology standardization effort (ISO 9000 quality standards) in which step-by-step implementation was never intended. However, that standard was not designed for joint implementation by two or more companies—it was a standard for individual companies to implement alone. By contrast, CPFR is meant for joint deployment by two or more organizations and, therefore, exhibits direct network effects. Consequently, flexibility in CPFR procedures is much more likely than in ISO 9000 standards to hinder standardization in practice. (A similar situation was seen in the case of EDI standards [Damsgaard & Truex, 2000]).

On the other hand, the standards lens helps explain how and why the nine-step CPFR methodology broke down. As already explained, Specialty Superstores could not automatically produce the order forecasts required by the nine-step methodology because of its huge number of store-SKU combinations. For many retailers, however, a more fundamental problem was that they did not have the incentive to provide order forecasts to suppliers (Holmstrom et al., 2002). Retailers do not need to forecast orders for their own immediate purposes (although they do need to forecast sales). If not for concerns about the timeliness of suppliers' deliveries, most retailers would not forecast orders but instead would place orders as the need arose. (Even so, some retailers prefer not to place orders, requiring suppliers to make replenishment decisions based on retailers' point-of-sale data.) By contrast, suppliers benefit from having information about retailers' future orders, because that information helps them schedule production efficiently. The nine-step CPFR process did not adequately balance these divergent needs of suppliers and retailers. Greater flexibility in the methodology balanced those needs better. By explaining the evolution of the CPFR methodology toward flexibility in terms of greater balance, the standards lens adds analytic value to the methodology lens.

The technology lens on CPFR highlights important differences between tools like CPFR software and technical infrastructure issues (i.e., deployment alternatives and integration with back-end systems). Standardization of data and message formats is necessary for widespread adoption of CPFR, and since vendors have adopted VICS standards, a dominant design for CPFR software is likely to emerge through market processes. However, the technology lens shows that tool standardization is not sufficient for CPFR to become a standardized business practice because of nonstandardized elements introduced by voluntary technical infrastructure choices.

On the other hand, by focusing on the matching problem and (in the case of CPFR) direct network effects from interdependent use, the standards lens explains precisely how and why non-standardized infrastructure elements threaten CPFR's prospects for adoption. Consider the situation of Consumer Products Manufacturer (CPM), a company that supplies Specialty Superstores and several of its competitors. CPM collaborates with about six different partners and, by the law of averages, finds itself forced to use every major CPFR software package on the market as well as numerous low-tech approaches, such as e-mailed spreadsheets. Even when two of CPM's customers use the same CPFR package, they use it in different ways. One retailer might provide order forecasts; the other may not. In addition, the retailers' forecast data might reflect different time periods and different levels of granularity (SKU, SKU-distribution center combination, SKU-store combination). CPM's IS specialists are required to develop and maintain a variety of special-purpose programs to extract data sent by retailers and load it into CPM's systems. (Considerable manual effort is devoted to this task every week.) Once that is done, CPM's supply chain analysts must review and analyze separately each retailer's data to account for differences in data quantity and quality. Although CPM believes that the benefits of collaborating with its customers outweigh the costs of nonstandard technologies and business processes, Specialty Superstores told us that other similar suppliers declined to participate in CPFR, because in the absence of Specialty's ability to supply an order forecast, the suppliers' perceived costs of performing CPFR exceeded their potential benefits.

In short, the four lenses overlap to a certain degree. However, each lens provides unique and important insights in the complex case of CPFR. Consequently, they are best used together.

DISCUSSION AND CONCLUSION

Our comparison of the four perspectives on the IS innovation of CPFR suggests three conclusions. First, there is value in combining both the emergent and the formal design/selection perspectives on the development of standards. The process by which formal standards evolve through experimentation in practice can enhance insights derived from a focus on the official efforts of standards bodies such as the industry association VICS. Similarly, as suggested by Fomin and Keil (1999), there is value in combining a focus on both the development/evolution of standards and on their adoption and diffusion. Implicit or explicit choices made during standards development can have important consequences for the extent of standards diffusion and its assimilation by adopting organizations.

Second, all four conceptual lenses shed light on the case of CPFR. Although not all IS innovations might be describable in all four ways, our analysis suggests that methodologies face different standardization challenges than technologies do. IS methodology standards face challenges around customization; IS tool standards face particular issues around integration with processes and infrastructures and around potential divergence between tool standardization and infrastructure standardization. Although the ultimate goal of standards theory is to account for all standardization efforts, it is important not to ignore differences in what is standardized that could produce variations in standardization processes and outcomes.

Finally, the case of CPFR raises an interesting question about what it takes for complex IS innovations to achieve standardization in practice. CPFR appears to be not yet standardized at any level of abstraction—philosophy, methodology, or technology. But what if it were standardized at only one level or two? Could CPFR be standardized in practice without standardization at every level? This, we believe, is an interesting question for future research on the standardization of IS innovations.

ACKNOWLEDGMENTS

This work was conducted as part of the Bentley Invision Project funded by Bentley College. We gratefully acknowledge our colleagues, Jane Fedorowicz (Principal Investigator), Janis Gogan, Amy Ray, Cathy Usoff, and Chris Williams.

REFERENCES

Alter, S. (2005). Architecture of sysperanto: A model-based ontology of the IS field. *Communications of the ACM, 15*(1), 1-40. Retrieved June 6, 2005, from http://www.cais.isworld.org

Anderson, P., & Tushman, M. L. (1990). Technological discontinuities and dominant designs: A cyclical model of technological change. *Administrative Science Quarterly, 35*(4), 604-633.

Barratt, M. (2004). Unveiling enablers and inhibitors of collaborative planning. *International Journal of Logistics Management, 15*(1), 73-90.

Barratt, M., & Oliveira, A. (2001). Exploring the experiences of collaborative planning initiatives. *International Journal of Physical Distribution & Logistics, 31*(4), 266-289.

Bruce, R., & Ireland, R. (2002). *What's the difference: VMI, co-managed, CPFR?* [white paper]. Retrieved June 1, 2004, from http://vccassociates.com/articles.asp

Brunsson, N., Jacobsson, B., & Associates. (2000). *A world of standards.* Oxford, UK: Oxford University Press.

Caldwell, B., Stein, T., & McGee, M. (1996). Uncertainty: A thing of the past. *Information Week,* 46.

Cargill, C. F. (1989). *Information technology standardization: Theory, process, and organizations.* Maynard, MA: Digital Press.

Clark, T. H., & Stoddard, D. B. (1996). Interorganizational business process redesign: Merging technological and process innovation. *Journal of Management Information Systems, 13*(2), 9-28.

Damsgaard, J., & Truex, D. (2000). Binary trading relations and the limits of EDI standards: The procrustean bed of standards. *European Journal of Information Systems, 9*(3), 142-158.

David, P. A., & Greenstein, S. (1990). The economics of compatibility standards: An introduction to recent research. *Economic Innovation and New Technology, 1*, 3-41.

de Vries, H. (1999). *Standardization—A business approach to the role of national standardization organizations.* Dordrecht, The Netherlands: Kluwer Academic Publishers.

de Vries, H. (2005). IT standards typology. In K. Jakobs (Ed.), *Advanced topics in IT standards & standardization research.* Hershey, PA: Idea Group Publishing (forthcoming).

Fisher, M. L. (1997). What is the right supply chain for your product? *Harvard Business Review,* 105-116.

Fomin, V., & Keil, T. (1999). Standardization: Bridging the gap between economic and social theory. *Proceedings of the International Conference on Information Systems*, Charlotte, NC.

Frankel, R., Goldsby, T. J., & Whipple, J. M. (2002). Grocery industry collaboration in the wake of ECR. *International Journal of Logistics Management, 13*(1), 57-72.

Greenwood, R., Hinings, C. R., & Suddaby, R. (2002). Theorizing change: The role of professional associations in the transformation of institutionalized fields. *Academy of Management Journal, 45*(1), 58-80.

Harrington, H. J. (1998). Performance improvement: The rise and fall of reengineering. *The TQM Magazine, 10,* 69.

Harris, J. K., Swatman, P. M. C., & Kurnia, S. (1997). Efficient consumer response (ECR): A survey of the Australian grocery industry. *Proceedings of the ACIS 1997—8th Australasian Conference on Information Systems*, Adelaide, Australia.

Harvard Business School. (1995). *Procter & Gamble: Improving consumer value through process redesign*. Cambridge, MA: Harvard Business School.

Heard, E. (1994, August). Quick response: Technology or knowledge? *Industrial Engineer,* 28-30.

Heracleous, L., & Barrett, M. (2001). Organizational change as discourse: Communicative actions and deep structures in the context of information technology implementation. *Academy of Management Journal, 44*(4), 753-778.

Holmstrom, J., Framling, K., Kaipia, R., & Saranen, J. (2002). Collaborative planning forecasting and replenishment: New solutions needed for mass collaboration. *Supply Chain Management: An International Journal,* 7(3/4), 136-145.

Howard, G. S., Bodnovich, T., Janicki, T., & Liegle, J., et al. (1999). The efficacy of matching information systems development methodologies with application characteristics—An empirical study. *The Journal of Systems and Software, 45*(3), 177-195.

Iacono, S., & Kling, R. (2001). Computerization movements: The rise of the Internet and distant forms of work. In J. Yates, & J. Van Maanen (Eds.), *Information technology and organizational transformation: History, rhetoric, and practice* (pp. 93-136). Thousand Oaks, CA: Sage Publications.

Iivari, J., Hirschheim, R., & Klein, H. K. (2001). A dynamic framework for classifying information systems development methodologies and approaches. *Journal of Management Information Systems, 17*(3), 179-218.

Iversen, E. J. (2000). Standardization and intellectual property rights: Conflicts between innovation and diffusion in new telecommunications systems. In K. Jakobs (Ed.), *Information technology standards and standardization: A global perspective* (pp. 80-101). Hershey, PA: Idea Group Publishing.

Johnson, M. (2004). Business-to-business electronic commerce standards [Presentation]. Waltham, MA: Bentley College.

Katz, M. L., & Shapiro, C. (1985). Network externalities, competition, and compatibility. *American Economic Review, 75*(3), 424-440.

Koch, C. (2002). It all began with drayer. *CIO Magazine, 15*(20), 1.

Lee, S. C., Pak, B. Y., & Lee, H. G. (2003). Business value of B2B electronic commerce: The critical role of inter-firm collaboration. *Electronic Commerce Research and Applications, 2,* 350-361.

Mathews, R. (1994a). CRP moves toward reality. *Progressive Grocer, 73,* 43-46.

Mathews, R. (1994b). CRP spells survival. *Progressive Grocer, 73,* 28-30.

Nandhakumar, J., & Avison, D. E. (1999). The fiction of methodological development: A field study of information systems development. *Information Technology & People, 12*(2), 176-191.

Orlikowski, W. J. (2000). Using technology and constituting structures: A practice lens for studying technology in organizations. *Organization Science, 11*(4), 404-428.

Orlikowski, W. J., & Gash, D. C. (1994). Technological frames: Making sense of information technology in organizations. *ACM Transactions on Information Systems, 12*(2), 174-207.

Orlikowski, W. J., & Iacono, C. S. (2001). Research commentary: Desperately seeking "IT" in IT research—A call to theorizing the IT artifact. *Information Systems Research, 12*(2), 121-134.

Palmer, J. W., & Markus, M. L. (2000). The performance impacts of quick response and strategic alignment in specialty retailing. *Information Systems Research, 11*(3), 241-259.

Raghunathan, S. (1999). Interorganizational collaborative forecasting and replenishment systems and supply chain innovation. *Decision Sciences, 30*(4), 1053-1071.

Ramiller, N. C., & Swanson, E. B. (2003). Organizing visions for information technology and the information systems executive response. *Journal of Management Information Systems, 20*(1), 13-50.

The RIS News Glossary. (2004). RIS News. Retrieved May 21, 2004, from http://www.risnews.com/Glossary/glossary.html

Romanow, K. (2004). *Collaboration roadmap: Success depends on process, not technology.* Frontline Solutions. Retrieved January 29, 2004, from http://www.frontlinetoday.com

Ross, J. (2003). Creating a strategic IT architecture competency: Learning in stages. *MIS Quarterly Executive, 2*(1), 31-43.

Sandoe, K., Corbitt, G., & Boykin, R. (2001). *Enterprise integration.* New York: John Wiley & Sons.

Seifert, D. (2003). *Collaborative planning, forecasting, and replenishment: How to create a supply chain advantage.* New York: AMACOM.

Soderstrom, E. (2002). Standardising the business vocabulary of standards. *Proceedings of the SAC 2002,* Madrid, Spain.

Swanson, E.B. (2002). Talking the IS innovation walk. *Proceedings of the IFIP WG8.2 Working Conference on Global and Organizational Discourse about Information Technology,* Barcelona, Spain.

Swanson, E.B., & Ramiller, N.C. (1997). The organizing vision in information systems innovation. *Organization Science, 8*(5), 458-474.

Truman, G. E. (2000). Integration in electronic exchange environments. *Journal of Management Information Systems, 17*(1), 209-244.

Verity, J. (1997). Collaborative forecasting: Vision quest. *Computerworld, 31,* 12-14.

VICS. (2004a). Collaborative planning, forecasting and replenishment (CPFR®). Retrieved June 15, 2004, from http://www.cpfr.org

VICS. (2004b). CPFR technical specification. Retrieved June 15, 2004, from http://www.cpfr.org

VICS. (2004c). CPFR® deployment scenarios. Retrieved June 15, 2004, from http://www.cpfr.org

VICS. (2004d). Supply chain initiatives. Retrieved June 15, 2004, from http://www.cpfr.org

Wynekoop, J. L., & Russo, N. L. (1993). System development methodologies: Unanswered questions and the research-practice gap. *Proceedings of the Fourteenth International Conference on Information Systems*, Orlando, FL.

Zhu, K. (2004). The complementarity of information technology infrastructure and e-commerce capability: A resource-based assessment of their business value. *Journal of Management Information Systems, 21*(1), 167-202.

This work was previously published in The Journal of IT Standards & Standardization Research, 4(1), edited by K. Jakobs, pp. 24-42, copyright 2006 by IGI Publishing, formerly known as Idea Group Publishing (an imprint of IGI Global).

Section V
Cases and Projects

Chapter XII
Market Response to ISO 9000 Certification of Software Engineering Processes

G. Keith Fuller
University of British Columbia, Canada

Ilan Vertinsky
University of British Columbia, Canada

ABSTRACT

A very large proportion of software projects are deemed to be failures. In most business sectors, this situation would be dealt with by improving quality assurance processes, frequently including certification of the business processes to standards such as ISO 9000. However, in the field of software engineering, there is controversy over whether or not certification of software engineering processes is a cost-effective response to quality issues. The value of certification in software engineering is examined in the present research by applying event-study methodology to examine the market response to announcements of certification of software engineering processes. The findings support the hypothesis that certification of software engineering processes leads to increased profits for companies that are primarily focused on developing products. Subsequent exploratory analysis suggests that the knowledge of the certification may leak out to the marketplace before the official announcement.

INTRODUCTION

Today, software development can be a risky business. Far too often the software development process fails. One way to reduce the costs associated with software development may be to modify the software engineering processes of a company so that they can be certified as meeting standards such as those specified in the ISO 9000 (International Organization for Standardization) series. A number of studies has argued that such certification should reduce the costs associated with software development (Kuilboer & Ashrafi, 2000; Stelzer, Mellis, & Herzwurm, 1997; Terziovski, Samson, & Dow, 1997), and a great deal has been written on how to apply these standards to software engineering (Schmauch, 1994; Stelzer et al.; Yang, 2001). A consulting specialization has even developed to help companies attain certification to these standards. There remains an open question, however, as to whether or not the time and money spent on certification yields benefits that exceed the costs of certification.

Certification should be effective in improving software engineering processes because it not only requires companies to adopt quality assurance processes, but it also puts in place a set of procedures to ensure that the improved processes are consistently followed in order to maintain the certification. Hence, it is reasonable to expect that adoption of quality assurance processes in software development should reduce the rate of project failures. Having more reliable software engineering processes that reduce the rate of project failure will, in turn, reduce software development costs. For projects that would not actually fail without the certification, costs should still be reduced because fewer resources will be expended on correcting errors introduced during the software design, development, and implementation processes. Certification may also act as a means to assure potential customers that a company's software development processes will produce better quality products (Anderson, Daly, & Johnson, 1999), thereby leading to increased sales.

The above arguments support the supposition that certification should increase profits by either reducing costs or by increasing sales. If certification is expected to contribute to profits, obtaining it will increase the value of the certifying firm. In this chapter, the event-study methodology is used to test whether the expected net benefits from the certification of U.S. companies are reflected as changes in the market valuation.

LITERATURE REVIEW OF EVENT STUDIES

The event-study methodology is well established, having been in use for over 30 years. In 1969, an early study sampled stock prices at monthly intervals in order to examine abnormal returns associated with stock splits (Fama, Fisher, Jensen, & Roll, 1969). By 1980, event studies had become well established, and simulation techniques (still based on a monthly sampling interval) were used to validate the methodology and to evaluate different techniques for carrying out the studies (Brown & Warner, 1980). Two important findings were that the use of a market model based on a least squares regression is a valid means of predicting normal stock prices, and that the methodology is relatively insensitive to sample size. This work was extended in 1985 to use daily instead of monthly returns. Although the use of daily returns violated normality assumptions, the methodology was shown to be robust with respect to these violations (Brown & Warner, 1985). The same study showed the methodology to be valid with sample sizes as small as five companies, although it was important that the stocks were regularly traded, with stocks from the New York Stock Exchange (NYSE) producing much better results than those from the American Stock Exchange (AMEX) because of the frequency of trading.

The event-study methodology has been extensively used in the fields of finance and marketing. Use of the methodology to look at certification announcements is not unique. In a study in which the methodology has been applied to the more general case of ISO 9000 registration in the United States across all business sectors, it was found that while small firms reaped significant abnormal market returns from certification (especially after the

Maastricht Treaty in 1992), there was no evidence to support the hypothesis for larger companies (Docking & Dowen, 1999). A subsequent study based on Portuguese companies across all business sectors that attained ISO 9000 certification also found statistically significant market response to the announcement of certification, independent of the size of the company (Beirao & Cabral, 2002). Two recent event studies examined the consequences of ISO 9000 certification in Spanish firms (Martinez-Costa & Martinez-Lorente, 2003; Nicolau & Sellers, 2002). The first Spanish study found significant abnormal returns on the day of the announcement, whereas the other did not. All of these event studies spanned most of the industrial sectors in the country of interest.

Several event studies have also been carried out on the effect of prestigious awards for quality management on the market's expectations of future profits. These studies have yielded conflicting results. One study that looked at a variety of quality awards found that the market did show statistically significant abnormal returns around the time of the announcement of the awards (Hendricks & Singhal, 1996). This finding was challenged in a follow-up article that criticized the original work based on the sample size and on the effect of a small number of what could be considered outliers in the results (Adams, McQueen, & Seawright, 1999). A similar study that looked at American companies that had won the Malcolm Baldridge National Quality Award failed to find any significant abnormal market returns around the time of the announcement of the award (Przasnyski & Tai, 1999). However, two other similar studies of American firms that had won major quality awards did find significant abnormal returns (Ramasesh, 1998; Soteriou & Zenios, 2000). While these studies looked at companies that had won awards for quality management, rather than looking at companies that had attained certification of the quality of their processes, both the awards and the certifications have in common the fact that they provide external validation of the quality of business processes.

Overall, prior research supports the supposition that certification of business processes can yield abnormal market returns, at least in some cases. It has been suggested that future work should focus on a single sector since the market response to certification may vary across industrial sectors (Soteriou & Zenios, 2000). To our knowledge, there is no research concerning the specific case of software engineering processes. The current study extends prior research by isolating the sector to software engineering. Studying the market response to ISO 9000 certification of software engineering is of particular interest because the certification is often resisted by practitioners in the field, with the justification being that software engineering is different from general manufacturing because its processes are inherently more creative, and hence more difficult to put into procedures. The event-study methodology provides insight into the value assigned by the marketplace to certification in this sector. The high failure rate of projects, and the costs incurred, makes the market response a valuable objective measure to complement the practitioner's subjective point of view.

METHODOLOGY

The event-study methodology starts by identifying suitable study companies that have announced that they have been certified and the date of the announcement. The daily stock prices for each company being studied are collected for a period before and after the announcement. The period of time running from approximately 8 months before the announcement through approximately 6 days before the announcement is used as a baseline, and is referred to as the estimation period. The period running from 5 days before the announcement through 5 days after the announcement is referred to as the evaluation period. Various subsets of the evaluation period will be examined to determine if the actual stock prices are significantly different from what would be expected if the announcement had no effect, and these will be referred to as event windows.

For each of the studied companies, an index is constructed by selecting a group of similar companies. The stock prices for each of the certified companies and their corresponding index are collected for the estimation period and the evaluation period. An index price is calculated as the sum of the stock prices for the companies comprising the index for each day in the estimation period and the evaluation period. The performance of the index and the studied company during the estimation period is used to develop a market model based on a least squares regression, which is then used to predict what the studied company's daily stock price would have been during the evaluation period if the announcement had no effect on the market evaluation of the company's worth, and hence its stock price.

The predicted returns (percent daily change in price) in the evaluation period are then compared to the actual returns. The daily difference between the two is measured for each of the studied companies. Since the null hypothesis is that the announcement of certification will have no effect on stock price, the daily difference between predicted and actual returns for each company is referred to as the *abnormal return* (AR). The ARs are then averaged across all studied companies for each day in the evaluation period (average abnormal returns or AARs) and accumulated over the days in each event window being studied (cumulative average abnormal returns or CAARs). The significance of the CAAR for each event window is then tested.

The most common technique used to test the significance of the CAARs is a parametric test. This test is premised on the assumption that the predictive model, built using the daily stock prices of the index companies and the respective studied companies over the estimation period, will form an accurate prediction of what the stock prices would be in the event window if the announcement of the certification has no effect. The standard deviation of the abnormal returns in the event window is assumed to be the same as the standard deviation of the abnormal returns in the substantially longer estimation period. The formula used in the parametric test is given in Appendix 1.

A key decision in the research design is the choice of the event window. The event window is the period of time around the event during which stock prices are tested to determine if they are significantly different from what would be expected, with all other things being equal. This study considers an event window that consists of the day of the announcement and the day after. This choice of event window assumes that the announcement is a "crisp" event, meaning that the certification has not come to the notice of the marketplace prior to the announcement, and that once the announcement is made, the information is rapidly disseminated to the marketplace.

The following sections will describe how the event-study methodology and the tests have been applied to data from the American software industry.

DATA COLLECTION

In 2001, American companies that had attained ISO 9000 certification of their software design and development processes between 1990 and 1999 were identified using the database of certification registrations provided at the Web site http://www.qualitydigest.com. The search of the database was limited to companies in the United States that had indicated that their business falls primarily into the standard industrial classifications (SICs) 7371, 7372, and 7373. Respectively, these correspond to custom software development, packaged software development, and systems integration. The database search produced 225 records of certification. The elimination of duplicate records and records pertaining to companies that were not publicly traded on either the NYSE or the NASDAQ Exchange yielded 23 companies of interest. (Limiting the choice of companies to those traded on either the NYSE or the NASDAQ is a conventional means of ensuring that the stocks are regularly traded. The daily stock prices were manually reviewed to confirm that they were regularly traded.)

The historical daily closing stock prices for each of these companies was obtained from http://finance.yahoo.com for a period ranging from 199 days before the announcement of the certification through 5 days after the announcement.

In order to create an index for each of the certified companies, the Yahoo Finance classification of each company's sector and industry was determined, and a random sampling of 10 additional companies from the same classification was made. The historical stock prices for each index's component companies were obtained using the same procedure that was used for the certified company. As discussed above, the indices' component companies' daily closing prices were summed to calculate the daily index price. The historical stock prices of the certified companies and their indices were next split into an

Table 1. Market model parameters calculated from data obtained from http://www.qualitydigest.com in 2001

	Company	Ticker	alpha	beta	R	
1	Abbott Laboratories	ABT	-3.519	0.139	0.981	***
2	ADP	ADP	17.001	0.0467	0.259	
3	Analog Devices Inc.	ADI	10.537	0.03738	0.606	***
4	Beckman Coulter	BEC	16.967	0.03754	0.094	
5	CIBER Inc.	CBR	-2.404	0.155	0.735	***
6	Honeywell Inc.	HON	18.11	-0.07737	0.270	
7	IBM	IBM	15.868	-0.01831	0.272	
8	L3	LLL	42.965	-0.0189	0.727	***
9	Lockhead Martin	LMT	-0.0261	0.296	0.885	***
10	MSC Software Corp.	MNS	13.306	-0.01603	0.203	
11	Nortel Networks	NT	-4.552	0.273	0.962	***
12	Raytheon Training Inc.	RTN	14.509	0.103	0.594	***
13	Schlumberger	SLB	-0.921	0.22	0.964	***
14	Unisys Corp.	UIS	-10.876	0.175	0.822	***
15	Varian Medical Systems	VAR	-13.092	0.197	0.771	***
16	Adc Software	ADCT	-1.331	0.06685	0.712	***
17	Direct Insite Corp.	DIRI	-1.233	0.03851	0.853	***
18	Evans and Sutherland	ESCC	-13.053	0.188	0.732	***
19	Intergraph Corp.	INGR	1.839	0.09086	0.778	***
20	Mapics Inc.	MAPX	26.181	-0.0424	0.060	
21	Mentor Graphics	MENT	1.775	0.06934	0.237	
22	Novell	NOVL	12.176	-0.02888	0.202	
23	OAO Corp	OAOT	-3.4	0.0259	0.804	***

*** Correlation above .5 threshold

estimation period ranging from 199 days before the announcement through 6 days before the announcement, and an evaluation period ranging from 5 days before the announcement through 5 days after the announcement.

A market model for each of the certified companies was next formed by conducting a linear regression analysis of the daily index prices against the corresponding certified company's daily prices using SPSS 11 software. The results of this analysis are shown in Table 1. The quality of the market models ranges from excellent to poor, as shown by the Pearson's correlation coefficients (Rs) ranging from 0.981 through 0.060. Since the purpose of developing the market models is to use them to predict the expected stock price of the certified companies in the absence of the certification, there is no justification to using companies whose stock prices cannot be effectively modeled. For this reason, companies for which the market model has an R of less than the arbitrary value of 0.5 were eliminated. This yields a set of 15 certified companies for which acceptable models were developed.

Event studies, particularly with small sample sizes, are likely to be sensitive to the presence of outliers. That is, the inclusion of a company that has returns that are significantly outside the typical range can swing the results of the entire group. To check for outliers, the abnormal returns in the evaluation period were plotted as shown in Figure 1. Two companies were identified as outliers. These companies were eliminated from the sample, reducing the number of studied companies to 13.

The selected 13 companies were rank ordered by their net revenue (as obtained from Edgar Online, http://www.edgar-online.com/), as shown in Table 2. The companies were then divided into two groups: seven small companies with net revenue below $2 billion, and six larger companies with net revenues above $2 billion. (All prices are in U.S. dollars.)

Besides looking at different sizes of companies, it was decided to also look at the companies broken into groups based on whether they were primarily engaged in the provision of services or products. The 13 companies divided into two

Figure 1. Daily abnormal returns for each company

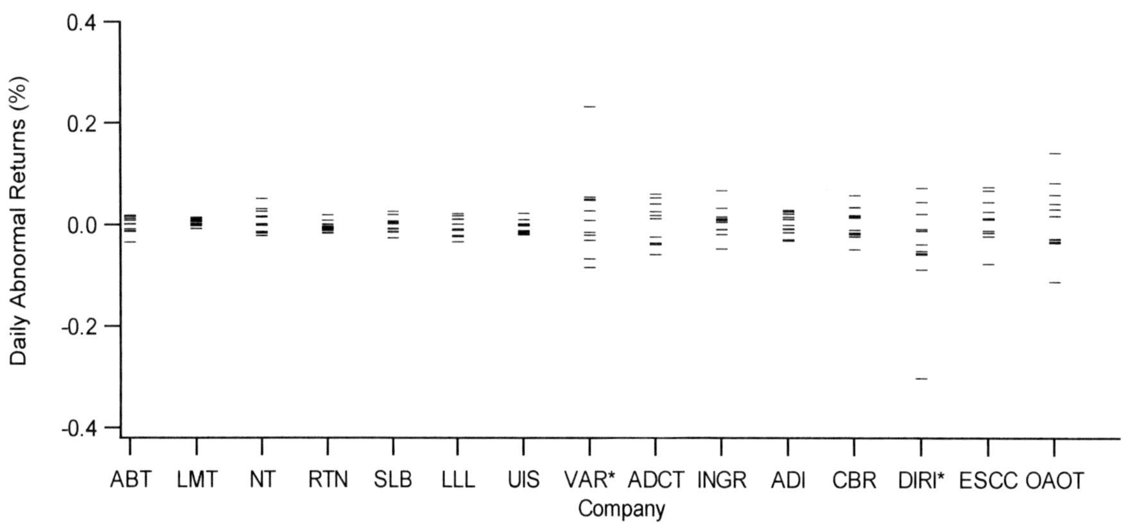

* Companies with abnormal returns that are considered outliers

Table 2. Company net revenues in year of certification

Company	Date	Net Revenue ($US)
LMT	1992	16,030,000,000
NT	1996	11,919,000,000
RTN	1994	10,098,000,000
ABT	1991	6,876,588,000
SLB	1993	6,705,000,000
UIS	1994	2,877,000,000
ADCT	1998	1,547,383,000
INGR	1993	1,050,277,000
LLL	1998	1,037,000,000
ADI	1994	773,474,000
ESCC	1992	148,594,000
CBR	1995	120,151,000
OAOT	1997	29,738,000

Table 3. Company membership in groups based on size and focus

Company	Small Companies	Producing Products
ABT		✓
ADI	✓	✓
CBR	✓	
LLL	✓	✓
LMT		✓
NT		✓
RTN		✓
SLB		
UIS		
ADCT	✓	
ESCC	✓	✓
INGR	✓	
OAOT	✓	

groups: six companies providing services, and seven companies providing products. Refer to Table 3 for a listing of which companies fall into which group based on size and on whether they focus on the provision of products or services.

The first analysis applied the event-study techniques, as described in the earlier section, to test the research hypothesis that announcement of the certification of software engineering processes leads the marketplace to increase its estimation of future earnings. This analysis was carried out on all 13 of the companies taken together, on groups of larger and smaller companies, and on the companies grouped by whether they are primarily engaged in the production of products or the provision of services. The event window that was tested was formed by the day of the announcement and the day after [0,1]. The analysis failed to reject the null hypothesis that there is no difference in the abnormal returns before and during the event window for the composite group

Table 4. Z statistics calculated using parametric test of the null hypothesis that certification makes no difference in cumulative average abnormal returns in the event window formed by the day of the announcement of certification and the day after for all companies and for companies grouped by size and output

Group	Event Window [0,1]
All Companies	1.085
Small Companies	1.027
Large Companies	0.387
Service Companies	0.147
Product Companies	2.018*

* significant at .05 level of confidence

of all companies, or for either of the groups of larger or smaller companies, or for the companies that are primarily engaged in the provision of services. However, the companies engaged in the provision of products showed a statistically significant positive market response for the event window [0,1] ($z = 2.018$; $p < .05$). That is to say, the parametric test was able to reject the null hypothesis for the companies that focus on the production of products.

The specific *z* statistics for each of the tests of the a priori hypothesis described above are given in Table 4.

Having failed to reject the a priori null hypothesis for the companies taken as one group, or for the companies grouped by size, the study shifts from confirmatory research to an exploratory search for any event window that exhibits a CAAR that would have reached statistical significance if it had been proposed a priori. This phase of the study starts by repeating the analysis for all possible event windows in the evaluation period, for the aggregate group of companies, for the small and large companies, and for the companies that provide products and services (61 event windows x five groups of companies for a total of 305 tests).

The combined data set showed no suggestions of significant cumulative average abnormal returns for any of the event windows. Similarly, the data sets composed of the smaller and larger companies also showed no evidence of positive abnormal returns in any of the event windows. The data set composed of the companies that were primarily concerned with the provision of services also showed no significant abnormal returns.

Companies that focused on the provision of products yielded parametric *z*-test statistics above the .05 level of significance for event windows [-2,1], [-2,2], [-2,3], and [-1,1] in addition to [0,1], which was noted in the confirmatory results section ($z = 2.23$, 2.01, 2.04, 2.12 respectively).

DISCUSSION

The significant confirmatory test result for the companies engaged in producing products is the most important finding in this study. This result indicates that the marketplace believes that the announcement of certification is a harbinger of improved future revenues for these companies. Why should the marketplace make this interpretation for companies that are primarily engaged in the production of products and not companies engaged in the provision of services? Recall that improved revenues could stem from reduced costs, or from improved sales. Improved quality will impact both costs and revenues for companies that develop products. Cost reductions would not only include reduced fixed costs to develop the product, but also variable support costs that would increase as volumes of sales increase. Sales would also be expected to benefit from improved quality, which could be a key product differentiator.

On the other hand, companies that are purveyors of services may expect to see increased sales due to third-party assurances of quality, but they are often shielded from the effects of costs by the terms of their contracts. Contracts may be either fee for service, fixed price, or some combination of the two. The profit from a fixed-price contract is directly affected by incremental costs from problems unforeseen at the time that the contract is formed. A typical example of this sort of incremental cost in software development is the cost of finding and fixing bugs in the software during the test phase. In fee-for-service contracts, the client retains closer control of the project and also bears the majority of the risk of incremental costs. The closer the contract is to a fixed price, the more the profit margin may be expected to be sensitive to costs due to quality issues. Fixed-price contracts, however, are seldom delivered as initially specified since the user requirements usually change during the course of the contract, and change requests expand the scope (and cost) of the contract. The increased costs in fixed-price contracts accrued to quality problems may then be masked by these changes in scope. This leads to a situation where service providers are more likely to see benefits from increased quality through increased sales rather than from reduced costs. If the marketplace sees announcements of certification as being significant for product developers and not for service providers, then it suggests that the marketplace is anticipating a greater effect on reduced costs than for increased sales.

Before summarizing the exploratory results, it is necessary to consider how the results may be interpreted. Performing a single test on the data from a group of companies based on an a priori hypothesis may yield results that are considered confirmatory. Once the test is repeated on varying event windows for a group of companies, however, we have moved to a more exploratory line of inquiry, which is looking for any event window showing results of interest.

Besides the change from confirmatory to exploratory, there is also a statistical issue that is introduced when we start repeating the analysis for any possible event window in the evaluation period. In this case, the parametric test has been applied to all possible event windows (61) for each group of companies (five) within the evaluation period surrounding the event. Since the significance level of each individual test is expected to yield a false positive in 1 out of 20 cases (alpha = .05), repeating the test on this number of varying subsets of the same data can be expected to yield approximately 15 false positive results for 300 tests. The risk of false positive results stemming from multiple tests is commonly addressed in statistics by applying a correction to the significance level using a variant of the Bonferroni correction (for example, dividing the significance value by the number of tests undertaken). The parametric test results cited in the exploratory portion of the "Results" section could be significant if the test that yielded each result had been the only such test performed on the data. However, the approach actually taken was to look at all possible event windows for each group of companies within the evaluation period surrounding the event. Applying a Bonferroni correction, the parametric test results of the exploratory analyses are not significant and fail to confirm the research hypothesis. However, the results are still of interest. Without Bonferroni correction, the exploratory parametric test results appear to support the confirmatory results in that the companies that produce products are the only group of companies that have any event windows with z statistics that exceed the critical value for a .05 level of confidence. All of these event windows include the day of the announcement, and there is a suggestion that windows that open one or two days prior to the announcement may also be significant. This may indicate that there is leakage of the certification announcement to the marketplace prior to the official announcement.

CONCLUSION

To conclude, the market assessment of ISO 9000 quality certification of software engineering processes supports the hypothesis that the certification does correspond to increased profitability in those companies engaged in developing products. Exploratory analysis of the data suggests that leakage of the certification announcement takes place.

The findings should be of use to companies engaged in producing products that are looking for a business justification for certification, and also to companies looking to market consulting services geared toward preparing companies engaged in software engineering for a certification audit. It would be useful if the study could be replicated with a different sample as this may allow confirmation of the exploratory findings that the announcement of certification leaks out to the marketplace a day or two in advance of the official announcement. It would also be interesting to explore the qualitative perceptions of the marketplace to certification of software engineering processes to determine when market observers first become aware of certification efforts and results, and to determine whether the market sees certification as being a glass half full or a glass half empty in the sense that it portends improved productivity or signals hitherto unseen problems in quality.

REFERENCES

Adams, G., McQueen, G., & Seawright, K. (1999). Revisiting the stock price impact of quality awards. *Omega, 27*(6), 595-604.

Anderson, S. W., Daly, J. D., & Johnson, M. F. (1999). Why firms seek ISO 9000 certification: Regulatory compliance or competitive advantage. *Production and Operations Management, 8*(1), 28-43.

Beirao, G., & Cabral, J. A. S. (2002). The reaction of the Portuguese stock market to ISO 9000 certification. *Total Quality Management, 13*(4), 465-474.

Brown, S. J., & Warner, J. B. (1980). Measuring security price performance. *Journal of Financial Economics, 8*, 205-258.

Brown, S. J., & Warner, J. B. (1985). Using daily stock returns: The case of event studies. *Journal of Financial Economics, 14*, 3-31.

Docking, D. S., & Dowen, R. J. (1999). Market interpretation of ISO 9000 registration. *Journal of Financial Research, 22*(2), 147-160.

Fama, E. F., Fisher, L., Jensen, M. C., & Roll, R. (1969). The adjustment of stock prices to new information. *International Economic Review, 10*(1), 1-21.

Hendricks, K. B., & Singhal, V. R. (1996). Quality awards and the market value of the firm: An empirical investigation. *Management Science, 42*(3), 415-436.

Kuilboer, J. P., & Ashrafi, N. (2000). Software process and product improvement: An empirical assessment. *Information and Software Technology, 42*, 27-34.

Martinez-Costa, M., & Martinez-Lorente, A. R. (2003). Effects of ISO 9000 certification on firms' performance: A vision from the market. *TQM & Business Excellence, 14*(10), 1179-1191.

Nicolau, J. L., & Sellers, R. (2002). The stock market's reaction to quality certification: Empirical evidence from Spain. *European Journal of Operational Research, 142*(3), 632-641.

Przasnyski, Z. H., & Tai, L. S. (1999). Stock market reaction to Malcolm Baldridge National Quality Award announcements: Does quality pay? *Total Quality Management, 10*(3), 391-400.

Ramasesh, R. V. (1998). Baldridge Award announcement and shareholder wealth. *International Journal of Quality Science, 3*(2), 114-125.

Schmauch, C. H. (1994). *ISO 9000 for software developers*. Milwaukee, WI: ASQC Quality Press.

Soteriou, A. C., & Zenios, S. A. (2000). *Searching for the value of quality in financial services* (Working Paper). Philadelphia: University of Pennsylvania, The Wharton Financial Institutions Center.

Stelzer, K., Mellis, W., & Herzwurm, G. (1997). A critical look at ISO 9000 for software quality management. *Software Quality Journal, 6*(2), 65-79.

Terziovski, M., Samson, D., & Dow, D. (1997). The business value of quality management systems certification: Evidence from Australia and New Zealand. *Journal of Operations Management, 15*(1), 1-18.

Yang, Y. H. (2001). Software quality management and ISO 9000 implementation. *Industrial Management and Data Systems, 101*(7), 329-338.

APPENDIX 1: THE PARAMETRIC EVENT-STUDY TEST

The standard parametric test for an event study is a *t*-test of the null hypothesis that there is no significant difference between the abnormal returns in the event window and in the estimation period. This is calculated as:

$$t := \frac{\text{EW\_CAAR}}{\text{EP\_s\_AAR} \cdot \sqrt{\text{EW\_Return\_Days}}}$$

where EW_CAAR is the cumulative average abnormal return in the event window, EW_Return_Days is the number of days in the event window, and EP_s_AAR is the standard deviation of the average abnormal returns in the estimation period (assumed to be the same in the event window), which is calculated as:

$$\text{EP\_s\_AAR} := \sqrt{\frac{\sum(\text{EP\_AAR} - \text{EP\_MAAR})^2}{\text{EP\_Return\_Days} - 1}}$$

where EP_AAR is the average abnormal return in the estimation period and EP_MAAR is the mean of the abnormal returns averaged across all companies in the estimation period. Due to the very large number of degrees of freedom, the test statistic can be treated as following an N(0,1) distribution.

Chapter XIII
The Value of Web Design Standards for Mobile Computing

Matt Germonprez
University of Wisconsin—Eau Claire, USA

Michel Avital
University of Amsterdam, The Netherlands

Nikhil Srinivasan
Case Western Reserve University, USA

ABSTRACT

The multiple and ever-evolving standards that govern mobile computing result in multilayered heterogeneous environments of mobile devices and services. Thus, as mobile computing becomes more prevalent, it is important that designers build systems that support as many unique, in-use, and user-defined characteristics as possible. This study explores the related effects of two existing standardized technologies: hypertext markup language (HTML) and cascading style sheets (CSS). Furthermore, whereas we investigate the impact of the CSS standard in the context of computing in general and mobile computing in particular, we also focus on two emerging roles of this standard: device independence and usability. Our findings suggest that the application of the CSS standard can improve data delivery across independent devices with varied bandwidth and resource availability, thereby providing device independence and improved usability respectively. We demonstrate that through their effect on device independence and usability, CSS plays an important role in the evolution, expansion, and openness of mobile computing.

INTRODUCTION

We often find that information systems, especially those used in mobile computing, represent an ad hoc collection of data, services, and devices that have been created and recreated by users on an as-needed basis. It is difficult for designers to predict how exactly applications will be used in mobile computing because it is the user who ultimately defines on the fly what applications and configurations are used in which contexts. As mobile computing becomes more prevalent, it is important that designers build systems that support these unique, in-use, and user-defined characteristics. This research aims to extend our knowledge about how standardized interface design can be used by designers to aid and improve mobile computing.

Mobile computing has gained increasing importance as systems and devices continue to move toward an assemblage of distributed Internet-based services that support the exchange and sharing of data and processes. Technologies such as cellular telephones, personal digital assistants (PDAs), and Web portals are used daily in mobile computing. These technologies allow for user expressiveness or customization around such issues as computing style, program preferences, and aesthetic layout (Pask, 1971). Therefore, mobile computing implies a dynamic environment that supports a vast array of varying interfaces, contexts, and automation (Abowd & Mynatt, 2000; Gabriel, 2002). In this chapter we explore the expressiveness within computing in general and mobile computing in particular, and propose ways to use existing technology standards to enhance this characteristic.

Many dimensions describe mobile computing, including the convergence of technologies, the relinquishment of personal data for the receipt of particular services, and the blurring boundaries between physical and virtual spaces (Abowd & Mynatt, 2000). We investigate and propose solutions for achieving two aspects of mobile computing: device independence and usability (Lum & Lau, 2002). Device independence is the separation of data and presentation in support of the movement of data between technologies and heterogeneous computing systems. Device independence describes any data-device relationship that is capable of being replaced without affecting data, services, workflow, or personal computing style. Device independence provides users access to the same data, where the data are unbound from a particular device, irrespective of the device from which the request originates. Whether passing data between devices or across new versions of the same devices, mobile computing and its support of user-designed expressive systems require that we consider device independence.

Usability refers to issues such as the delivery of data to a variety of devices with unpredictable screen sizes and technological capabilities as well as the rapid delivery of data over heterogeneous networks. We contend that usability is a user-centered issue that affects the success of computing in general and mobile computing in particular. Shneiderman (2000) argues that usability has become an important issue for computing research. In particular, he suggests that broad technical variety has a direct impact on usability and that speed in the distribution of data across devices must be considered. With that, we explore usability in the context of the technical variety available within mobile computing through three measures of computing and network efficiency.

We use published design standards (cascading style sheets [CSS] and hypertext markup language [HTML]) and their ability to separate requested data and their presentation to both achieve device independence and improve usability. Accessing and using the same data regardless of platform on a PDA, a laptop, and a telephone demands the separation of data and presentation, and we need only to revisit existing and emerging roles of the published interface design standard of CSS to realize these aspects of mobile computing. Since the debut of markup languages, we have long had the ability but not the motivation to separate the data and presentation. Mobile computing has provided the motivation. The use of published standards is addressed by Abowd and Mynatt (2000), who call for evaluative research of existing technologies in the design of mobile computing, and Weiser (1991, 1993) who notes that research on mobile computing must ultimately be functional for both designers and users. In addition, our work extends the writing of Lie and Saarela (1999) that looked at the early role of CSS with HTML by illustrating benefits gained from their combination.

Using available standards to deliver content presents service providers with a dilemma: Employing proprietary technologies may help in developing unique services that in turn distinguish them from the pack, but at the same time, employing standards is the key for interoperability and true mobility. The dilemma stems from the fact that standardization relies on agreement between competitors to develop common specifications at the expense of limiting the extent to which service providers derive value from proprietary technologies. However, the interface standards of HTML and CSS work at a broad level where few proprietary technologies exist. They provide an efficient and simple way to deliver content to a suite of mobile devices at high levels of usability without the complexities of producing proprietary interface design solutions.

Next, we define the role that device independence and usability play in mobile computing. Then, we look at how device independence can be provided through CSS. Through an illustrative case, we show how published design standards lead to the creation of device independence and subsequently mobile computing. We then illustrate how CSS also provides improved usability for mobile computing. Finally, we discuss how the CSS standard provides a readily available technology for application designers in the development and implementation of mobile computing.

DEVICE INDEPENDENCE AND USABILITY FOR MOBILE COMPUTING

Mobile computing is a sociotechnical phenomenon where devices, services, and data are integrated into our daily lives (Weiser, 1991). Table 1 illustrates three research themes and eight dimensions that help define the context of mobile computing (Abowd & Mynatt, 2000; Avital & Germonprez, 2003) and the underlying scope of our study. The table represents research and practical areas where mobile computing can be improved and expanded through existing and emerging technologies that support the specific dimensions. For example, the 802.11 suite of wireless standards supports universal capture and access in the form of improved service mobility.

The ability to access data across a variety of devices, that is, device independence, is a prerequisite of mobile computing. Device independence provides users access to the same data, where the data are unbound from a particular device, irrespective of the device from which the request originates. This emphasis on device independence for mobile computing is timely as several researchers have been calling for the development of multidevice, adaptive user interfaces (Grundy & Yang, 2002) and location-sensitive systems (Abowd & Mynatt, 2000) in support of this requirement. We focus on device independence in the explanation of mobile computing because of its extended use in computing today and its potential development through existing standards and technologies. We also extend the thinking of device independence, illustrating that it can be provided not only through intelligent systems, but also through the application of CSS standards. Using CSS for device independence is a more simplified approach than the "tour guide" systems developed by Cheverst, Mitchell, and Davies (1998) or the decision engine of Lum and Lau

Table 1. Research themes and computing dimensions for mobile computing

Research Themes (Abowd & Mynatt, 2000)	Computing Dimensions (Avital & Germonprez, 2003)	Description
Universal Capture and Access	Mobility of Services	Services are portable for anytime, anywhere computing using wireless communication technologies.
	Expanding into Virtual Spaces	Virtual presence, virtual work, and virtual relationships break the limiting boundaries of physical places.
	Proliferation of Task-Specialized Devices	Preconfigured devices and service agents operate in the background to carry out delegated and specialized tasks.
Context Awareness	Information Transparency	Personal information is available across traditional boundaries in return for specialized services.
	Device Independence	***Data are unbound from technology devices.***
	Computing Automation	Autonomous systems are less dependent on human management.
Natural Interfaces	Tailoring and Customization	Devices and services are modified in the context of use according to an individual's specifications or preferences.
	Convergence of Technologies	There are the integration and unification of previously distinct digital technologies.

(2002). These studies placed a specific emphasis on object recognition and context identification through specific devices. In this study, however, we investigate how to create device independence with little real-time awareness by any device. Instead, we propose CSS as a means of achieving device independence through otherwise passive and standardized design rules.

Faced with rapidly changing technologies and social demographics, it is important to understand usability for mobile computing in the support of technical variety and network access capabilities as well as the ability to accommodate a variety of users with different capabilities (Shneiderman, 2000). Usability in our context refers to issues such as the delivery of data to a variety of devices with unpredictable screen sizes and technological capabilities as well as the rapid delivery of data over heterogeneous networks. We believe that by improving effective data delivery, interface standards play an important role in the usability of mobile computing. While CSS does not convert information, it can seamlessly deliver the same information to multiple devices. In this light, usability is operationalized as the efficiency at which data are presented across devices (Abowd & Mynatt, 2000; Calongne, 2001) in the form of page sizes, CPU (central processing unit) utilization, and rendering times. The efficiency of data presentation is a quantifiable measure that can be used to define usability and provide designers with a more concrete performance metric (Calongne; Lum & Lau, 2002; Shneiderman, 1992).

CASCADING STYLE SHEETS FOR DEVICE INDEPENDENCE AND USABILITY

Cascading styles sheets are used for controlling the presentation of data on the Internet. The development of standards for the presentation of data began in 1978 with the development of the standard generalized markup language (SGML) standard, which was eventually published by ISO (International Organization for Standardization) in 1986. The commonly used HTML standard emerged from a single SGML definition in an effort to reduce the broad scope of SGML into a more limited interface design standard. Despite its popularity, HTML has drawbacks as an interface design standard, which includes the vagaries of the standards and elements used in its specification. In addition, despite the advantages that newer standards such as CSS provide, their adoption has been slow. In the remainder of this section, we briefly review HTML and CSS to illustrate the advances that have taken place in design standards. We then conclude with a discussion of other SGML descendents—XML (extensible markup language), XSL (extensible style language), and XSLT (extensible style language transformation)—as alternative design technologies that may provide similar solutions to the ones presented in this chapter.

Hypertext Markup Language

HTML is the language most commonly used to display data on the Internet, or the standard of published documents on the Web, as described by its maintainers, the World Wide Web Consortium (W3C).[1] It is a structural markup language that describes how the different elements of a document are related. The data are formatted through tags, which control presentation characteristics such as font sizes, color, and layout. A drawback of HTML is its inability to characterize data semantically. Instead, specific data about an organization, a new product, or the weather are embedded in general and data-unspecific HTML tags of <table> and <H1>.

Figure 1. CSS example

```
p.title{
margin-left: +10em;
color: #1f1f1f;
font-weight: bolder;
font-size: medium;
font-family: Verdana, Helvetica;}
```

Cascading Style Sheets

CSS is a complementary interface design standard to HTML. The use of style sheets allows data within an HTML document to be separated from their presentation. Effectively combining the use of CSS with HTML is dependent on the existing structure of HTML (Figure 1).

In the illustration, CSS provides formatting to title paragraphs (p.title) in the form of margins, coloring, and font styling. The combination pro-

Figure 2. CSS changes the layout specification but not the presentation

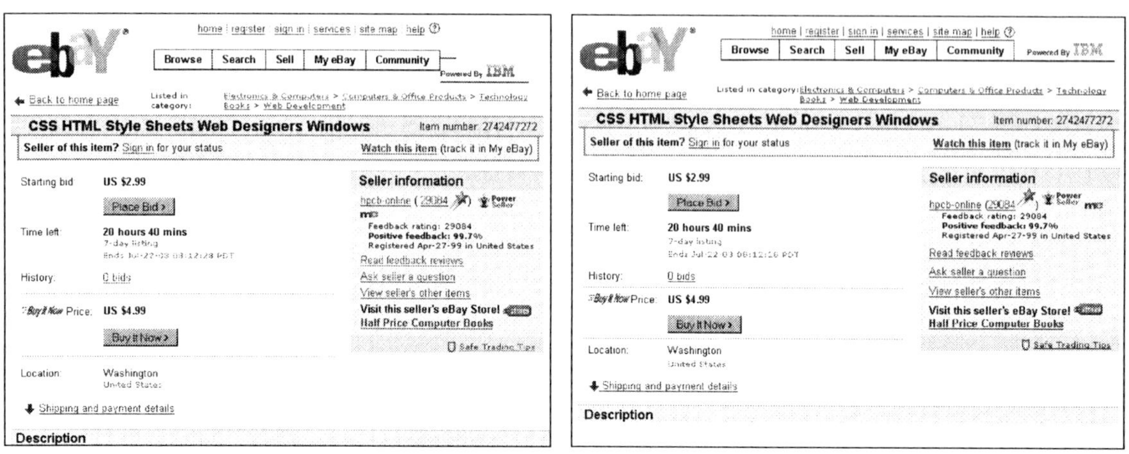

eBay page with only HTML

eBay page with HTML and CSS

vides support for device independence because the presentation and device-specific data are described by the CSS (Lie & Saarela, 1999) and not by the HTML. From a user perspective, style sheets have a negligible impact on how data are displayed on the same device. Figure 2 shows how CSS handles the display of data on the same device (a desktop Web client). In the example, CSS was applied to the HTML data page generated by an eBay auction. The definition of the margins and tables were pulled out of the HTML and placed in CSS.

XML, XSL, and XSLT

XML has emerged as a powerful language used to describe people (e.g., FOAF [friend of a friend]), frameworks (e.g., RDF [resource description framework]), and data (e.g., SOAP [simple object application protocol]). In all of these cases, system administrators are empowered to define the tags in use to provide the desired descriptions. In the following sample of code (Figure 3), XML is used for a Google search. In this case, XML takes the form of SOAP where the elements are in part standardized (Envelope, Body) and in part created by Google (doGoogleSearch).

This SOAP code can be rendered to a computer screen through XSL formatting (XSL-FO) or CSS, and can be transformed into other markup (e.g., see HTML) through XSLT. In any case, XML can be rendered to a screen or transformed into something else that can also be rendered.

Figure 3. XML example

```
<SOAP-ENV:Envelope>
<SOAP-ENV:Body>
<doGoogleSearch>
<key type="string">Qx+Nz/1QFHKJc4smLr</key>
<q type="string">cleveland</q>
<maxResults type="int">2</maxResults>
</doGoogleSearch>
</SOAP-ENV:Body>
</SOAP-ENV:Envelope>
```

It is very likely that similar efficiencies to computing and mobile computing in particular can be obtained through the use of XML, XSL, and XSLT. However, determining the most technically or cost-efficient technologies for achieving this goal is outside the scope of this chapter. Instead, we present two widely implemented technologies (HTML and CSS) as illustrations of how two widely accepted interface design standards can be used to improve mobile computing.

RESEARCH DESIGN

Cascading style sheets were expected to improve device independence by reducing the number of images, the amount of tabling, and the redundancy of presentation rules. Specifically, cascading style sheets were used to unbind the presentation of data from any single device, thereby contributing to device independence. The CSS standards were also expected to have an impact on usability through an improvement of both network and computer efficiencies. This study investigated the effect of design standards on device independence and on usability. The expected findings aim to extend the current understanding of device independence and usability, and indirectly enhance our understanding of their established effect on mobile computing performance (Figure 4).

Device independence was illustrated in more detail through the redesign of one Web site, which shows how CSS improves the presentation of data across devices, therefore supporting device independence (Arrow 1 in Figure 4). This provides our first research question.

- **Question 1:** Can CSS be used to promote device independence?

In addition, through CSS, page sizes were consistently reduced by roughly 50%. For example, using the CSS standard, the aforementioned eBay page (Figure 2) was reduced from 20Kb to 11Kb.

Figure 4. Research model: The effect of cascading style sheets on device independence and usability

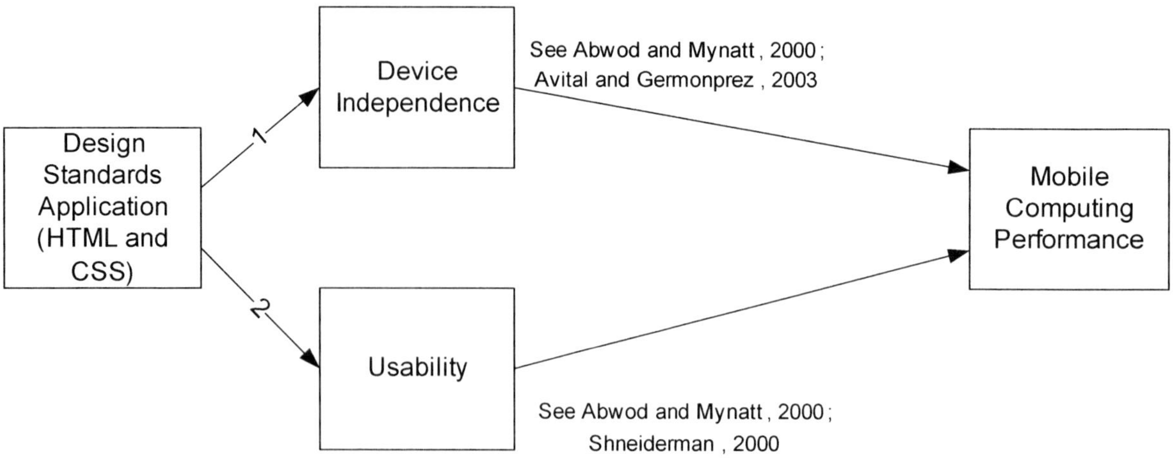

We believe that this reduction depends on the ratio between the size of the text and the amount of formatting done through HTML. An HTML page with little layout information would have lower page size. The reduction in page size is likely to result in usability improvements as represented by three measures (Arrow 2 in Figure 4). This is expected based on the modularity that CSS provides to the Web page. This provides a set of three research questions concerning usability.

- **Question 2a:** What impact does CSS and the subsequent reductions of page size have on the amount of time it takes for a client to display data?
- **Question 2b:** What is the impact of CSS and the subsequent reductions in page size on CPU resources consumed by clients?
- **Question 2c:** What is the impact of CSS and the subsequent reductions in the page size on CPU resources consumed by servers?

The study was based on the analysis of the application of CSS to three Web pages. Although it may appear small, the sample was sufficient for the examination of the impact of CSS on the display and delivery of data, and subsequently provided a reasonable answer for the underlying research questions. We applied a two-step approach. First, we used a descriptive explanation in an illustrative case that demonstrates how CSS can be used to support and extend device independence across computing environments (Question 1). Second, we used a statistical relevance explanation in an experiment that illustrated how CSS improves usability through three efficiency variables: the rendering time of the client application, the client CPU utilization, and the server CPU utilization (Questions 2a-2c). All statistical relevance was determined through multiple tests on each data point. In this case, validity was provided through multiple tests of each Web page ($N = 50$ in this study), not through the number of pages tested ($N=3$ in this study). Following a power assessment, we believe that additional cases would not strengthen our descriptive explanation nor improve our statistical power.

RESEARCH FINDINGS

Impact of CSS for Device Independence

In this section, we answer the first question and explore the effect of CSS standards on device

Figure 5. Site through desktop using HTML tables

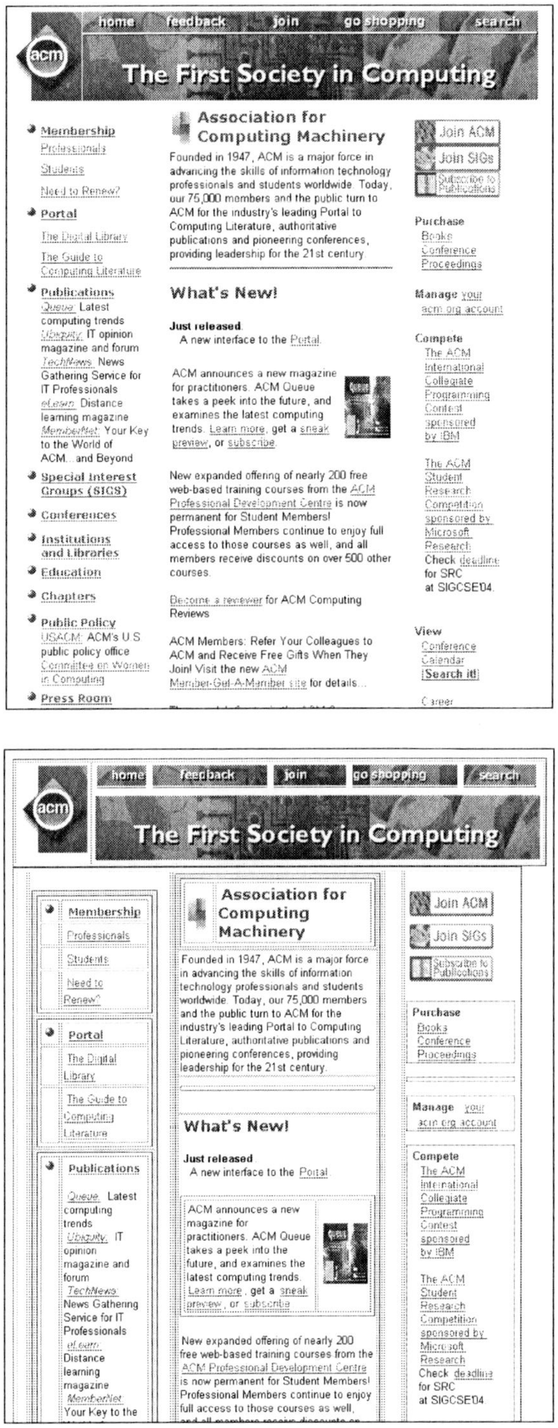

independence. One of the most powerful characteristics of CSS is the capacity to manage data presentation. Generally, so far, the most common presentation vehicle has been HTML tables. Specifically, data are presented in a series of tables, embedded tables, and cells. In Figure 5, the Association of Computing Machinery (ACM) data are presented using HTML tables. In the figure, the left image is the ACM Web site at a desktop view; the right image is the same Web page with the tables highlighted. The colors of the table borders are reflective of nesting order: Outer tables are blue, second-level tables are green, and third-level and deeper tables are red.[2]

As shown, the page design relied heavily on HTML tables, a technique that works well only when displaying data on a desktop. However, when table structures are accessed through a smaller device such as a PDA, the presentation of data cannot adapt to the smaller interface space. Figure 6 shows a segment of the same site accessed through a PDA.

As seen in Figure 6, the exact presentation was maintained in the PDA, even though it provides a smaller window through which the same data are viewed. Both vertical and horizontal scrolling

Figure 6. Site through PDA using HTML tables

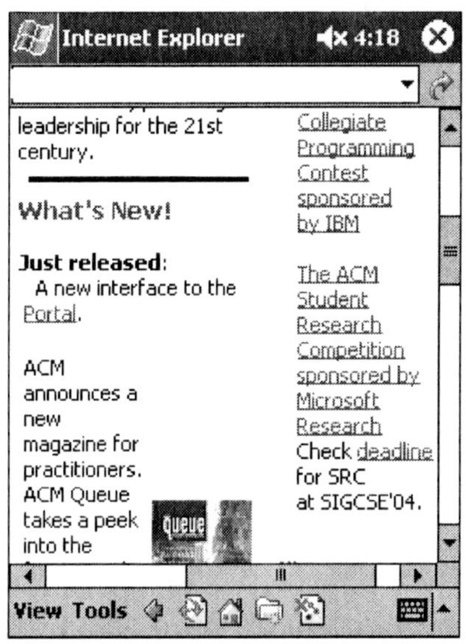

Figure 7. Web site similarity using HTML and CSS

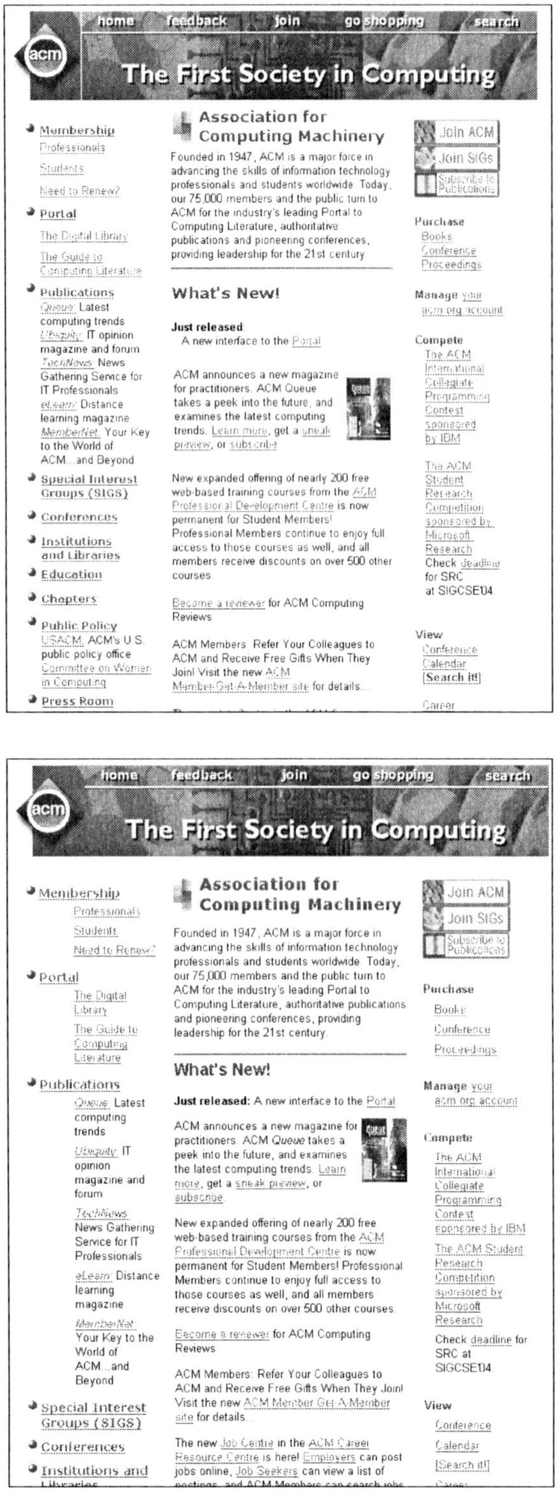

was required as the presentation of data maintains its large screen design. In essence, presentation through HTML tables treats the PDA as a small lens that is moved up and down and left and right across the larger canvas of a desktop interface. In this instance, the system did not scale across devices and failed to provide support for a broad suite of devices needed for mobile computing. In the case of presentation through HTML tables, device independence was not supported through the natural interface of a PDA or any small-screen device.

The same site was restructured to remove many of the HTML tables. Using the interface design standard of CSS, presentation occurred through HTML markers, not tables. Instead, the data were laid out using HTML divisions (<DIV>) and the CSS properties of margins, padding, and positioning. Figure 7 shows the presentation of the site using CSS. The left image is the earlier HTML-based Web page and the right image is the page redesigned using HTML and CSS.

The differences that CSS provide via the desktop were visually insignificant. However, a closer look at the site through a PDA shows that the data were rendered on the fly in order to ac-

Figure 8. Web Site through PDA using CSS

commodate the new device. Figure 8 shows the same Web page displayed through a PDA after being redesigned using HTML and the interface design standard of CSS.

By replacing the HTML tables with CSS, the presentation became entirely vertical without any horizontal navigation. CSS presented the data sequentially by removing the majority of HTML tables that, in the previous example, fixed the width of the site. In this context, the site is presented equally well on a large space like a desktop and the small space of a PDA. In fact, the presentation of the data is more accommodating to the natural interface of the PDA. In separating the data and presentation, we were able to extend device independence to the mobile computing environment without using multiple design templates or altering the intelligence of any device. Designers may be overdesigning conditional interfaces for multiple devices (e.g., if device A, use style A, etc.), and engineers overengineering devices to handle the multitude of interfaces possible for any one device. Instead, and following the first research question, we argue that device independence is promoted in part through the otherwise simplified and passive technology of CSS.

Impact of CSS for Usability

In this section, we answer the second set of questions and explore the effect of CSS standards on usability. As mentioned, an improvement provided by CSS was a reduction in page size. The impact of reduced page size was examined on three usability outcomes. First, following Question 2a, we investigated the role that reduced page size played in determining the amount of time required for a client to render an interface. Rendering time is an important issue in usability research and mobile computing as users move across devices with varying levels of processing power. Second, servers and clients use computing resources to deliver and display data. Following Questions 2b and 2c, we investigated the effect of page size with respect to the reduction of resources that servers used in distributing data and the decrease in resources clients used to display them. Resource utilization is an important usability issue in mobile computing as users place an increasing burden on clients and servers to deliver an increasingly expansive set of data and services to devices with increasing technical variety.

In the design of the empirical experiment to investigate these questions, the data server and the clients were run on the same device, disconnected from the main network to avoid the interference of network parameters. Caching was also disabled among clients and within the server because complete page loads from the server, not local copies, were required. Finally, all interface design code (i.e., the CSS) was embedded into the HTML and not in a separate file. This feature was designed to illustrate that CSS can reduce page size without simply splitting a file into two parts. The experiment made use of two desktop Web browsers: Netscape and Internet Explorer. We designed the study with two clients in order to examine if the different rendering engines of the clients showed similar improvements in rendering times. Additionally, the clients were treated independently and not as a combined representation of a Web client because rendering times and client CPU utilization were shown to be significantly different between the two.

The data used and served from the server to the clients were three Web pages that were used in their original HTML format and then reformatted using HTML and CSS. The sites were chosen based on their use of tables for formatting. Their selection was arbitrary in that there are an untold number of Web sites to choose from; however, we felt that these three could be used to describe our findings as well as any other site. The sample size of three pages was sufficient because the technique used to reformat the table-based HTML was identical in all cases, not leading to a deeper descriptive explanation of CSS and HTML. Additionally, the redesigned Web pages were visually the same as

Table 2. Tested Web sites and reduced page size

Page Size	Association of Computing Machinery	eBay	Los Alamos National Laboratory
HTML Only	26Kb	20Kb	21Kb
HTML and CSS	13Kb	11Kb	11Kb

the originals (see Figures 2 and 7). As expected, size was reduced significantly, virtually to half in our cases, using CSS across all three pages (Table 2).[3]

With regard to the three efficiency variables, rendering time was measured from the start of the page load to the completion of the last character on the page, providing an accurate capture of the load time for the page. The measurement of CPU utilization for both client and server was done using a performance monitor application. Fifty samples were taken to gather rendering times and resource utilization on the client and server (N=50) for every Web site.

Table 3 shows the results of the t-tests that suggest improved efficiencies from an HTML page to an HTML-CSS formatted page. The table represents percentage improvements in CPU load times (for both the client and the server) and the rendering times when using the HTML-CSS reformatted pages. Significance indicates that HTML-CSS reformatted pages statistically improved the associated measures. With respect to Question 2a, regarding the impact of CSS and the subsequent reduction of page size on the amount of time it takes for a client to display data, rendering time was significantly reduced in all cases for both the desktop and PDA. With respect to Questions 2b and 2c, regarding the impact on CPU resources, the effects of CSS in reducing the CPU utilization was mixed in both the client and the server.

The values represent percentage improvements in CPU load times (for both the client and the server) and the rendering times when using the HTML-CSS reformatted pages. Significance indicates that HTML-CSS reformatted pages statistically improved the associated measures.

In the experiment, each page (in both its original and redesigned state) was loaded into two popular Web browsers. As seen, there was no significant improvement in the utilization of client or server resources due to the reduction of page size for the ACM Web page. It is unclear as to why this was the case. The initial rendering times of the ACM page were faster than those of Los Alamos or eBay and the reduced rendering time using CSS was simply not enough to provide significance. These two sites, however, showed an improvement in the CPU utilization for both the client and server processes.

DISCUSSION

Cascading style sheets have been shown to play an emerging role in mobile computing. We believe these findings can also be applied to more traditional computing as well. Our study illustrated that there are multiple benefits to using CSS and HTML, two of which include device

Table 3. Results of CSS on resource utilization and rendering time

	Netscape			Internet Explorer		
	ACM	eBay	Los Alamos	ACM	eBay	Los Alamos
CPU Load: Client	(9.0%)	13.9%**	12.2%**	4.7%	30.9%**	10.8%*
CPU Load: Server	8.0%	18.9%*	35.7%**	(1.5%)	21.7%*	16.2%*
Rendering Time: Desktop	33.0%**	36.8%**	18.2%**	30.0%**	63.4%**	47.6%**
Rendering Time: PDA				41.3%**	N/A	79.6%**

* Indicates significance $p<.05$
** Indicates significance $p<.001$
( ) Indicates a decrease in performance

independence and improved usability. In the case of device independence, we were able to focus on one specific characteristic of mobile computing (Avital & Germonprez, 2003) and illustrate how this standard provided improvements across devices. With usability, we demonstrated advantages not only for mobile computing but computing in general. These advantages included lower bandwidth needed to deliver content (lowering page size), reduced CPU utilization (both client and server), and improved rendering times, all of which represent improvements in any computing domain.

In mobile computing, device independence and usability play a significant role in determining how data are delivered and displayed across various devices. As users are capable of surveying and selecting data that suit their current needs, system designers must be aware that the delivery and display of data must be as streamlined and dynamic as possible. Serving data to an unknown suite of devices including desktops, handhelds, and wireless telephones increases the need for device independence and efficiency in the delivery and rendering of data. CSS has been shown to alleviate both of these concerns by improving usability and device independence.

HTML and CSS fit into a broader possible space of technologies and standards that provide improved usability and device independence. Usability can likely be improved through XML-based data structuring standards such as RSS (real simple syndication). RSS relies on data feeds to present users with structured content without the accompanying HTML. RSS has been shown to improve the accessibility of Web sites by reducing the size of content and summarizing the information on the Web page (Belkas, Garofalakis, & Stefanis, 2006). What we proposed and demonstrated addresses a small portion of what comprises mobile computing. How additional design standards, whether CSS, XML, or XSL-FO, can be used to improve aspects of tailoring and customization or device automation has yet to be determined. In addition, much of what was described in this chapter may be possible through other more efficient or complementary technologies. For example, numerous Web sites rely on different pages served to different devices based on browsers, operating systems, and connection speeds (Lum & Lau, 2002). Technologies such as caching, data distribution networks, and object identification provide improved data delivery and rendering speeds. In any case, CSS is capable of complementing or augmenting these alternate technologies in the effective delivery and display of data. Fine-tuning which technologies are best for which aspects of mobile computing is an ongoing inquiry of future research. What researchers in this domain should ultimately strive for is a better use of mobile devices in displaying all types of information, not just markup-based information. How and which technologies best achieve these ends is an ongoing forum for debate, discussion, and research.

The impact of CSS on mobile computing should not be underestimated. As mobile computing devices proliferate and become used on a mass scale, networks are going to become burdened with data traveling between providers and the users. Providers have the opportunity to improve the efficiency of their servers and network connections by making the data that they serve richer in experience and smaller in size. Additionally, providers must maintain device independence as their data and services are consumed across various devices. The use of CSS as a standard for developing data on the Web for mobile computing devices will allow providers to take advantage of the multifaceted nature of this design standard.

CONCLUSION

We explored the utilization of two existing standards, HTML and CSS, and the impact of the latter on device independence and usability. The findings suggest that CSS improves network efficiencies to a large degree, and it is also an important component for delivering data across clients with varied bandwidth, resource availability, and technical variety. As devices, particularly in mobile computing, are expected to vary in bandwidth and resource availability, improvements in network efficiencies may prove valuable in the future design, implementation, and use of such systems. Our research supports a line of inquiry that can benefit both mobile computing as well as standards research. This research acknowledges that there are characteristics that describe mobile computing and design standards that can support and improve these characteristics.

ACKNOWLEDGMENT

An earlier version of this chapter, entitled "The Impacts of the Cascading Style Sheet Standard on Mobile Computing," was published in the *Journal of IT Standards and Standardization Research* (2006, Vol. 4, Issue 2, pp. 55-69).

REFERENCES

Abowd, G., & Mynatt, E. (2000). Charting past, present, and future research in ubiquitous computing. *ACM Transactions on Computer-Human Interaction, 7*, 29-58.

Avital, M., & Germonprez, M. (2003). *Ubiquitous computing: Surfing the trend in a balanced act.* Proceedings of the 2003 Workshop on Ubiquitous Computing Environment, Cleveland, OH.

Belkas, A., Garofalakis, J., & Stefanis, V. (2006). *Use of RSS feeds for content adaptation in mobile Web browsing.* Paper presented at the Proceedings of the 2006 International Cross-Disciplinary Workshop on Web Accessibility (W4A) Building the Mobile Web: Rediscovering Accessibility? Edinburgh, UK.

Calongne, C. M. (2001). Designing for Web site usability. *The Journal of Computing in Small Colleges, 16*, 39-45.

Cheverst, K., Mitchell, K., & Davies, N. (1998). *Design of an object model for a context sensitive tour guide.* Proceedings of the Conference on Interactive Applications of Mobile Computing, Rostock, Germany.

Gabriel, Y. (2002). Essay: On paragrammatic uses of organizational theory. A provocation. *Organization Studies, 23*, 133-151.

Grundy, J., & Yang, B. (2002). *An environment for developing adaptive, multi-device user interfaces.* Paper presented at the Fourth Australasian User Interface Conference, Adelaide, Australia.

Lie, H. W., & Saarela, J. (1999). Multipurpose Web publishing using HTML, XML, and CSS. *Communications of the ACM, 42*, 95-101.

Lum, W., & Lau, F. (2002). A context-aware decision engine for content adaptation. *IEEE Mobile Computing, 1*, 41-49.

Pask, G. (1971). A comment, a case history, and a plan. In J. Reichardt (Ed.), *Cybernetics, art and ideas* (pp. 76-99). London: Studio Vista.

Shneiderman, B. (1992). *Designing the user interface: Strategies for effective human-computer interaction* (2nd ed.). Reading, MA: Addison-Wesley.

Shneiderman, B. (2000). Universal usability. *Communications of the ACM, 43, 84-91.*

Weiser, M. (1991). The computer for the 21st century. *Scientific American, 265*, 94-104.

Weiser, M. (1993). Some computer science issues in ubiquitous computing. *Communications of the ACM, 36*, 75-84.

Chapter XIV
Developing Country Perspectives on Software: Intellectual Property and Open Source. A Case Study of Microsoft and Linux in China

Xiaobai Shen
University of Edinburgh, UK

ABSTRACT

This chapter looks at implications of the emerging global intellectual property (IP) regime for developing countries (DCs) and their attempts to improve their technological capabilities. It further highlights the new perspectives for DCs opened up by the emergence of non-proprietary (open source/free) software, such as Linux. A case study of the battle between Microsoft and Linux in China is used to explore the dilemmas faced by China in determining what IP regime (strict or weak) to adopt, and the threats and opportunities that either may pose for indigenous technology development. Based on the case analysis, the chapter criticizes the simplistic polarized views that have been presented of the implications of the global IP regime and of the potential of non-proprietary software. It explores some of the complex considerations about the interplay between technology strategy and IP protection for China and discusses the policy implications for China and other DCs.

INTRODUCTION

There has been increasing pressure on developing countries to adopt intellectual property (IP) regimes alongside the development of the globalization process, under which major technology holders in the developed world seek to strengthen technology protection in developing countries (DCs) and other emerging markets. Conventional modernization thinking suggests that a strong IP regime will promote growth and the development of technological capabilities in DCs by attracting foreign investment and technology transfer. An alternative, critical current, with its roots in dependency theory (Colman & Nixson, 1994), highlights the threats to DCs from simply adopting Western IP regimes. A similar dichotomy of views has emerged in debates about the respective advantages of proprietary and non-proprietary (e.g., open source) software—with the latter being held out as providing important opportunities for DCs and an alternative strategy to dependence upon the Western multi-national enterprises (MNEs) that have achieved near-monopoly positions worldwide in some software market segments. A case study of the recent history of Linux and Microsoft in China calls into question these simplistic polarized accounts (Dutt, Kim, & Singh, 1994) of the choices before DCs. It highlights the complex interplay between IP and innovation and the balance of considerations DCs face in designing their IP regime and strategies to promote indigenous capabilities in the key area of software.

This paper first seeks to explicate the impact of IP issues on indigenous and exogenous technology development in DCs. It focuses on the implications for strategies in the software industry, partly because IP issues in the software industry are subject to considerable concern even in developed countries, and because software has been considered a more promising field for DCs to exploit than other technologies. Specifically, it explores some of the key issues that a DC like China may need to consider in establishing its own IP regime while promoting indigenous technology development and use.

The second concern of the paper is to discuss how China may exploit opportunities from the emergence of non-proprietary software (open source/free software) that is protected by general public license (GPL)—COPYLEFT in contrast with COPYRIGHT.

A case study of the battle between proprietary software (the Microsoft operating system) and non-proprietary software (the Linux operating system) in China is used to explore some of the complex considerations about the interplay between technology strategy and IP protection and their implications for a developing country like China.

The case has potentially broader significance, to the extent that the outcomes of the contest between Microsoft and Linux in China may upset the current global technology trajectory and pose significant challenges to the current international IP regime—in particular, in software protection—and thus may have an impact on software development globally.

CONTEXT OF IP PROTECTION AND A DILEMMA FOR DCS

In a world characterized by rapid technological development, we find that most advanced technologies are created by corporations in the developed world. Western governments—increasingly concerned about their corporations protecting their technological know-how from being acquired and copied by others—are therefore keen to strengthen and extend current IP regimes. With the rapid development of globalization, protecting Western technologies in emerging markets and DCs becomes imperative. These powerful pressures are represented by trade-related IP rights (TRIPS) and other international agreements. DCs are under intense pressure to adopt the same IP regimes.

The mainstream modernist thesis, also espoused by some experts from developing countries, argues that improving IP protection will benefit developing countries, assisting technology development and therefore economic growth by promoting foreign investment and by encouraging indigenous innovative activity. With increasing globalization and the growing role of MNEs, we find increasing acceptance of the expansion of the international IP regime to the developing world under the name of global harmonization.

Most DCs, concerned with attracting foreign technology and investments, have become committed to improving the IP protection system in their own nations. Countries like China, India, Brazil and Mexico, and so forth, have been significantly building up their IP protection systems in a rather short period of time. However, today their IP protection practices are still far from achieving the level demanded by MNEs and Western governments, and many of them still appear on the top of the list of worst IP violation nations by the United States (U.S.). In most cases, the shortcomings stem not from public policy and the promulgation of IP laws, but derive instead from problems in IP enforcement and practice. The obstacles to enforcement lay with those who are involved in operating the political and economic mechanisms, but may also have deeper and broader roots in cultural traditions (for example, in populations that might well not endorse the rules set by the current global IP regime). The combination of these factors is manifested in weak and ineffective IP mechanisms, including lack of effective enforcement and prosecution; resources constraints, such as lack of finance, training, qualified professionals; the lack of an infrastructure/facilities for handling IP data and cases; and the lack of experience in dealing with all IP-related routine tasks.

These difficulties with the current international IP framework are by no means limited to DCs. During the last 20 years or so, as the level, scope, territorial extent and role of IP protection have expanded at an unprecedented pace, concerns have arisen in the developed as well as the developing worlds (Drahos & Braithwaite, 2002). There has been some evidence that IP rights may pave the way for monopoly and constrain competition, which may discourage rather than encourage innovation, as originally intended. Moreover, the cost of IP rights applications and, in particular, the enormous legal costs in cases of infringement, mean that IP protection is largely favorable for richer and bigger players. Many small players are likely to lose out. With the expansion of IP rights protection both locally and internationally, the cost of handling data and going through legal procedures is increasing dramatically, especially when dealing with international cases. On top of this, correct processing of IP rights applications and ensuring the quality of patents registered may become onerous.

These considerations suggest that, in general, the current IP protection system is favorable for the larger players and for those with more knowledge of the current IP system: for MNEs (mainly based in developed countries) owning and exercising IPR over a large number of technologies (and thus for the governments and national economies of those countries). None of these circumstances would seem to prevail in developing countries. A key question that needs to be answered concerns whether implementing the current international IP regime can benefit developing countries.

We must also examine sceptically the arguments that IP protection will encourage innovative activity and promote foreign investment. The consequences of applying the international IP regime in DCs are still unproven. When we examine the experience of perhaps the most successful DCs, the "four little Asian tigers," we find that they accumulated their technological capabilities initially through reverse engineering and imitation, facilitated by their domestic context of weak IP regimes. The fact that China has attracted large foreign investment is clearly not because of its IP protection record. The UK

Commission on Intellectual Property Rights, in its comprehensive general report on these very issues, observes that developing countries are "second comers" in a world that has been shaped by the "first comers" (CIPR, 2002). It is not a question of whether the international IP regime is a good or bad thing: DCs have to live with the international IP regime and its likely trend of development by understanding it and finding a way to utilize it for their own good. While much discussion to date has been rather general regarding whether DCs should dilute or delay enforcement of the global IP regime (Kumar, 2002), this paper instead argues for a more active (and proactive) strategy, adjusting the IP regime around specific goals. Different strategies may be needed for individual regions, countries, sectors and technologies to achieve these purposes.

In recent years, international IP protection has expanded to cover technologies like information technology, including software. Some acute problems have arisen. The example of software reveals controversies about whether current IP regimes provide the incentives for innovation and creativity, which are the underlying rationale for IP protection.

IP AND PROPRIETARY SOFTWARE

The U.S.-based MNE, Microsoft, has been subject to much attention, often critical, on the grounds that it has been exercising monopolistic power and threatening the development of the whole software industry. Recent anti-trust actions accuse Microsoft of exploiting its control over its proprietary operating system software and other industry standard platforms, withholding interface information and similar connectivity data on a timely basis in order to maximize market share and price and, crucially, achieve leverage and strategic advantage in the market for IT applications. Microsoft's opponents warn that these will have a detrimental effect on the productivity of the industry and endanger the future of the Internet as a competitive and productive facility for American commerce[1]. Although the bulk of popular commentary and opinion is critical of Microsoft, alternative accounts are also eminent. Some (Rothstein, 2001; Buckley Jr., 1999; Lee & McKenzie, 2001) argue that Microsoft has merely done what a good corporation should do and attribute its success to the effective management of its technologies. Campbell-Kelly (2003) points out that Microsoft's share of the software market is not more monopolistic than a number of other fields (e.g., microprocessor and memory chips). Some even portray Microsoft as a visionary and innovator whose success in software development has contributed to the improvement of people's lives (Elkin, 2000). Regardless of whether the charges against Microsoft are eventually confirmed, the criticisms and the responses reflect the tensions surrounding the software industry and the salience of the substantial monopoly Microsoft has enjoyed. Many still doubt whether the final outcome of anti-trust actions will solve the problems of fair competition, arguing that the proposals to split the company will neither weaken the power of Microsoft nor help reduce software prices for customers.

Microsoft's success is, of course, only partly a result of the current IP regime, resting more upon secrecy and success in continually innovating rather than, for example, formal IP measures such as patenting, licensing and copyright. The other element is Microsoft's success in establishing its offerings as the *de facto* industry standard platforms, which, by offering interoperability between complementary products, can mobilize powerful network externalities (Cowan, 1992) and help in sustaining that position through the erosion of existing standards and migration to new standards (Williams, 1999).

A number of problems have been detected, raising concerns about the practicality and effectiveness of the current IP regime with respect to software protection. It has been recognized that copyright protection for software interfaces will tend to over-reward the established software developer, thereby blocking the development of competing and complementary products, and making users unable to benefit fully from their own investments and limiting the product choices available to them (Samuelson, 1995). Despite that, software patenting has currently become fashionable; the number of patents is increasing dramatically in the U.S. We see on the one hand, difficulties arising from the lack of available prior-art against which to examine software applications. The rapid growth in the number of patent applications submitted has involved much patented software that is of low quality. On the other hand, we find unduly broad software patents blocking others from new software development (Stallman, 1999). Individuals and companies incur considerable time and money costs to determine how and/or whether to conduct research without infringing upon others' patent rights, or to defend their own patent rights against others (Office of Industry Liaison, 2000).

These problems do not match the claims for which IP frameworks were originally designed. Moreover, the cost burden incurred by individuals and companies for software protection is likely to increase in line with the expansion of the data, information exchange infrastructure and facilities required, and the judicial costs incurred when disputes occur. All these costs will eventually fall onto the consumers. In reality, because of the high cost of judicial procedure, even small players who could afford to apply for software protection might not be able to afford to take up disputes when the outcome is uncertain. On top of that, there is a question whether the public will benefit from such a system, in terms of the cost they have to bear, not to mention additional restrictions on access to knowledge and information.

There has always been a problem for the current IP regime to effectively protect software because of the unique nature of software technology. The software industry is very different in many ways compared with other leading industries.

In the last two decades, the changes taking place in the software industry have been colossal. Until relatively recently, software development was expensive and only a few people and companies could undertake to produce it, because computing equipment was expensive and because it remained labor intensive (Friedman, 1989). Now, with the development of the PC and its relatively lower costs, many people can collaborate in software development. Many take up software development as a means of living; others get involved simply for leisure or as a hobby. The usage of software has thus been widened incredibly, from merely operating a machine to running a specialized network such as banking, and widely accessible Internet-based services that are gradually becoming a necessity of our everyday life. Software has been used in almost every industry. Software varies in terms of cost of development, scale of usage and the level of importance to and impact on our human society. Policies for the software industry, therefore, are of great significance.

As far as many developing countries are concerned, software could prove to be the industry that will bring opportunities for leapfrogging—catching up with or overtaking capabilities in developed countries. Above all, the software industry is labor intensive, with non-material products that can be delivered across the Internet and which does not require massive investment in fixed plant capacity or infrastructure (Steinmueller, 2001). Once a particular product is developed, manufacturing (copying) and transport are cheap.

However, the disadvantages are mountainous, too. For example, as latecomers, developing countries do not own their core software for computing technologies such as operating systems, which form the basis for the software industry and for running networks, like the Internet. DCs have to buy in this software and pay thereafter for upgrades. This is particularly an issue in relation to computer operating systems; if a country is dependent on a proprietary operating system, its software developers find themselves effectively operating in a technological terrain developed, controlled and further innovated by the foreign platform software developers, even where their products are geared for local users.

No doubt, intellectual property systems, such as patents, have contributed to growth in the early economic development of the world. The great success of the IP mechanism is attributed to its role in encouraging domestic innovation by regulating and providing economic incentives for the dissemination of information. At least one part of its industrial and economic success owes to a democratization of access to intellectual property (Khan, 2002).

However, we have to be aware that the situation in the developed world then was very different from DCs now. IP systems were established gradually in individual countries in the West, in line with domestic needs at the time. The process of adopting other countries' IP provisions and harmonizing between them, and leading ultimately to the agreement to establish international conventions, has a different story. Discrimination against foreign IP rights was common practice, which allowed latecomers a window of opportunity to exploit foreign technologies without constraint. The U.S. can be seen as a good example, which enjoyed a weak IP regime, during which it experienced a boom in technological development as well as economic growth (Drahos & Braithwaite, 2002).

Whether the implementation of IPRs leads to technology development and industrialization in developing countries is very much in question.[2] There is a series of strategic issues for DCs to decide upon, drawing lessons from experiences in developed countries. Individual countries may need to design/adopt their IP system to meet local circumstances and progress at a pace that can best promote indigenous technological development and benefit their own economies.

A NEW PROSPECT FOR DCS: NON-PROPRIETARY SOFTWARE WITH GPL—COPYLEFT

The emergence of non-proprietary software and software based on open interoperability standards is projected as opening important potential opportunities, not least for DCs (CIPR, 2002; UNCTAD, 2002; Noronha, 2002). One important example is provided by open source/free software. Software developed with a general public license (GPL) creates the freedom for people to copy, study, modify and redistribute software. It forbids anyone to forbid others to copy, study, modify and redistribute the software. The GNU/Linux operating system[3] software is such a product; initially developed by a Finnish student and taken over by a community of software developers cross the world who value software sharing and volunteer to work on the system. Non-proprietary software is portrayed as offering important advantages over proprietary solutions.

First, Linux is cheap—cheap enough even for people in developing countries to buy. Second, the right to access, to study and to modify means that people, wherever they are, including developing countries like China, India, and so forth—can access the same code.[4] It means that programmers in different places can modify the system to meet specific requirements; to fix a problem, enhance a program or adapt the program to local needs. This is of crucial significance, since application developers, though seeking to anticipate all the ways in which others may use their software and under what local conditions, can only anticipate these to a limited degree. Therefore, modification is necessary for the improvement of existing products. However, in contrast, the source code of proprietary software, like Microsoft's operating systems, is kept secret, preventing others from scrutinising the software and modifying it.

Behind non-proprietary software and GPL is a powerful ideology that defines itself in counter-position to the dominant profit-oriented corporate culture of IT.[5] GPL is designed to restrict the emergence of monopoly and encourage collaboration. It values wide engagement and knowledge sharing, and presents itself as best protecting the interests of the public and consumers. There is strong commitment to these principles amongst a community of technical enthusiasts. Some technical specialists have argued that open-source projects can produce better-quality software than traditional corporate R&D (Raymond, 1998).

A distinctive characteristic of open-source non-proprietary software projects, when compared with traditional corporate software development projects, is the way intellectual property rights are handled. One key innovation in open source has been the GNU general public license (Stallman, 1999), which has made it possible to legally improve and adopt software developed by others, facilitating continuous improvement. COPYLEFT can be seen as a mechanism that uses the legal system to protect software development in a way contrasted to COPYRIGHT. The COPYLEFT mechanism provides access to source code, keeping the development path open and therefore facilitating learning, improvement and integration with other systems. Some have further argued that the COPYLEFT mechanism can overcome difficulties experienced in the commercial appropriation of research and development, and thereby enhance technological development.

However, controversies remain whether Linux is sufficiently secure or dependable, with suggestions that lack of adequate testing/control and open access to code make it more vulnerable to attack by hackers and viruses. Others point to the weaknesses of non-proprietary software in terms of commercially related management, from financing to marketing strategies (Broersma, 2002). Even the powerful underpinning ideology might be eroded with the commercialization of Linux.[6] Though open-source software may not be able to replace proprietary ones, it can certainly offer an alternative for developing countries and other actors to exploit. Some high-profile corporations have recently decided to shift to open-source software as a means to reduce their escalating computing budgets. In Europe, the European Commission is encouraging government agencies to consider adopting open-source environments.[7] Other governments in developing countries are also looking to this alternative, to reduce the outward capital flow to proprietary software developers like Microsoft.

CASE STUDY: LINUX DEVELOPMENT IN CHINA IN BATTLING WITH MICROSOFT[8]

Microsoft Dominance in PC Software Market

In China, Microsoft Windows operating systems and Office products have dominated the desktop market. However, due to rampant piracy, the revenue from sales to individual users is still small. Before the recent price reductions across the world market, the cost of Microsoft Windows and Office licensed software was substantial in comparison to the low prices of PC hardware made in China. Thus, many PCs, especially non-branded PCs sold in the Chinese market, do not have operating software pre-loaded. Most individual users tended to install a pirated copy of Microsoft Window systems. A study released by the U.S. Business Software Alliance and Software & Information Industry Association suggests that 95% of China's newly installed business software in 1998 was pirated,[9] though, as we see below, this figure overlooks the enormous differences between the personal and small-firm markets and larger corporations and public organizations.

China's entry into WTO has made the government tighten regulations and policing in the software market. Since the onset of economic reforms, China has placed on the statute books a series of laws to protect intellectual property that sought to bring China in line with international practices. However, popular support for such protection is largely lacking. Government regulations may have an effect on large organizational users, but little influence on users in the home/SME market. Pirated copies of Microsoft Windows systems can still be readily purchased 'under the counter' from popular markets in China. Like any other pirated version of software, Microsoft Windows XP costs around five to 15 yuan RMB each (approximately $1-$2 or lower), with some regional variations. The cost of pirated software is in effect determined by the cost of copying a CD, having nothing to do with the real market value of the specific product. Compared to this, buying a licensed Microsoft Windows operating system plus Office suite in China used to cost about 3,800 RMB ($400-$500), equivalent to a couple of months' salary for an average office worker (Smith, 2000).

Microsoft initially took a tough stance to combat piracy in China alongside making large investments in China—which is considered by Bill Gates to be the second- or third-largest PC market in the world. In May 1998, Microsoft succeeded in suing two Chinese companies, Beijing Hai Si Da Science and Technology Development Company, and Beijing Min (an investment consulting company), for copyright violations. The victory was perhaps the first foreign-related IP case since the foundation of the PRC. The two Chinese companies paid Microsoft a total compensation of some 790,000 yuan (RMB) (about $95,000 U.S. dollars) and made public apologies in designated publications (People's Daily, 1999).[10]

Microsoft's victories were welcomed by media commentators, governments and MNEs in the developed world, but hardly helped improve its public image in China. Regardless of Microsoft (China)'s claim that the basic goal is to raise the awareness of people about IPR and to educate Chinese consumers to respect the value of software,[11] its market position and high product-pricing policy were viewed critically by Chinese software developers and the wider population. When another court case in which Microsoft was seeking 1.5m yuan (about $180,000 U.S.) ended with Microsoft's defeat, all present in the court cheered. In 1999, the former general manager for Microsoft (China), Ms Wu, came to be seen as a heroine after she resigned, publicly criticizing Microsoft's pricing policy in China. As we see later, Microsoft subsequently embarked on a rather different strategy.

Emergence of Linux in China

Linux began to be recognized in China in 1998, and it gained momentum from 2000 (Marketing News, 2002). Since then, its development has speeded up significantly. Chinese have been involved in Linux software development at both operating system and applications levels. Some have suggested that the free software/open-source philosophy might resonate better with China's socialist culture, appealing effectively to the Chinese as "the peoples' software."

Linux has had a particular attraction to the Chinese government. That Linux source code is open provides some reassurance about security in the face of concerns that Microsoft Windows contained a 'back door.' It also reduces the government's fear of becoming dependent on Microsoft. Using Linux-based systems has significant economic advantages for the nation. The Linux operating system for a desktop PC costs only 98 yuan RMB, compared to more than 3000 yuan RMB for Microsoft OS. The Ministry of Information Industry has therefore established an open source alliance to bolster its support for Linux systems (Marketing News, 2002).

From the outset, the attraction of Linux went far beyond governmental circles, and as in the West, included technical specialists. There was growing enthusiasm amongst software developers.

Xteam took the lead in developing a Chinese Linux operating system. It was set up in February 1998 as the first Chinese Linux R&D group in China, by a group of computer technology enthusiasts with limited resources. In only a year's time, on 1 March 1999, the first Chinese Linux distribution —XteamLinux Chinese version 1.0—was released into the market. In the following four months on the market, it sold 100,000 copies. A year later, on 1 March 2000, the company released XteamLinux Chinese version 3.0, which sold 40,000 copies in less than two weeks. In October the same year, Xteam announced a Linux Server—(XteamServer 3.0), a Linux operating system for servers specially designed for Internet services and users such as Internet data centres, e-commerce companies and SMEs. Subsequently, a range of products, Xteam severMonitor, Xteam LogAnalyzer and Xteam WebMail, all came to the market in that year.[12] XteamLinux 4.0 was also released in July 2001, which has two versions, one for desktop PC and one for server administrators.[13]

At the same time, with the backing of the Chinese government, a state-owned software company, CS&S, and Institute of Software in the Chinese Academy of Sciences started investing in developing a Chinese Linux operating system. Together, these two organizations possessed perhaps the strongest talent pool in software development in China. They both were involved in a decade-long national project, COSIX, developing a Chinese operating system based on UNIX. To speed up the pace of the indigenous operating system development and to meet the world standards, in the late 1990s the government extended the focus also to include the adaptation of available Linux to the Chinese system, despite some scepticism. The government attitude was that to invest might not bring the expected reward, but to not invest will definitely miss the opportunity.[14]

In September 1999, CS&S released COSIX32 Linux V1.0. Later, several versions of the Linux operating system were released into the market, including Linux operating systems for desktop, server, mail server, plug-in network computer, and so forth. Linux 3.1 was the latest version as of the beginning of 2003, and succeeded with other Chinese Linux OS products in winning the Beijing government's bid for desktop operating systems in 2002. Because of CS&S' strong position in China, its Linux development has a firm financial, political and organizational backing.

Compared to Xteam and CS&S, Red Flag was a late comer in the Chinese Linux OS market. After several basic versions of a Chinese Linux operating system were developed by the software research institute of the Chinese Academy of Sciences, the institute decided to establish a commercial company—Red Flag Software Co. Ltd.—to develop this into a commercial product. Later, the Ministry of Information Industry became a key investor. Two months after its birth in August 1999, its first product, Redflag Linux Server version 1.0, became commercially available. In March 2000, with Compaq, it lunched Redflag Linux ISV. Other new releases that year included Redflag Linux Desktop 2.0 and IBM S/390-based Redflag Linux Operating Systems. Redflag Linux embedded solutions including STB, PDA and thin client, and Redflag Linux Server version 2.0 in simplified Chinese, traditional Chinese and English.

Since its birth, Redflag has demonstrated a strong strategy based on establishing alliances with domestic and overseas partners, and focusing on providing a wide range of services. As early as April 2000, half a year after its birth, the company, in close collaboration with Intel and the China Software Association, established the Linux Technical Support Centre. It had a full range of Redflag Linux solutions showcased at the Comdex exhibition. Shortly afterwards, it reached an agreement with Hong Kong-based Sunwah Group to establish training/technical support/sales centres in Southeast Asia. In a very short period of time, Red Flag Software Co. Ltd. has gained a considerable position in the market. Its Chinese Linux OS for desktop machines has prevailed over its competitors, including Microsoft, to win a number of large contracts with PC vendors, regional governments and ministries. In 2002, the company was about to break even and is expecting to be in profit by the end of 2003.[15]

The software research institute of Chinese Academy of Sciences understood very well that a Chinese Linux Office product is necessary for the success of Linux OS in China. It approached Kingsoft, China's flagship office software maker, to convince it to make a Chinese Linux office suite. But Kingsoft failed to respond, on the grounds that this was an unknown future market. Thus, the software research institute of the Chinese Academy of Sciences was forced to develop a product based on the available Open Office. While establishing Red Flag to commercialize the Chinese Linux OS, it also invested in a new company, RedFlag Chinese 2000 Software Co. Ltd., established at the end of 2000. Its Chinese Office products for Linux and for Windows have already sold well in the Chinese market. The price of its products is only 400-500 RMB, one-tenth of Microsoft Office's price in China.[16]

Though Kingsoft did not respond to the software research institute's early call, it has been keeping an eye on Linux development in China. Soon after the Chinese Linux OS market took off, it began to develop a Chinese Office suite for Linux.[17] Before the entry of Microsoft, Kingsoft had been for a long period a dominant player in the Chinese word-processor market with its earliest word processing system, WPS, which was based on DOS. However, its position was severely weakened after the introduction of Microsoft Windows and Office into the Chinese market. Here again, software piracy helped Microsoft to build a user base and establish itself as the *de facto* standard. Very quickly, with little cost, users moved onto Windows and Microsoft Office.

Kingsoft's experience in collaborating with Microsoft also turned out to be complex. In 1994, to work with the Chinese language, Microsoft and Kingsoft reached an agreement to make each other's RTF formats compatible. This enabled Microsoft users to read WPS format and vice versa. However, several years later, when Microsoft had developed its own Chinese version of Office products and became dominant in the Chinese market, it eliminated this bridge with Kingsoft WPS products (eNet, 2002). When Kingsoft lunched its WPS Office 2002, Microsoft also introduced the Chinese version of OFFICE XP. Microsoft claims it will keep consolidating its share of the Chinese office software market (currently over 90%).[18] After this bitter lesson, Kingsoft has determined to fight back to regain its territory. Its Linux project follows two sets of developing plans. One is porting its existing Windows-based product range onto Linux, while the other is to be based on the available Open Office. Both are expected to be released into the market in 2004.[19]

The Role of the Chinese Government

The Chinese government policy of giving domestic products priority in public sector procurement might have been the incentive for Kingsoft to eventually decide to develop the Chinese Office for Linux, given the fact that pirated products account for the overwhelming share of the Chinese software market. Even though Microsoft products dominate the Chinese home/SME desktop market, only a small percentage of products are legally purchased. Most of these legitimate users are from business organizations and central and local governmental organisations. Public administration probably represents the most reliable customers, and its procurement of software in China is estimated at $30 billion U.S. this year. Its significance is increased in that government attitudes toward the choice of operating system might effectively influence the purchasing choices made by business communities. The e-government project is seen as having a pivotal role in shaping the future profile of the software market in China.

According to a report, Liu He, deputy director of the State Informatization Office at the State Council (China's cabinet), announced that China will build unified platforms for their e-government project during the 10th five-year plan (2001-2005) (Business Weekly, 2002). This attracts great attention from the entire IT industry. Many believe that this "unified platform" for the Chinese e-government program is likely to be based on the Linux operating system. However, officials from the State Informatization Office claimed that the Chinese government holds a neutral attitude towards the operating platforms. Some officials in the Ministry of Information Industry endorse the government's relatively "hands-off" attitudes toward software development (compared, for example, to the government policies for telecommunications).[20]

Both Linux-based software developers and Microsoft have been eyeing the e-government project. Chinese Linux-based software developers are not only competing with Microsoft but also with each other for the e-government projects. Many media reports show that these companies, including foreign Linux dealers, have been quite successful in bidding for deals with central and local government (Lei, 2002).

Under the new threat from Linux, on 27 February 2003, Microsoft signed a contract with the Chinese government to allow controlled access to source code and technical information (following similar recent agreements with India, Russia, NATO and UK governments).[21] On Bill Gates' recent visit to Beijing, he met with Chinese President Jiang Zemin, and hinted that China would be privy to all, not just part, of the source code the government wishes to inspect (Gao, 2003).

Dynamics Between Players Involved

Linux-based software developers also team up with PC makers, in particular, the strongest players in the Chinese PC market, to pre-load the Linux operating system. Red Flag has signed contracts with many Chinese Original Equipment Manufacturers, including Founder, Great Wall and TCL, to adopt Linux (Lei, 2002). Great Wall Computer, one of China's biggest PC makers, has already shipped 200,000 desktop computers loaded with the Linux operating system, which looks much like Windows though it cannot yet match all of Microsoft's features (Smith, 2000).

The Linux operating system for servers has gained an even stronger position in the Chinese market. Computer companies, including IBM (China) and Hewlett-Packard (China), sell servers containing RedFlag's version of the OS (Kanellos, 2002). The National Ministry of Science, the Ministry of Statistics and the National Labour Unit have all adopted Linux to run their servers. Linux has been applied in sectors such as finance, telecommunications, transport, education and railway, and so forth (Smith, 2000).

The market share of Linux-based desktop applications in China is still small compared to the server market. To challenge foreign products' dominance, with the backing of Chinese central and local governments, domestic office software companies have been actively involved in forming a united front by adopting common standards. On 12th October, a contract was signed amongst more than 10 Office software developers, including Kingsoft, Jinghua Net, Eastsoft, Chuanzhi, Chinese Information Association, and so forth, to launch "the Alliance for Office Software Standards," which is based on the Guangdong Provincial Alliance for Office Informatization and E-government Software. The Alliance's goal is to standardize domestic Office software and in this way squeeze the market share of Microsoft Office products (Nanfang Dushi Daily, 2002).

Evans Data Corp.'s Chinese Developer Survey released in July 2002[22] suggests that Chinese developers favor Linux and are strongly confident about using Linux for mission-critical applications. A significant number of programmers in China are beginning to develop applications for Linux or are planning to do so in coming years.[23]

The study also suggests that the use of Linux as the primary host OS may jump by 175% next year, although the current usage levels of the Linux operating system is still low. According to Evans Data, the level of confidence in open-source operating systems was also greater in China compared to North America. The survey shows 77% of respondents in China found the environment to be "very supportive" for building mission-critical applications to run on Linux, while in North America, only about 58% of those taking part in the survey said Linux was sturdy enough to be used for mission-critical applications.[24]

The Chinese Linux market is expected to grow rapidly, despite a small initial base, from $4.5 million U.S. in 2001 to $26.8 million U.S. by 2006 (Zhao, 2003). An enormous increase in demand for technical expertise in Linux systems is expected in the next 5 years, with estimates in the order of 1,200,000 recruits expected in China. Many software training centres have been running courses on Linux. Thiz Software has established a Linux open-source software training association, aiming to provide training for up to 100,000 people by setting up 75 management centres and 450 training centres across the country in the next two years.[25]

The shift towards Linux is not without problems—for example, in relation to standardization of Chinese Linux versions. One particular issue concerns the weakness of links and participation by Chinese Linux developers with the community of Linux developers.[26] As well as implying that Chinese developers are not contributing fully to the further development/improvement of 'global' Linux technology, it raises the question of how Chinese Linux developers can keep up with the rapid release of new Linux versions and standards.

One extremely interesting phenomenon is that many powerful MNEs, like IBM, Hewlett Packard and Sun Microsystems, join forces in making significant investments in China to promote open-source software; on the other hand, China has found the software lock-in effect is much greater than anticipated. For example, in many government offices, computers that came preinstalled with Linux OS and Office software have to be equipped additionally with Microsoft Windows OS and Office to be able to exchange documents with outside.[27]

CASE ANALYSIS

Clearly, there are lessons to be learned from the case study in terms of IP protection in China. First, IP rights protection is certainly needed. The case shows that rampant piracy, to some extent, hinders indigenous software development.

In the face of the rapid development and uptake of computing and Internet technology, the PC has increasingly become a household item, and demand for software has been increasing dramatically. In the software industry, most software products tend to wither and die. Products like operating systems and personal productivity tools/Office products such as word processors, spreadsheets, and so forth require a critical mass of users and inter-operability or other network effects to survive and to sustain their position in the market. These products exhibit strong network externalities (Cowan, 1992). In other words, the more users a software product acquires, the more valuable it becomes to each user. These circumstances tend to favor the emergence and dominance of *de facto* standards (Samuelson, 1995).

Proprietary products like Microsoft Windows, developed in the West, have prices that are only acceptable in the developed world. In China, Microsoft products are disproportionately more expensive than equivalent indigenous software, and only a few individuals can afford to buy them. China was not in a position to insist on all citizens using legal software. Only rampant piracy has made Microsoft products so popular (Williams, 2002). This arguably has had a long-term effect of entrenching their standards, as

Microsoft proprietary formats become established as data exchange formats. In addition, users have acquired the skills to use these products, and it becomes difficult for them to switch to different products later on. Therefore, we can argue that Microsoft has benefited from illegal copying in terms of establishing it as the *de facto* standard. Microsoft's large market share in turn presents longer-term obstacles for indigenous software developers.

Equally, given its lack of advanced technology in the early stages of economic development, it can be argued that piracy has greatly helped a developing country like China promote the uptake of computing and Internet.

In order to meet domestic demand and promote indigenous technological development in the software industry, both "affordable product prices" and "accessible source code" are crucial.

The price of products is a major issue for DCs, considering the goal of promoting the widest access to computing and the Internet, and the economic and social benefits presumed to flow from this (and conversely, the need to avoid excessive IT costs widening the 'digital divide'). With the enormous population/potential market, relying completely on expensive exogenous proprietary software products hardly seems an attractive solution for a country like China. For example, doubling the uptake of PCs in China from the 2000 figure of 2.2 for every 100 people would require 28 million PCs. If all of these used the Microsoft software package, the total cost at $300 each would involve more than $8 billion, mainly going to Microsoft in license fees.[28] It is certainly worthwhile for China to make substantial investments to develop indigenous proprietary products or non-proprietary products like Linux rather than contribute to the profits of a foreign company.

"Accessible source code" is equally crucial. Most advanced software was developed in developed countries. The functions and standard technical features and application models of those products are mainly designed for such markets and based on the presumptions about customer requirements/demands in advanced economies. The application of such software in developing countries with very different contexts often requires modifications. Proprietary software, the source code of which is kept secret, leaves little scope for technological participation by developing countries. Given the nature of fast-changing software products, the requirement of frequent upgrades makes DCs' dependence on these technologies even worse. Conversely, local developers have advantages in developing products for domestic markets, in terms of their better understanding of local needs.

What China needs, however, is not a simple uniform adoption of the global IPR regime, but a more selective implementation—that may need to change over time—depending upon particular circumstances. Simply emphasizing a temporary "weak IP regime" or "strict IP regime" is certainly not helpful in the software sector. The Chinese case shows that software piracy on the one hand allowed many Chinese people to use expensive foreign developed products. On the other hand, the lax regime helped foreign products like Microsoft Windows and Office to achieve a dominant position in the Chinese market. We saw how in the advanced economies IP regimes developed 'organically' in this way, depending on the balance between national and external knowledge bases, changing over time: increasing IP rights as the base of knowledge requiring protection grew. China needs a similarly pragmatic IP strategy that recognizes its different needs at different stages and under different circumstances.

In terms of balancing different needs and interests for developing countries, the current international IP regime seems to be far from appropriate. These problems cannot be solved by only focusing on IP issues. Many other measures need to be adopted and strategies explored to compensate the inappropriate IP regime.

The case study highlights new perspectives around the resort to non-proprietary software like Linux. The Linux community of software developers across the world, with its culture of cooperation and openness, provides an extremely valuable opportunity for software developers in developing countries to learn, to study and to communicate with their counterparts. Overall, when we consider a range of important issues like technological innovation and creation, and knowledge sharing in the communities and technology localization, emphasizing consumers' interests and choices, and so forth, Linux indeed opens up important opportunities for developing countries to explore.

However, the case study also reveals a potential problem in the commercialization of the Linux products—in particular, Linux operating systems for desktop machines. All Chinese Linux OSs are offered at a rather low price, currently only about 98 RMB per copy, and can also be downloaded free. To form a sustainable business model for suppliers, a critical mass of users (and paying users) is needed. However, in a country like China, in which Microsoft Windows has already established a dominant position, (partly through piracy), it is extremely difficult for individual companies to challenge this situation and build an alternative base without other measures.

In this context, Chinese government support is perhaps crucial. At the same time, Chinese developers have to cooperate and align themselves with global players in the software industry as well as the PC industry.

Since China is operating within the world political and economic order, China has to comply with international agreements (e.g., the WTO, TRIPS, etc.). China has to work with other developing countries to engage proactively in further improvement of the current international IP regime and address the interests of developing countries.

We can see there is an ethical issue: How far should an IP system go to protect the private interests? Views may vary between different cultures and traditions. The case study reveals the unsupportive attitudes of Chinese people towards software IP protection. The phenomenon might be derived from Chinese socialist thinking as well as a cultural tradition that values public over private and respects learning highly. Software for computing and accessing the Internet are commonly regarded as tools for learning and knowledge sharing—the public interest. It is difficult for them to accept that high prices ought to be secured to protect the interest of a private corporation like Microsoft, whose wealth far exceeds that of many economies in the third world.

The case study also confirms our view that the improvement of IP protection is not just a question of getting the legislature to install all the relevant laws and mechanisms to prosecute IP violations. It rests in addition on processes of social development, depending on peoples' awareness and understanding of the importance of IP protection. It will not be successful without people's consent and commitment to improving IP protection. The more people—software users—have been motivated to curb software piracy to help the development of the domestic software industry, the more chance that the problem of software piracy would wane away. Encouraging users' participation in decision-making and promoting dialogue between software developers and users will also help establish an effective software protection system, along with a system that strikes a balance between protecting the interests of software developers and users.

We need, finally, to look at competition and its interaction with IP regimes. Because of the existence of Linux and its growing support in China, Microsoft has been forced to move from its early position narrowly focused on enforcement of IPR towards a more mutual position, culminating in signing the GSP agreement with the Chinese government. A stronger Linux base in China will help maintain competition and might well help drive down Microsoft's high prices and encourage greater disclosure of proprietary knowledge. It changes the conditions in ways that may allow more mutual and beneficial outcomes to emerge. Under these circumstances, the presence of Microsoft in China can be more beneficial for the Chinese software industry. Microsoft's investment and its operations in China help foster both technical and managerial manpower and knowledge about the global state of the art, which are helpful for the Chinese software industry as a whole. Chinese companies (e.g., Kingsoft) are maturing by learning from dealing with successful corporations like Microsoft.

CONCLUSION

Under strong pressure from developed countries to improve IP protection, a developing country like China has to have a clear vision about how best to do this and to consider carefully the implications for the nation's development before installing such an IP regime. Software IP protection is an important case, and one in which there have been many concerns—even in developed countries—regarding whether the current software IP protection promotes or suppresses competition and technological innovation, and whether it protects the interests of the public, individual and commercial consumers and IT producers.

The case study demonstrates that curbing software piracy is necessary for China. To date, rampant piracy has allowed private citizens to benefit from advanced products (e.g., Microsoft) that would not otherwise be affordable. However, it makes the price difference between cheap domestic and expensive foreign products irrelevant, and paradoxically results in domestic products, which are often later comers than foreign ones, having lower market share. Policy goals of curbing software piracy and improving IP protection might need to be integrated with broader national strategies, including economic and technological development goals (for example, measures to create more rapidly a critical mass of users for domestic products and/or non-proprietary software) and even social goals (for example, to reduce IT prices and minimize the risk of creating a new digital divide).

Given the economic and social benefits of allowing as many people as possible to access the Internet and computing, national governments will wish to keep the price of widely required software as low as possible. We can see in the software industry that the prices of proprietary products do not necessary reflect the R&D and manufacturing costs. Network externalities and other factors that favor lock-in and monopolistic trends mean that the prices of these products can be manipulated by the owner, who can adopt different strategies to maximize income streams/profits (for example, by not providing 'longitudinal compatibility' between successive upgrades, thereby forcing users to buy the latest version [Williams, 1999]). Thus, it is important while improving software IP protection to have a national strategy that can strike a balance between protecting the software owners' interests and the interests of the public. For instance, this could be achieved by a strategy of pursuing a large market though low and locally affordable prices—especially for mass-consumer products, such as operating systems, word processor, e-mail and Internet communication tools.

While reinforcing software IP protection, it might also be necessary to have a national strategy of promoting competition as well as facilitating collaboration; for example, in agreeing on interoperability standards to help build the 'critical mass' of users and mobilize network effects in favor of indigenous software development. The establishment of *de facto* standards and platforms also creates markets for complementary products. There may also be technical and economic benefits from having technological diversity (e.g., increasing competition, reducing monopoly/oligopoly effects, reducing lock-in and providing more flexibility in responding to future challenges).

As discussed above, installing an IP protection mechanism in a country is not only a matter of having written laws and litigation institutions. It also involves a gradual learning process within IP-related communities, as well as wider society. The establishment of an IP regime and/or making it work effectively is a social project. Merely adopting a top-down dictatorial approach and heavy-handed measures to crack down on piracy is unlikely to be successful. Instead, users (and distributors, etc.) need to understand the importance of IP protection for the society and the importance of their role in actively participating in establishing an effective IP regime, which would benefit the nation and the people. This will be better achieved in a consensual manner than by imposing a punitive regime.

As a developing country, China has to proactively address the problems of the current international IP regime. At the same time, under high pressure to comply with the current international IP regime from developed countries, China must also adopt other measures and strategies to protect domestic industries and to curb the digital divide from its early stage.

Our analysis suggests that Linux can be a driving force for breaking the Microsoft monopoly and moreover, provide an opportunity for China to develop its indigenous software industry and avoid the many shortcomings of proprietary software like Microsoft Windows and Office products. The major challenges facing China are the dominance of Microsoft products in the Chinese market backed up by Microsoft's sophisticated finance, R&D and development strategies. To win the battle, a wider collaboration and alliances amongst domestic software developers is needed, together with stronger support from domestic users and public and private procurement. In these circumstances, Chinese government backing is particularly crucial. For example, the Chinese government's decisions on the choice of operating system for their e-government project may well play an extremely critical role in deciding the final victory of the battle.

The battle between Linux and Microsoft can be seen as of crucial national importance in the search to develop an IP regime that will benefit a developing country like China by balancing the protection of the interests of producers and users, and of private and the public; to create an IP regime which will ultimately encourage innovation and global connectivity as well as localization. This national project requires detailed strategies, a good organization and fine-tuning for the development of the battle in every stage.

There may be lessons for other DCs. It could be argued that the case of China is somewhat exceptional; certainly its sheer size gives China a number of advantages—for example, in terms of bargaining power with foreign MNEs, since China's domestic market will potentially become such a large proportion of the global market. Smaller economies may be in a weaker position in terms of influencing global corporations and technological trends. However, with the globalization of technology (in a context of escalating development costs), strategies to promote national technology development have given way to strategies to promote national leadership in global technology development (see Molina, 1994). This point also applies to China—which must integrate its technology strategies with global development—though the parameters are rather different. The case of open source and Linux illustrates these dynamics well.

For example, India, slightly smaller in population but with its successful entry into the global software market, exhibits some similarities with the Chinese case, in terms of developing a significant Linux community and in its strategic negotiations with Microsoft (Noronha, 2002). However, we can also find a range of initiatives across the smaller South Asian economies. For example, in Malaysia, government policy since 2002 has given powerful support for open-source software (Noronha, 2002; Smith, 2003).[29] The other feature is the promotion of Asian collaboration on open-source software; for example, through the International Open Source Network being developed by the Asia-Pacific Development Information Programme (APDIP, 2003).[30]

Similar developments to promote open source have been taking place across the globe, in Europe as well as DCs. Microsoft's concern about government policies requiring open-source solutions underpins its support for the Initiative for Software Choice, launched in 2002 by The Computer Technology Industry Association (in which Microsoft is the biggest software backer). The Initiative for Software Choice campaigns against the adoption of public policies requiring Linux and other open-source software on the grounds that this might impair choice and quality of software (Broersma, 2002).

These developments are still ongoing. Their outcomes may be of great significance. The stance and strategy of the Chinese government in complying with international IP protection regimes and developing its own national IP regime are crucial for China and may also provide an exemplar and a further stimulus for improving the IP system in the world. The Chinese government's software industry strategy (and especially its decision regarding which systems will be adopted as the platform for e-government systems) might tip the balance in favor of Linux in China and could have considerable impact on the world software market. And, moves towards non-proprietary software in the developed economies may, conversely, open up prospects for DCs.

REFERENCES

APDIP. (2003). IOSN International Open Source Network Asia-Pacific Development Information Programme. *Proceedings of the CICC/NECTEC Asian OpenSource Symposium*. Retrieved May 2, 2003, from www.asiaosc.org

Awang, H. (2002). Open Source Software as a catalyst in narrowing digital divide. *Jaring Internet Magazine,* August. Retrieved May 2, 2003, from *www.magazine.jaring.my/2002/august/index2.html?content=stay7.html*

Broersma, M. (2002). Group campaigns against open source. *CNETAsia*, August 14, 2002. Retrieved May 2, 2003, from *http://asia.cnet.com/newstech/systems/0,39001153,39072261,00.htm*

Buckley Jr., W.F. (1999). Whose Side Are We On?. *National Review*, June, Vol. 51, 23, 74.

Business Weekly. (10/09/2002). *Windows or Linux? Or Both?* Retrieved from www.chinadaily.com.cn/bw/2002-09-03/85731.html

Campbell-Kelly, M. (2003). *From Airline Reservations to Sonic the Hedgehog: A History of the Software Industry*, MIT Press.

CIPR. (2002). *Integrating Intellectual Property Rights and Development Policy—Report of the Commission on Intellectual Property Rights*, London: Commission on Intellectual Property Rights. Retrieved from www.iprcommission.org.

Colman, D., & Nixson, F. (1994). *Economics of Change in Less Developed Countries*, 3rd edition. New York and London: Harvester Wheatsheaf.

Cowan, R. (1992). High technology and the economics of standardization. In Dierkes, Meinolf and Hoffmann, Ute (Eds.), *New Technology at the Outset: Social Forces in the Shaping of Technological Innovations* (279-300), Frankfurt/NY: Campus/Westview.

Drahos, P., & Braithwaite, J. (2002). *Information Feudalism: Who Owns the Knowledge Economy?*. Earthscan.

Dutt, A. K., Kim, K.S., & Singh, A. (1994). *The State, Markets and Development—beyond the Neoclassical Dichotomy* (pp. 3-21). Cheltenham: Edward Elgar.

Elkin, T. (2000). Monopoly ruling won't destroy Microsoft's image, experts say. *Advertising Age*, *10*(71), 15, 4.

eNet. (06/17/2002). How could Kingsoft challenge the domination of Microsoft?. Retrieved June 17, 2002, from *www.cfan.net.cn/news/soft-news/0617/1.htm*

Financial Times. (02/05/2003). A meeting of Microsoft's minds. *Research and Development*, 11.

Friedman, A. (1989). *Computer Systems Development: History, Organisation and Implementation.* Wiley, Chichester.

Gao, K. (2003). China to view Windows code. *News.Com, Tech News First.* Retrieved February 28, 2003, from *http://news.com.com/2100-1007-990526.html*

Isinolaw Research Centre. (2001). Intellectual Property. Retrieved from www.isinolaw.com/jsp/ip/main.jsp?LangID=0

Kanellos, M. (06/05/2002,). Microsoft gets diplomatic in China. *News.com.* Retrieved from http://news.com.com/2102-1001-932927.html

Khan, B.Z. (2002). Intellectual Property and Economic Development: Lessons from American and European History. Commission on Intellectual Property Rights. *Proceedings of the Conference on How IPRs could Work Better for Developing Countries and Poor People.*

Kumar, N. (2002). Intellectual Property Rights, Technology and Economic Development: Experiences of Asian Countries. *RIS Discussion Paper*, 25-2002.

Lee, D.R., & McKenzie, R.B. (2001). Technology, Competition, and Antitrust Enforcement. *USA Today Magazine*, *3*(129), 2670, 26.

Lei, J. (10/11/2002). ChinaByte Report: E-government project—an opportunity for Linux. Retrieved from http://it.sohu.com/88/40/article17114088.shtml.

Lewis, O. (2001). China: Protecting the Knowledge-based Economy and Intellectual Property Rights. Country Panel IV. *Proceedings of the Harvard Asia Business Conference.*Retrieved from www.fas.harvard.edu/~asiactr/haq/200102/0102a001cp5.htm.

Marketing News. (11/22/2002). The emergence of Linux in the fore of opportunities and challenges. Retrieved from http://it.sohu.com/14/80/article204508014.shtml

Molina, A. (1994). Understanding the emergence of a large-scale European initiative in technology. *Science and Public Policy*, *21*(1), 31-41.

Nanfang Dushi Daily. (10/30/2002). The battle amongst Office software.

Noronha, F. (2002). Openness and sharing make Linux attractive. *Tribune,* 3 June 2003. Retrieved May 2, 2003, from www.school.net.th/linux/news/tribuneindia/

Office of Industry Liaison. (2000) Intellectual Property. The University of Western Ontario. Retrieved from www.uwo.ca/industry/intellectual/#IP

People's Daily. (6/4/2001). Kingsoft, Microsoft Contend for Supremacy in China. Retrieved from http://english.peopledaily.com.cn/200107/04/eng20010704_74170.html

People's Daily. (12/18/1999). Microsoft Loses Suit Against Beijing Company. Retrieved from http://english.peopledaily.com.cn/19991218T105.html

Raymond, E. (3/2/1998). The Cathedral and the Bazaar. *Peer Review Journal on Internet*, 3(3). Retrieved from www.firstmonday.dk/issues/issue3_3/raymond/

Rothstein, E. (2001). Wronging Microsoft. *Commentary, 9*(112), 2, 46.

Samuelson, P. (1995). Software Compatibility and the Law. *Communications of the ACM, 38*(8), 15-22.

Smith, C.S. (06/07/2000). Fearing Control by Microsoft, China Backs the Linux System. *The New York Times*. Retrieved from www.hartford-hwp.com/archieves/55/424.html

Smith, I.W. (2003). The Status of Open Source and Free Software in Malaysia, especially Government/Policy. *Proceedings of the CICC/NECTEC Asian OpenSource Symposium*.Retrieved May 2, 2003, from www.asiaosc.org

Smith, I.W. (2003a). Asia Opensource Symposium, Thailand Day 2. *Proceedings of the Asian OpenSource Symposium*. Retrieved May 2, 2003, from www.asiaosc.org

Stallman, R. (5/16/1999). Saving Europe from Software Patents. 20 :11 UTC, http://features.linuxtoday.com/news_story.php3?ltsn=1999-05-16-003-05-NW-LF.

Stallman, R. (5/17/1999a). 15 Years of Free Software, 17 :23 UTC, http://features.linuxtoday.com/news_story.php3?ltsn=1999-03-17-003-10-NW-LF.

Steinmueller, W.E (2001). ICT and the possibilities for leapfrogging by developing countries. *International Labour Review, 140*(2), 193-210.

Taylor, P. (1999). *Hackers: Crime in the Digital Sublime*. London and NY: Routledge.

UNCTAD. (2002). E-Commerce and Development Report 2002. *Proceedings of the United Nations Conference on Trade and Development (UNCTAD).*

Williams, R. (1999). ICT standards setting from an innovation studies perspective. Proceedings of the SIIT '99 1st IEEE Conference on Standardisation and Innovation in Information Technology. In Jakobs and Williams (Eds) *Standardisation and Innovation in Information Technology* (pp. 251-262). Piscataway: IEEE.

Williams, S. (9/26/2002). Profits from piracy. *Salon*. Retrieved from http://salon.com/tech/feature/2002/09/26/piracy_unlimited/print.html

Zhao, C. (2003). OSS Movement of Redflag. *Proceedings of the Asian Open Source Symposium.* Retrieved May 2, 2003, from www.asiaosc.org

ENDNOTES

1 Silicon Valley Software Industry Coalition, letter to United States Department of Justice, Antitrust division, August 23, 1996, www.softwareindustry.org/issues/docs-htm/svsic-doj.html.

2 "A Summary report of the conference on How IPRs could work better for developing countries and poor people," *Sustainable Development*, Vol. 70, No. 1, Thursday 28 February 2002, www.iprcommission.org/papers/pdfs/conferences/conference_minutes.pdf.

3 According to Stallman (1999a), it should be called GNU/Linux, as this is the operating system bundled with the Linux kernel.

4 Software comes in two forms, one readable only by computers and the other readable by people. The form that the computer can read is what the computer runs. This form is called a binary or executable. The form that a human can read is called source code. It is what a human programmer creates, and is translated by another computer program into the binary or executable form. Modern software is so complex that it is not possible to determine how a programme functions simply by examining executable code and working backwards. In proprietary software, you may buy the executable code, but you do not gain access to source code. You cannot, therefore, scrutinise the operation of the programme. This prevents the purchaser from amending the software, and restricts scrutiny over other aspects of the software relating, for example, to dependability and security aspects.

5 This can be traced back to the computer enthusiast 'Hacker' sub-culture that arose from the earliest days of commercial computing, notably in universities offering computing courses, inspired by utopian visions of the benefits that might flow from wide access to computer systems and information (Taylor, 1999).

6 There have been some complaints about the high cost of supporting services.

7 Financial Times 22/01/2003, p17.

8 Except for the referred sources, much of the material comes from the interviews carried out by the author with these Chinese companies and organisations, including Red Flag, CS&S, RedFlag Chinese 2000, Kingsoft and the Institute of Software of CAS, in April 2003.

9 Zdnet News, online article, "A year Ago: Microsoft sues Chinese firm for software piracy," 20/11/2000, http://news.zdnet.co.uk/story/0,,t269-s2082655,00.html.

10 Online source, "Microsoft Loses Suit Against Beijing Company," http://english.peopledaily.com.cn/19991218T105.html

11 Tian Fengchang, corporate attorney for Microsoft in Beijing, ibid.

12 http://www.Xteamlinux.com.cn.milestone.php.

13 http://www.Xteamlinux.com.cn.milestone.php

14 Author's interview with a manager in CS&S on 25 March 2003.

15 Author's interview with Liu Bo, CEO of RedFlag, on 23 March 2003.

16 Author's interview with Professor Sun, one of the key founders for Redflag, in the Institute of Software, Academia Sinica (Chinese Academy of Sciences) on 25 March 2003.

17 Telephone interview with a technical manager in Kingsoft Beijing office on 28 March 2003.

18 Online article, "Kingsoft, Microsoft Contend for Supremacy in China," People's Daily, 4/6/2001, http://english.peopledaily.com.cn/200107/04/eng20010704_74170.html.

19 Telephone interview with a technical manager in Kingsoft Beijing office on 28 March 2003.

20 Author's interview with an official in Ministry of Information Industry (21 October 2003).

21 *Sina S&T*, 28.02.2003, online news, http://tech.sina.com.cn/s/n/2003-02-28/1420168775.shtml

22 The company questioned 700 Chinese software developers during May 2002 (Paul Hales, "Chinese Developers like Linux—the people's software," 18/06/2002; www.theinquirer.net/?article=4501).

23 Evans Data found 27% of the developers polled in China said they were currently (2002) writing Linux applications and 66% said they would do so in 2003. ("Chinese Developer is ready for Linux application development," 15/07/2002, www.linuxmax.net/print.php?ArticleID=813.)

24 ibid.

25 www.thizlinux.com.cn/edu_01.htm

26 Author's interviews with RedFlag, Kingsoft, CS&S and the Institute of Software of CASS.

27 *Sina S&T*, 28.02.2003, online news, http://tech.sina.com.cn/s/n/2003-02-28/1420168775.shtml

28 Though we should consider that these IT acquisition costs may be outweighed by the economic benefits of more efficient business (UNCTAD 2002).

29 In 2002, following representations from PIKOM, the industry association (Association of Computer and Multimedia Industry), the Malaysian Minister for Energy Communications and Multimedia announced

that the government will begin to implement OSS within its departments and in schools. Unofficial procurement policy supports LINUX and other initiatives, including Komnas 2020, a low-cost LINUX-based P.C. (Smith, 2003; Awang, 2002).

30 The International Open Source Network is a United Nations Development Programme centre of excellence, established by the Asia-Pacific Development Information Programme, and bringing together government and international bodies practitioners and other players from 24 Asia Pacific countries, around a vision that "Developing countries in the Asia-Pacific Region can achieve rapid and sustained economic and social development by using affordable yet effective Open Source ICT solutions for bridging the digital divide" (APDIP, 2003). Networking is important, not just for knowledge sharing, but because of the need to develop standards and enhance the LINUX core as well as developing national versions of LINUX (Smith, 2003a).

This work was previously published in The Journal of IT Standards and Standardization Research, 3(1), edited by K. Jakobs, pp. 21-43, copyright 2005 by IGI Publishing, formerly known as Idea Group Publishing (an imprint of IGI Global).

Chapter XV
COPRAS: Encouraging ICT Research Projects to Produce More Tangible Standardization Results

Bart Brusse
ConTeSt Consultancy, The Netherlands

ABSTRACT

Between early 2004 and 2007, the cooperation platform for research and standards (COPRAS) deployed a series of activities to improve the interfacing process between ICT research and standardization. This included the conclusion of standardization action plans with a series of projects in the EU-funded Sixth Framework Programme and the development of a set of generic standardization guidelines supporting future ICT research projects in their interfacing with standardization. COPRAS's results show that cooperation and cross-fertilization between research and standardization is perceived as increasingly important by projects, and also demonstrate how direct and indirect support mechanisms are able to increase the amount of tangible results ICT research will be able to contribute to ongoing standards work. However, results also show that structurally improving the research and standards interfacing process will not be possible unless all parties to the process, including the research and standards communities and the administrators of the research programmes, take additional, and continuous, action.

INTRODUCTION

ICT research that is carried out under European framework programmes is often closely connected to standardization. Projects addressing technical or scientific issues generally produce results that can be used to develop a new standard, to improve an existing one, or to anticipate a future standard. Moreover, experience teaches that even projects that do not primarily set out to develop standards may contain elements supporting ongoing or new standardization processes.

The ICT standards world, however, is a dynamic environment with several hundred standards bodies, industry consortia, trade organizations, and others operating in the same arena on a national, regional, or global level. Consequently, this can turn the process of finding the right standards organization(s) into a complicated and time-consuming activity for ICT research projects and thus can block cross-fertilization between ICT research and standardization. This is because the standards community will not receive contributions at the earliest possible point in time, and research projects may miss out on the latest developments and state of the art in standardization. As a consequence, windows for standardization opportunities will be missed.

These are the issues that were addressed by the cooperation platform for research and standards (COPRAS), a project initiated by the European Committee for Standardization (CEN), the European Committee for Electrotechnical Standardization (CENELEC), and the European Telecommunications Standards Institute (ETSI), together with the World Wide Web Consortium (W3C) and The Open Group. Between early 2004 and early 2007, COPRAS supported a large variety of ICT research projects in the Sixth European Framework Programme in their interfacing with standards organizations, and, in doing so, also managed to generate a set of guidelines and tools that will assist future projects in their efforts in passing output though standards processes. However, as experience in COPRAS also shows, its achievements can only be seen as a first step toward improving research and standards interfacing in general that can only be effective in a framework of measures.

This chapter will provide an overview of the objectives COPRAS set out with and the methodology it deployed to achieve its targets. Subsequently, the work that was carried out during the 3 years in which the project was active will be described and the conclusions emerging from it will be discussed and translated into recommendations for future work in this area.

OBJECTIVES AND METHODOLOGIES

The two core objectives that COPRAS set out with at the start of its activities can be described as follows.

- To provide the ICT projects it directly addressed with individual support, allowing them to establish cooperation with standards organizations relevant to their work
- To develop a set of tools and guidelines that could provide future ICT research projects with similar levels of support, thus structurally improving the research and standards interfacing process

When looking at the first of these two objectives, it was COPRAS's aim to develop so-called standardization action plans (SAPs) for those ICT research projects in the first calls of the Sixth Framework Programme, which, after a close analysis of their planned activities and results, could be expected to benefit considerably from receiving direct support in their standards-related activities. The aim of these plans was to specify and structure the steps a project should take to arrange interfacing with standards organizations, establishing their prime target in the quickest and most effective way so that the process of building a constituency among the relevant participants to these targeted standardization processes, necessary for the adoption of their contributions, could start as early as possible. As there were several hundred ICT research projects in the first calls of the Sixth Framework Programme, the target was to develop these plans for at least 8% of them.

In order to manage the process of selecting the projects, four methodological steps were defined that could be deployed in a cyclical way.

1. Gathering of information from all projects in the initial calls of the Sixth Framework Programme in order to determine the standardization potential of their output
2. Analysis of the information gathered and, where applicable, clustering this information around specific themes or areas
3. Selection of projects that are expected to benefit most from having standardization action plans developed for them
4. Development of standardization action plans followed by the approval and execution of the plans by the projects

One of the ultimate goals for COPRAS was to demonstrate that improving interfacing between research projects and standards organizations that would be triggered through the execution of the standardization action plans would actually lead to an increased number of tangible contributions to (ongoing) standards processes, and consequently increase the qualitative and quantitative output of standards organizations as well. The target was to generate a minimum of six tangible contributions to ongoing standardization work; this could concern technical specifications, but, for example, could also be the establishment of a new constituency or contributions to increase the deployment of an existing standard.

As a second goal, COPRAS aimed to use the experience it would build up by conducting the process of improving research and standards interfacing for a group of selected projects as the basis for a set of generic standardization guidelines that would assist projects in their research and standards interfacing processes, also beyond COPRAS's own life span. In order to optimize these guidelines, a four-step methodological process was adopted.

1. Preparation of a first document version of a set of generic standardization guidelines after the first standardization action plans had been developed, and the distribution of this document to ICT research projects in the last calls of the Sixth Framework Programme
2. Gathering of feedback from projects that received the standardization guidelines with respect to their appreciation and usage of the document
3. Development of an interactive platform version of the standardization guidelines including more tools and allowing differentiation between different types and levels of research and standards interfacing information and/or guidelines
4. Preparation of a second release of the standardization guidelines as well as an upgrade of the interactive platform following the feedback received

Work Performed and Results Achieved

At the earliest possible moment, COPRAS launched the information gathering process addressing projects in the first calls of the Sixth Framework Programme, inquiring into their standards-related activities, plans, and expected results. For that purpose, it sent out an information package on its objectives as well as on the benefits the cooperation with COPRAS could bring, and invited project members to fill out a questionnaire on their standardization intentions and requirements. The response rate of more than 50% that this questionnaire generated not only provided COPRAS with a massive amount of information on the standardization requirements of the projects, but also indicated that research and standards interfacing is an issue for many projects in ICT research programmes.

In the subsequent analysis of the information that was received, it showed that many

Table 1. COPRAS's results against targets in Steps 1 to 4

Call	Number of Projects	Step 1				Step 2-3		Step 4	
		Addressed		Responding		Selected & Invited		SAP Development	
		Target	Result	Target	Result	Target	Result	Target	Result
1	176	176	164	> 70	92	>14<18	40	>14<18	16
2	111	111	107	> 44	55	>9<11	41	>9<11	26
1 & 2	287	287	271	> 115	147	>23<29	81	>23<29	38

more projects than originally expected could be identified as being likely to benefit from support from COPRAS. A closer look at many of them was therefore necessary in order to prevent an amount of projects that COPRAS would not be able to handle. Meetings were therefore arranged with several of the projects to obtain a more precise understanding of the standards issues they planned to address, and all of them were invited to participate in kick-off meetings where their standardization objectives were discussed with representatives from standards organizations relevant to their projects.

The kick-off meetings were attended by some 40 different ICT research projects as well as by representatives from about 15 different standardization working groups, and proved to be a very effective instrument, not only for creating a clear perspective on ICT projects' standardization potential, but also for establishing contacts between relevant parties at an early point, thus allowing preparatory processes, heading toward projects' contributions to standards organizations, to start. The kick-off meetings marked the start of the development of standardization action plans for 38 projects across most of the ICT research areas covered by the Sixth Framework Programme. Some of these were individual plans, meaning that the plans were tailored to the standardization objectives of a single project, while others were clustered, meaning that the plans described the efforts of several projects toward one or more joint standardization targets. Taking into account this distinction, COPRAS developed in total 18 standardization action plans addressing 14 different ICT research areas and a variety of standards organizations such as CEN, ETSI, W3C, IP multimedia subsystem (IMS), IMS forum, Internet engineering task force (IETF), Java community process (JCP), home gateway initiative (HGI), and many others.

When looking at the overall result of the methodology deployed, it can be concluded that COPRAS was quite successful in achieving the results it anticipated, and even managed to outperform the targets it set itself. This is documented in more detail in Table 1.

The execution of the SAPs, describing projects' steps toward standardization results, and the support COPRAS would provide in this process was started in early 2005 when the first plans were completed and approved by the projects. This process was closely monitored and, where necessary, adjustments were made and additional assistance was provided. Although some projects started at a later point in time with the execution of their plans and thus have not been able to conclude their standards work yet, a considerable number of projects have already completed their plans successfully.

The results that this last step in the COPRAS methodological process generated can best be assessed by evaluating the number of tangible contributions to (ongoing) standards work it actually generated vs. the target of six that was set at the beginning of COPRAS's activities. As the overview in Table 2 shows, projects managed to generate impact in standardization processes across at least 10 different areas—already considerably more than originally expected, even though

Table 2. Impact of standardization action plans executed by ICT research projects

Project	Standardization Impact
Embedded Systems Cluster	Creation of a new working group within the Java Community Process (JCP) that adopted the HIJA project results as the basis for a new safety-critical standard for the Java programming language
SIMILAR	Promotion of UsiXML as a new standard in W3C
GRID Cluster	Establishment of a new technical committee in ETSI working toward a first set of specifications for new GRID standards
E-Learning Cluster	Dramatic increase of the number of units of learning produced using the IMS learning design standard specification
TALK	Creation of a constituency in W3C around the advanced research technologies developed within the project
POLYMNIA	Submissions to the W3C Semantic Web Deployment Working Group
TEAHA	Submission of several UPnP contributions to the HGI
CWE Cluster	Formalising the process for establishing a common architecture as a new industry reference for use in building collaborative working tools and applications
EUAIN	Creation of CEN/ISSS Workshop on Accessible Document Processing
MediaNet	Contribution of a reference architecture to IETF; contributions to the DSL (Digital Subscriber Line) Forum as well as to ETSI TISPAN on Video over IP (Internet Protocol)

several research projects are still in the process of working toward standardization results. It is therefore not unlikely that the standardization action plans may generate twice as many tangible results as originally targeted.

The fact that COPRAS from the start generated considerably higher response rates and better results than it initially targeted led to the conclusion that the need to improve research and standards interfacing in ICT was much higher than expected, at least on the side of ICT research projects. This raised the question whether confirmation of this assumption could be found on the side of the standards organizations as well, or, in other terms, it led to the question to what extent the main current focus areas for ICT standards organizations were actually addressed by research programmes.

Therefore it was decided to complete the activities carried out toward the ICT research projects in the first calls of the Sixth Framework Programme by means of a reverse mapping analysis. Contrary to defining standards organizations for projects to interface with, the reverse mapping process held the main standardization topics of 11 different organizations in the ICT Standard Board (the coordinating platform for ICT standardization in Europe) against the areas the projects addressed to determine to what extent the research programme was addressing today's main standardization issues. The results of this analysis, which can be found in Table 3, indeed show a remarkable overlap, again stressing the necessity to streamline research and standards interfacing, specifically in ICT.

In order to achieve the second of its main objectives, the development of a set of generic standardization guidelines and tools supporting research and standards interfacing in ICT, COPRAS spent considerable effort using, transforming, and translating the information and knowledge it generated working with research projects into deliverables addressing the more generic aspects of research and standards interfacing. As a first step, this resulted in a set of generic guidelines for projects interfacing with standards organizations, demonstrating the benefits of standardization to research projects, and helping them to determine whether or not an interface with standardization should be pursued. The document pointed out the most important milestones within a project life span where standardization should be considered

Table 3. Reverse mapping of current main standards areas against research projects' focus

Standards Organization	Number of Main Standardization Areas	Number of Areas Covered by at Least One Research Project	Overlap (%)
CEN/ISSS	13	10	76.9
CENELEC	10	5	50.0
ETSI	21	14	66.7
DVB	8	16	50.0
Ecma International	9	2	22.2
ERTICO	1	1	100.0
OASIS	19	19	100.0
OMG	7	7	100.0
RosettaNET	7	7	100.0
W3C	9	9	100.0
The Open Group	22	21	95.5
Total	**126**	**111**	**88.1**

and, for example, stressed the importance of interfacing with standards organizations at the beginning of a project's activities.

Upon their completion, the guidelines were distributed to ICT research projects that were on the verge of launching their activities with a dual purpose.

- To provide them with support in building the interface to standardization into their initial work plan
- To generate feedback from these projects on the appreciation and usage of the guidelines in order to be able to improve them

In order to be able to measure the effect of the standardization guidelines, it was decided to distribute the material to projects in two of the three last calls of the Sixth Framework Programme, one of which included projects that were about to start their activities (and hence could not use the guidelines to adjust their work plan and allocation of resources anymore), while the other included project consortia that were still putting together their proposals (and therefore were able to use the guidelines to adjust their work plan and allocation of resources).

To measure the impact of the guidelines, a feedback process, similar to the one used to gather information from projects in earlier calls of the Sixth Framework Programme, was launched. Taking into account the high feedback rate of 40%, this process not only confirmed that the vast majority of ICT research projects does foresee the need to interface with standards organizations at a certain point during its life span, but also pointed out that generic support mechanisms, such as the COPRAS standardization guidelines, can indeed have an impact, as demonstrated in Figure 1.

As shown, the percentage of projects that did not allocate any resources is relatively consistent across Calls 1, 2, and 4, whereas it suddenly decreases for Call 5. Taking into account that projects in this last group were the first that had the opportunity to use the standardization guidelines before submitting their project proposals, a positive impact can be assumed here. Further, the analysis of the response showed that the vast majority of projects that received or downloaded the standardization guidelines used (or planned to use) these during the course of their work.

Triggered by the positive feedback and good results COPRAS achieved with its initial set of

Figure 1. Allocation of ICT projects' resources to standards work in several calls

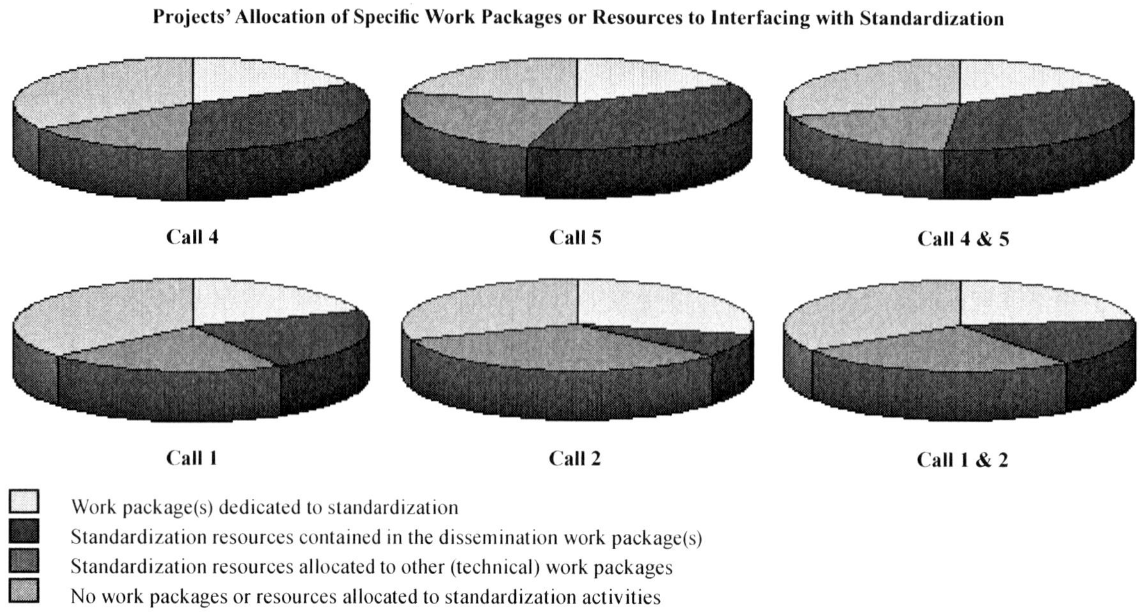

generic standardization guidelines, an interactive platform was built to improve the accessibility of relevant information on research and standards interfacing in ICT based on a set of frequently asked questions (FAQs). These questions focused on specific aspects of the standardization process, as well as on the interests of the different constituencies, so that stakeholders could easily find their way through the information. The platform was put online toward the close of the COPRAS project early 2007 in conjunction with an upgraded version of the standardization guidelines document. Further research and monitoring of the platform will be necessary to evaluate its impact and to determine the specific elements that will still have to be addressed to improve overall research and standards interfacing.

CONCLUSION AND RECOMMENDATIONS

When looking back at the work COPRAS has done and the results it managed to achieve, a first evaluation shows that the project outperformed most of the (aggressive) quantitative targets it set for itself. The number of standardization action plans it developed was considerably higher (13%) then the 8% that was planned, and the impact generated by ICT projects with respect to ongoing standards work will most likely prove to be twice as high as initially expected.

More important than the positive results itself is, however, the recognition—or better, the confirmation—that ICT research and standardization are very much interlinked and should proceed in parallel as much as possible in order to enable the cross-fertilization processes necessary to address the ever-increasing pace of progress on both sides. In this respect it is also important to realize that the impact COPRAS generated has not been limited to specific areas in ICT research or standardization. Standardization action plans involved projects across 14 out of the 18 research areas addressed, and involved standards activities in more than 20 different standards organizations on a European (e.g., CEN and ETSI) as well as on a global level (e.g., W3C, JCP, IETF, IMS, and

HGI). Moreover, as a result of the plans, technical specifications were submitted to standardization, new constituencies were built addressing emerging challenges, and the uptake of newly developed standards was promoted.

However, despite the good overall results, it should also be recognized that considerable and, more importantly, continuous improvement of the research and standards interfacing process remains necessary as, generally speaking, projects often remain too optimistic with respect to the standards work they will be able to complete and frequently have to adjust their plans due to lack of resources. Lowering ambitions in this respect to a more realistic level significantly increases the chance a project is able to complete all of its plans. Moreover, many additional issues will have to be addressed, for example, through future support activity or research, ranging from establishing a clearer view on the benefit of standardization for a research project to addressing the standardization gap that occurs at the end of a project's life span, when standards activities often cannot be continued as resources and time have run out.

Although the standardization guidelines were highly appreciated by ICT research projects, and demonstrably improved their cooperation with standards organizations, feedback also showed that many issues need to be addressed to further improve research and standards interfacing in future framework programmes. Examples of these are barriers created by confidentiality, IPR (intellectual property right) or membership issues, mapping research activities with standards work, or finding the standards and standards organizations most relevant to a project and contacting them. This not only means that additional COPRAS-like support activity directed at individual or clustered projects in future framework programmes will remain necessary, but also that the international standards community as well as the administrators of the ICT research programmes will have to focus on offering more information on standards and standardization processes, better marketing of the benefits of standardization (e.g., to SMEs [small and medium-sized enterprises]), addressing current barriers to participation in standards processes, enabling research projects to acquire additional resources for completing their standards work, and developing better tools to help projects in finding the standards and standards organizations that are most relevant to them.

Moreover, in order to capitalize on the initial impact COPRAS generated, these follow-up activities will be essential to seize the momentum created, otherwise there will be a severe risk that the first steps toward overall improvement of the research standards interfacing process will not get any maintenance, and the initial impact will fade away.

Compilation of References

Aben, G. (2002). *Report on the survey regarding the use of international standards.* Geneva, Switzerland: International Federation of Standards Users.

Abernathy, W. J., & Clark, K. B. (1985). Innovation: Mapping the winds of creative destruction. *Research Policy, 14*, 3-22.

Abernathy, W. J., & Utterback, J. M. (1978). Patterns of industrial innovation. *Technology Review, 80*(7), 40-47.

Abowd, G., & Mynatt, E. (2000). Charting past, present, and future research in ubiquitous computing. *ACM Transactions on Computer-Human Interaction, 7*, 29-58.

Access Board Standards. (2000). Electronic and Information Technology Accessibility Standards. Retrieved from www.access-board.gov/sec508/508standards.htm

Adams, G., McQueen, G., & Seawright, K. (1999). Revisiting the stock price impact of quality awards. *Omega, 27*(6), 595-604.

Adler, P., & Kwon, S. (2002). Social capital: Prospects for a new concept. *Academy of Management Review, 27*(1) 17-40.

Adolphi, H. (1997). *Strategische Konzepte Zur Organisation Der Betrieblichen Standardisierung.* Berlin, Germany: Beuth Verlag.

Ailleret, P. (1982). *Essai de la théorie de la normalisation.* Paris: Eyrolles.

Akerlof, G. A. (1970). The market for "lemons": Quality uncertainty and the market mechanism. *The Quarterly Journal of Economics, 84*(3), 488-500.

Alter, S. (2005). Architecture of sysperanto: A model-based ontology of the IS field. *Communications of the ACM, 15*(1), 1-40. Retrieved June 6, 2005, from http://www.cais.isworld.org

American National Standards Institute. (1998). *Procedures for the development and coordination of American national standards.* Washington, DC: American National Standards Institute.

Andersen, E. S. (1991). Techno-economic paradigms as typical interfaces between producers and users. *Journal of Evolutionary Economics, 1*(2), 119-144.

Anderson, P., & Tushman, M. L. (1990). Technological discontinuities and dominant designs: A cyclical model of technological change. *Administrative Science Quarterly, 35*(4), 604-633.

Anderson, S. W., Daly, J. D., & Johnson, M. F. (1999). Why firms seek ISO 9000 certification: Regulatory compliance or competitive advantage. *Production and Operations Management, 8*(1), 28-43.

ANEC: The European consumer voice in standardization. (n.d.). Retrieved from http://www.anec.org/

APDIP. (2003). IOSN International Open Source Network Asia-Pacific Development Information Programme. *Proceedings of the CICC/NECTEC Asian OpenSource Symposium*. Retrieved May 2, 2003, from www.asiaosc.org

Arthur, B. (1989). Competing technologies, increasing returns, and lock-in by historical events. *The Economic Journal, 99*, 116-131.

Arthur, W.B. (1996). Increasing returns and the new world of business. *Harvard Business Review, 74*(4), 100-109.

Association Française de Normalisation (AFNOR). (1967). *La normalisation dans l'entreprise*. Paris: Author.

Avital, M., & Germonprez, M. (2003). *Ubiquitous computing: Surfing the trend in a balanced act.* Proceedings of the 2003 Workshop on Ubiquitous Computing Environment, Cleveland, OH.

Awang, H. (2002). Open Source Software as a catalyst in narrowing digital divide. *Jaring Internet Magazine*, August. Retrieved May 2, 2003, from *www.magazine.jaring.my/2002/august/index2.html?content=stay7.html*

Axelrod, R. (1997). The dissemination of culture: A model with local convergence and global polarization. *Journal of Conflict Resolution, 41*, 203-226.

Balci, O. (1997). *Verification, validation and accreditation of simulation models.* Paper presented at the Winter Simulation Conference.

Baldwin, C. Y., & Clark, K. (2000). *Design rules: The power of modularity.* Cambridge, MA: The MIT Press.

Band, J. (1995). Competing definitions of "openness" on the NII. In B. Kahin & J. Abbate (Eds.), *Standards policy for information infrastructure* (pp. 351-367). Cambridge, MA: MIT Press.

Barratt, M. (2004). Unveiling enablers and inhibitors of collaborative planning. *International Journal of Logistics Management, 15*(1), 73-90.

Barratt, M., & Oliveira, A. (2001). Exploring the experiences of collaborative planning initiatives. *International Journal of Physical Distribution & Logistics, 31*(4), 266-289.

Barzel, Y. (1989). *Economic analysis of property rights.* Cambridge, MA: Cambridge University Press.

Barzel, Y. (2004). Standards and the form of agreement. *Economic Inquiry, 42*(1), 1-13.

Barzel, Y. (2005). Organizational form and measurement costs. *Journal of Institutional and Theoretical Economics, 161*(3), 357-373.

Beardsmore, H. (1986). *Bilingualism: Basic principles.* Clevedon, UK: Multilingual Matters.

Beirao, G., & Cabral, J. A. S. (2002). The reaction of the Portuguese stock market to ISO 9000 certification. *Total Quality Management, 13*(4), 465-474.

Belkas, A., Garofalakis, J., & Stefanis, V. (2006). *Use of RSS feeds for content adaptation in mobile Web browsing.* Paper presented at the Proceedings of the 2006 International Cross-Disciplinary Workshop on Web Accessibility (W4A) Building the Mobile Web: Rediscovering Accessibility? Edinburgh, UK.

Benbasat, I., Goldstein, D.K., & Mead, M. (1987). The case research strategy in studies of information systems. *MIS Quarterly, 11*(3), 369-386.

Benton, W. C., & Maloni, M. (2000). Power influences in the supply chain. *Journal of Business Logistics, 21*, 49-74.

Besen, S. M., & Farrell, J. (1994). Choosing how to compete: Strategies and tactics in standardization. *The Journal of Economic Perspectives, 8*(2), 117-131.

Biller, S., Bish, E. K., & Muriel, A. (2001). Impact of manufacturing flexibility on supply chain performance in the automotive industry. In J. Song & D. Yao (Eds.), *Supply chain structures: Coordination, information and optimization* (pp. 73-118). Boston: Kluwer Academic Publishers.

Bloemen, F. E. M., Van den Molegraaf, J. C. M., & Van Mal, H. H. (1992). Normalisatie vermindert levensduuruitgaven van chemische installaties aanzienlijk. *B& Id, 4*(8), 17-20.

Bonner, P., & Potter, D. (2000). *Achieving best practices in national standardisation: A benchmarking study of the national standardisation systems of Finland, Sweden, Denmark and Italy.* Helsinki, Finland: Ministry of Trade Industry.

Bouma, J. J., & Winter, W. (1982). *Standardization fundamentals.* Delft, The Netherlands: Nederlands Normalisatie-instituut.

Brady, R. A. (1929). *Industrial standardization.* New York: National Industrial Conference Board.

Bresnahan, T. F., & Chopra, A. (1990). The development of the local area network market as determined by user needs. *Economics of Innovation and New Technology, 1*, 97-110.

Bresnahan, T. F., & Greenstein, S. (1999). Technological competition and the structure of the computer industry. *Journal of Industrial Economics, 47*(1), 1-40.

British Standards Institution. (1970). *PD 3542: The operations of a company standards department.* London: Author.

British Standards Society. (1995). *PD 3542: 1995 Standards and quality management. An integrated approach.* London: British Standards Institution.

Broersma, M. (2002). Group campaigns against open source. *CNETAsia*, August 14, 2002. Retrieved May 2, 2003, from *http://asia.cnet.com/newstech/systems/0,39001153,39072261,00.htm*

Brown, J., &. Duguid, P. (1991). Organizational learning communities of practice: Toward a unified view of working, learning and innovation. *Organization Science, 2*(1), 40-57.

Brown, S. J., & Warner, J. B. (1980). Measuring security price performance. *Journal of Financial Economics, 8*, 205-258.

Brown, S. J., & Warner, J. B. (1985). Using daily stock returns: The case of event studies. *Journal of Financial Economics, 14*, 3-31.

Bruce, R., & Ireland, R. (2002). *What's the difference: VMI, co-managed, CPFR?* [white paper]. Retrieved June 1, 2004, from http://vccassociates.com/articles.asp

Brunsson, N., Jacobsson, B., & Associates. (2000). *A world of standards.* Oxford, UK: Oxford University Press.

Brynjolfsson, E., & Kemerer, C.F. (1996). Network externalities in microcomputer software: An econometric analysis of the spreadsheet market. *Management Science, 42*(12), 1627-1647.

Buckley Jr., W.F. (1999). Whose Side Are We On?. *National Review*, June, Vol. 51, 23, 74.

Burrows, J. H. (1999). Information technology standards in a changing world: The role of users. *Computer Standards & Interfaces, 20*, 323-331.

Business Weekly. (10/09/2002). *Windows or Linux? Or Both?* Retrieved from www.chinadaily.com.cn/bw/2002-09-03/85731.html

Buxmann, P. (1996). *Standardisierung betrieblicher Informationssysteme.* Wiesbaden, Germany: Gabler Verlag.

Buxmann, P., Weitzel, T., Westarp, F. von, & König, W. (1999). *The standardization problem: An economic analysis of standards in information systems.* Proceedings of the First IEEE Conference on Standardization and Innovation in Information Technology, SIIT '99, Aachen, Germany.

Caldwell, B., Stein, T., & McGee, M. (1996). Uncertainty: A thing of the past. *Information Week,* 46.

Calongne, C. M. (2001). Designing for Web site usability. *The Journal of Computing in Small Colleges, 16*, 39-45.

Campbell-Kelly, M. (2003). *From Airline Reservations to Sonic the Hedgehog: A History of the Software Industry*, MIT Press.

Caprile, M., CIREM Foundation, & Llorens, C. (2000). *Outsourcing und Arbeitsbeziehungen in der Automobilindustrie* (Tech. Rep.). Retrieved August 25, 2004, from http://www.eiro.eurofound.eu.int/2000/08/study/tn0008203s.html

Cargil, C. (2001). Why are we doing this? *IEEE Computer, 34*(10), 116-117.

Cargill, C. (1994). Evolution and revolution in open systems. *StandardsView, 2*(1), 3-13.

Cargill, C. (1997). *Open systems standardization: A business approach.* Upper Saddle River, NJ: Prentice Hall.

Cargill, C. (1997). Section 2: Sun and standardization wars. *StandardsView, 5*(4), 133-135.

Cargill, C. F. (1989). *Information technology standardization: Theory, process, and organizations.* Maynard, MA: Digital Press.

Cargill, C., & Bolin, S. (2004). Standardization: A failing paradigm. *Proceedings of the Standards and Public Policy Conference*, Chicago.

Catford, J. (1980). *A linguistic theory of translation.* Oxford: Oxford University Press.

Chau, P.Y.K, & Tam, K.Y. (1997). Factors affecting the adoption of open systems: An exploratory study. *MIS Quarterly, 21*(1), 1-24.

Cheverst, K., Mitchell, K., & Davies, N. (1998). *Design of an object model for a context sensitive tour guide.* Proceedings of the Conference on Interactive Applications of Mobile Computing, Rostock, Germany.

Chiesa, V., Coughlan, P., & Voss, C.A. (1996). Development of a technical innovation audit. *Journal of Production Innovation Management, 13*, 105-136.

Chow, G., Heaver, T. D., & Henriksson, L. E. (1995). Strategy, structure and performance: A framework for logistics research. *Logistics and Transportation Review, 31*(4), 285-308.

Chow, S. Y. (2001). The effects of UCITA on software component development and marketing. In G. T. Heineman & W. T. Councill (Eds.), *Component based software engineering: Putting the pieces together* (pp. 719-730). Boston: Addison-Wesley.

Christensen, C. M., Verlinden, M., & Westerman, G. (2002). Disruption, disintegration and the dissipation of differentiability. *Industrial and Corporate Change, 11*(5), 955-993.

Church, J., Gandal, N., & Krause, D. (2002). *Indirect network effects and adoption externalities.* University of Calgary, Department of Economics.

CIPR. (2002). *Integrating Intellectual Property Rights and Development Policy – Report of the Commission on Intellectual Property Rights*, London: Commission on Intellectual Property Rights. Retrieved from www.iprcommission.org.

Clark, D. C. (1995). Interoperation, open interfaces and protocol architecture. Retrieved from *http://www.csd.uch.gr/~hy490-05/lectures/Clark_inter operation.htm*

Clark, K. B. (1995). *Notes on modularity in design and innovation in advanced ceramics and engineering plastics* (Working Paper No. 95-073). Boston: Harvard Business School.

Clark, T. H., & Stoddard, D. B. (1996). Interorganizational business process redesign: Merging technological and process innovation. *Journal of Management Information Systems, 13*(2), 9-28.

Clarke, M. (2004). *Standards and intellectual property rights: A practical guide for innovative business.* London: NSSF.

Coase, R. H. (1937). The nature of the firm. *Economica, 4*, 386-405.

Coase, R. H. (1984). The new institutional economics. *Zeitschrift für die Gesamte Staatswissenschaft (Journal of Institutional and Theoretical Economics), 140*(1), 229-231.

Coase, R. H. (1988). The firm, the market, and the law. In R. H. Coase (Ed.), *The firm, the market, and the law* (pp. 1-31). Chicago: The University of Chicago Press.

Coleman, J. (1988). Social capital in the creation of human capital. *The American Journal of Sociology, 94*, S95-s120.

Colman, D., & Nixson, F. (1994). *Economics of Change in Less Developed Countries*, 3rd edition. New York and London: Harvester Wheatsheaf.

Cowan, R. (1992). High technology and the economics of standardization. In Dierkes, Meinolf and Hoffmann, Ute (Eds.), *New Technology at the Outset: Social Forces in the Shaping of Technological Innovations* (279-300), Frankfurt/NY: Campus/Westview.

Cowan, R., & Foray, D. (1997). The economics of codification and the diffusion of knowledge. *Industrial and Corporate Change, 6*(5), 595-622.

Cowan, R., David, P. A., & Foray, D. (2000). The explicit economics of knowledge codification and tacitness. *Industrial and Corporate Change, 9*(2), 211-253.

Cox, A. (2001). Managing with power: Strategies for improving value appropriation from supply relationships. *Journal of Supply Chain Management, 37*(2), 42-47.

Crain Communications. (2002, April). Car cutaways: An inside look at who supplies what on some of Europe's most important new models. *Automotive News Europe.*

Critchley, T. A., & Batty, K. C. (1993). *Open systems: The reality.* Hertfordshire, UK: Prentice Hall.

Cubine, M., & Smith, K. (2001). Lack of communication, standards builds barriers to paper e-commerce. *Pulp & Paper Magazine, 75*(2), 32-35.

Dale, B., & Oakland, J. (1991). *Quality improvements through standards.* Leckhampton, UK: Stanley Thornes Ltd.

Damsgaard, J., & Truex, D. (2000). Binary trading relations and the limits of EDI standards: The procrustean bed of standards. *European Journal of Information Systems, 9*(3), 142-158.

Dankbaar, B., & van Tulder, R. (1992). The influence of users in standardization: The case of MAP. In M. Dierkes & U. Hoffmann (Eds.), *New technology at the outset: Social forces in the shaping of technological innovations* (pp. 327-350). Frankfurt; New York: Campus.

Daoud, M. (1991). *The processing of EST discourse: Arabic and French native speakers' recognition of rhetorical relationships in engineering texts.* Los Angeles: University of California.

Darby, M. R., & Karni, E. (1973). Free competition and the optimal amount of fraud. *The Journal of Law and Economics, 16*, 68-88.

David, P. A. (1985). Clio and the economics of QWERTY. *The American Economic Review, 75*(2), 332-337.

David, P. A. (2000). *Path dependence, its critics and the quest for "historical economics"* [Working paper]. Stanford University, Department of Economics.

David, P. A., & Greenstein, S. M. (1990). The economics of compatibility standards: An introduction to recent research. *Economics of Innovation and New Technology, 1*(1), 3-41.

David, P.A. (1987). Some new standards for the economics of standardization in the information age. In P. Dasgupta, & P. Stoneman (Eds.), *Economic policy and technological performance* (pp. 206-239). Cambridge: Cambridge University Press.

Davis, C.K. (2000). Managing expert power in information technology. *Journal of Database Management, 11*(2), 34-37.

De Gelder, A. (1989). *Opstellen van normen.* Delft, The Netherlands: Nederlands Normalisatie-instituut.

De Vries, H. (2005). IT standards typology. In K. Jakobs (Ed.), *Advanced topics in IT standards & standardization research.* Hershey, PA: Idea Group Publishing (forthcoming).

De Vries, H. J. (1999). *Standardization: A business approach to the role of national standardization organisations.* Boston: Kluwer Academic Publishers.

De Vries, H. J. (2002). *Procedures voor ISO 9000:2000.* Delft, The Netherlands: NEN.

De Vries, H. J. (2007). Developing a standardization best practice by cooperation between multinationals. In K. J. O'Sullivan (Ed.), *Strategic knowledge management in multinational organizations*. Hershey, PA: Information Science Reference.

De Vries, H. J., & Slob, F. J. C. (2008). Building a model of best practice of company standardization. In J. Dul & T. Hak (Eds.), *Case study methodology in business research*. Oxford, UK: Butterworth Heinemann.

De Vries, H., Feilzer, A., & Verheul, H. (2004). Removing barriers for participation in formal standardization. In F. Bousquet, Y. Buntzly, H. Coenen, & K. Jakobs (Eds.), *EURAS Proceedings 2004* (pp. 171-176). Aachen, Germany: Aachener Beiträge zur Informatik.

Demsetz, H. (1967). Toward a theory of property rights. *The American Economic Review, 57*(2), 347-359.

Desruelle, D., Gaudet, G., & Richelle, Y. (1996). Complementarity, coordination and compatibility: The role of fixed costs in the economics of systems. *International Journal of Industrial Organization, 14*(6), 747-768.

Deutsches Institut für Normung (DIN). (2004). *Translator's manual I, (DIN-Normen), (German-English).* Unpublished manuscript.

Diesken, M., & Hoffmann, U. (1992). *New technology at the outset*. Frankfurt, Germany: Campus.

Dietzsch, A., & Esswein, W. (2001). Gibt es eine "softwarekomponenten industrie"? Ergebnisse einer empirischen untersuchung. In H. U. Buhl, A. Huther, & B. Reitwiesner (Eds.), *Information Age Economy: 5. Internationale Fachtagung Wirtschaftsinformatik 2001* (pp. 697-710). Heidelberg, Germany: Physika-Verlag.

Dioan, A., Madhavan, J., Domingos, P., & Halvevy, A. (2004). Ontology matching: A machine learning approach. In S. Staab & R. Studer (Eds.), *Handbook on ontologies* (pp. 385-403). Berlin, Germany: Springer.

Docking, D. S., & Dowen, R. J. (1999). Market interpretation of ISO 9000 registration. *Journal of Financial Research, 22*(2), 147-160.

Dokko, G., & Rosenkopf, L. (2004). *Social capital for hire? Mobility of technical professionals and firm influence in wireless standards committees* [work in process].

Dosi, G. (1982). Technological paradigms and technological trajectories. *Research Policy, 11*, 147-162.

Drahos, P., & Braithwaite, J. (2002). *Information Feudalism: Who Owns the Knowledge Economy?*. Earthscan.

Dranove, D., & Gandal, N. (1999). *The DVD vs. DIVX standard war: Network effects and empirical evidence of vaporware*. Tel Aviv, Israel: Tel Aviv University, Foerder Institute of Economic Research, Faculty of Social Sciences.

Düsterbeck, B., Hesser, W., Inklaar, A., & Vischer, J. (1995). Company standardization. In W. Hesser & A. Inklaar (Eds.), *An introduction to standards and standardization* (pp. 99-138). Berlin, Germany: Beuth Verlag.

Dutt, A. K., Kim, K.S., & Singh, A. (1994). *The State, Markets and Development – beyond the Neoclassical Dichotomy* (pp. 3-21). Cheltenham: Edward Elgar.

Dutton, W.H. (1981). The rejection of an innovation: The political environment of a computer-based model. *Systems, Objectives, Solutions, 1*(4), 179-201.

ECMA. (2002). ECMA 334 C# language specification. Retrieved September 19, 2005, from *http://www.ecma-international.org*

ECMA. (2002). ECMA 335 common language infrastructure (CLI). Retrieved September 19, 2005, from *http://www.ecma-international. org*

Economides, N. (1989). Desirability of compatibility in the absence of network externalities. *American Economic Review, 79*(5), 1165-1181.

Economides, N. (1991). Compatibility and the creation of shared networks. In M. E. Guerin-Calvert & S. S. Wildman (Eds.), *Electronic services networks: A business and public policy challenge* (pp. 39-55). New York: Praeger.

Economides, N. (1996). The economics of networks. *International Journal of Industrial Organization, 14*(6), 673-699.

Economides, N., & Himmelberg, C. (1995). *Critical mass and network size with application to the US fax market* (Discussion Paper No. EC-95-11). NYU Stern School of Business.

Economides, N., & Lehr, L. (1994). The quality of complex systems and industry structure. In W. Lehr (Ed.), *Quality and reliability of telecommunications infrastructure.* Hillsdale, NJ: Lawrence Erlbaum.

Economides, N., & Salop, S. C. (1992). Competition and integration among complements, and network market structure. *The Journal of Industrial Economics, 40*(1), 105-123.

Edwards, J. (1994). *Multilingualism.* London: Penguin.

Egyedi, T. (1996). *Shaping standardization.* Delft, the Netherlands: Delft University Press.

Egyedi, T. M. (2001). Why Java was—not—standardized twice. *Computer Standards & Interfaces, 23,* 253-265.

Eisenhardt, K.M. (1989). Building theories from case study research. *Academy of Management Review, 14*(4), 532-550.

Elkin, T. (2000). Monopoly ruling won't destroy Microsoft's image, experts say. *Advertising Age, 10*(71), 15, 4.

eNet. (06/17/2002). How could Kingsoft challenge the domination of Microsoft?. Retrieved June 17, 2002, from *www.cfan.net.cn/news/softnews/0617/1.htm*

Essers, J., & Schreinemakers, J. (1996). The conceptions of knowledge and information in knowledge management. In J. E. Schreinemakers (Ed.), *Knowledge management: Organization, competence and methodology* (pp. 93-104). Würzburg, Germany: Ergon.

Eveland, J.D., & Tornatzky, L.G. (1990). The deployment of technology. In L.G. Tornatzky, & M. Fleischer (Eds.), *The processes of technological innovation* (pp. 117-147). Lexington, MA: Lexington Books.

Fama, E. F., Fisher, L., Jensen, M. C., & Roll, R. (1969). The adjustment of stock prices to new information. *International Economic Review, 10*(1), 1-21.

Farrell, J. G. (1984). Installed base and compatibility: Innovation, product preannouncements, and predation. *The American Economic Review, 76*, 940-955.

Farrell, J., & Gallini, N. (1988). Second-sourcing as a commitment: Monopoly incentives to attract competition. *Quarterly Journal of Economics, 103*(4), 673-694.

Farrell, J., & Saloner, G. (1985). Standardization, compatibility & innovation. *Rand Journal of Economics, 16*(1), 70-83.

Farrell, J., & Saloner, G. (1986). Installed base and compatibility: Innovation, product preannouncements, and predation. *The American Economic Review, 76*(5), 940-955.

Fédération des Institutions Internationales Semi-Officielles et Privées Établies à Genève (FIIG). (1979). *The use of languages in organizations and in meetings.* Geneva, Switzerland: Author.

Feld, C.S., & Stoddard, D.B. (2004). Getting IT right. *Harvard Business Review, 82*(2), 72-79.

Fellner, K. J., & Turowski, K. (2000). Classification framework for business components. *Proceedings of the 33rd Annual Hawaii International Conference on System Sciences.*

Financial Times. (02/05/2003). A meeting of Microsoft's minds. *Research and Development*, 11.

FIPS 151. (1986). Proposed Federal Information Processing Standard for UNIX Operating System Derived Environments. *Federal Register, 51*(168).

Fisher, F. M., McKie, J. W., & Mancke, R. B. (1983). *IBM and the U.S. data processing industry: An economic history.* New York: Praeger.

Fisher, M. L. (1997). What is the right supply chain for your product? *Harvard Business Review*, 105-116.

Fleming, L., & Sorenson, O. (2001). Technology as a complex adaptive system: Evidence from patent data. *Research Policy, 30*(7), 1019-1039.

Fomin, V. (2001). *The process of standards making: The case of the cellular mobile telephony.* Jyväskylä, Finland: University of Jyväskylä.

Fomin, V., & Keil, T. (1999). Standardization: Bridging the gap between economic and social theory. *Proceedings of the International Conference on Information Systems*, Charlotte, NC.

Foray, D. (1995). Coalitions and committees: How users get involved in information technology (IT) standardization. In R. Hawkins, R. Mansell, & J. Skea (Eds.), *Standards, innovation and competitiveness* (pp. 192-212). Hants, UK: Edward Elgar Publishing.

Frankel, R., Goldsby, T. J., & Whipple, J. M. (2002). Grocery industry collaboration in the wake of ECR. *International Journal of Logistics Management, 13*(1), 57-72.

Fricke, M. (2004). *Information logistics in supply chain networks: Concept, empirical analysis, and design.* Stuttgart, Germany: Ibidem-Verlag.

Friedman, A. (1989). *Computer Systems Development: History, Organisation and Implementation.* Wiley, Chichester.

Furubotn, E. G., & Pejovich, S. (1972). Property rights and economic theory: A survey of recent literature. *Journal of Economic Literature, 10*(4), 1137-1162.

Furubotn, E. G., & Richter, R. (2001). *Institutions and economic theory: The contribution of the new institutional economics.* Ann Arbor: University of Michigan Press.

Furubotn, E., & Richter, R. (Eds.). (1991). *The new institutional economics: A collection of articles from the Journal of Institutional and Theoretical Economics.* Tübingen, Germany: Mohr.

Gabel, H.L. (1987). Open standards in computers: The Case of X/OPEN. In H.L. Gabel (Ed.), *Product standardization and competitive strategy* (pp. 91-123). Amsterdam: North Holland.

Gabriel, Y. (2002). Essay: On paragrammatic uses of organizational theory. A provocation. *Organization Studies, 23*, 133-151.

Gallagher, S., & Park, S.H. (2002). Innovation and competition in standard-based industries: A historical analysis of the U.S. home video game market. *IEEE Transactions on Engineering Management, 49*(1), 67-82.

Gallaugher, J.M., & Wang, Y.-M. (2002). Understanding network effects in software markets: Evidence from Web server pricing. *MIS Quarterly, 26*(4), 303-27.

Gandal, N. (1994). Hedonic price indexes for spreadsheets and an empirical test for network externalities. *RAND Journal of Economics, 25*, 160-170.

Gandal, N. (1995). Competing compatibility standards and network externalities in the PC software market. *Review of Economics and Statistics, 77*, 599-608.

Gandal, N., Kende, M., & Rob, R. (1997). *The dynamics of technological adoption in hardware/software systems: The case of compact disc players* (Working Paper No. 21-97). Sackler Institute of Economics Studies.

Gao, K. (2003). China to view Windows code. *News. Com, Tech News First.* Retrieved February 28, 2003, from *http://news.com.com/2100-1007-990526.html*

Garcia, L. (1993). A New Role for Government in Standard Setting? *StandardView, 1*(2), 2-10.

Gates, B. (1998). Compete, don't delete. *The Economist* (p. 19). Retrieved from http://www.economist.com

Geels, F. W. (2002). Technological transitions as evolutionary reconfiguration processes: A multi-level perspective and a case-study. *Research Policy, 31*(8-9), 1257-1274.

Gerst, M., & Bunduchi, R. (2005). Shaping IT standardisation in the automotive industry: The role of power in driving portal standardisation. *Electronic Markets, 15*(4), 335-343

Glaser, B.G., & Strauss, A. (1967). *The discovery of grounded theory: Strategies of qualitative research.* London: Wiedenfeld and Nicholson.

Goldenberg, J., Libai, B., & Muller, E. (2002). Riding the saddle: How cross-market communication can create a major slump in sales. *Journal of Marketing, 66*, 1-16.

Gouldner, A. W. (1954). *Patterns of industrial bureaucracy*. New York: The Free Press.

Gove, Ph. B. (1993). *Webster's third new international dictionary of the English language unabridged*. Cologne, Germany: Könemann Verlagsgesellschaft.

Greenstein, S. M. (1993). Did installed base give an incumbent any (measurable) advantages in federal computer procurement? *RAND Journal of Economics, 24*(1), 19-39.

Greenstein, S.M. (1997). Lock-in and the costs of switching mainframe computer vendors: What do buyers see? *Industrial and Corporate Change, 6*(2), 247-274.

Greenwood, R., Hinings, C. R., & Suddaby, R. (2002). Theorizing change: The role of professional associations in the transformation of institutionalized fields. *Academy of Management Journal, 45*(1), 58-80.

Grindley, P. (1995). *Standards, strategy, and policy: Cases and stories*. Oxford: Oxford University Press.

Grundy, J., & Yang, B. (2002). *An environment for developing adaptive, multi-device user interfaces*. Paper presented at the Fourth Australasian User Interface Conference, Adelaide, Australia.

Hanseth, O., & Braa, K. (2000). Who's in control: Designers, managers—Or technology? Infrastructures at norsk hydro. In C.U. Ciborra (Ed.), *From control to drift—The dynamics of corporate information infrastructures* (pp. 125-147). Oxford: Oxford University Press.

Harrington, H. J. (1998). Performance improvement: The rise and fall of reengineering. *The TQM Magazine, 10*, 69.

Harris, J. K., Swatman, P. M. C., & Kurnia, S. (1997). Efficient consumer response (ECR): A survey of the Australian grocery industry. *Proceedings of the ACIS 1997—8th Australasian Conference on Information Systems*, Adelaide, Australia.

Harvard Business School. (1995). *Procter & Gamble: Improving consumer value through process redesign*. Cambridge, MA: Harvard Business School.

Hatim, B. (2001). *Communication across cultures: Translation theory and contrastive text linguistics*. Exeter, United Kingdom: University of Exeter Press.

Heard, E. (1994, August). Quick response: Technology or knowledge? *Industrial Engineer*, 28-30.

Hein, O., Schwind M., & König, W. (2006). Scale-free networks: The impact of fat tailed degree distribution on diffusion and communication processes. *Wirtschaftsinformatik, 48*(4), 267-275.

Hendricks, K. B., & Singhal, V. R. (1996). Quality awards and the market value of the firm: An empirical investigation. *Management Science, 42*(3), 415-436.

Heracleous, L., & Barrett, M. (2001). Organizational change as discourse: Communicative actions and deep structures in the context of information technology implementation. *Academy of Management Journal, 44*(4), 753-778.

Hertwig, M., Mühge, G., & Tackenberg, H. (2002). The impact of e-business on the organization of the German automobile supply industry. *Zeitschrift für die Gesamte Wertschöpfungskette Automobilwirtschaft, 6*, 17-25.

Hesser, W. (2006). *Standardization within a company: A strategic perspective*. In W. Hesser (Ed.), *Standardisation in companies and markets* (pp. 177-215). Hamburg, Germany: Helmut-Schmidt-University Hamburg.

Hesser, W., & Inklaar, A. (Eds.). (1997). *An introduction to standards and standardization*. Berlin, Germany: Beuth Verlag.

Higgins, W. (2005). *Engine of change: Standards Australia since 1922*. Blackheath, Australia: Brandl & Schlesinger.

Hippel, E. v. (1986). Lead users: A source of novel product concepts. *Management Science, 32*(7), 791-805.

Hodgson, G. (1988). *Economics and institutions*. Cambridge: Polity Press.

Holmstrom, J., Framling, K., Kaipia, R., & Saranen, J. (2002). Collaborative planning forecasting and replenishment: New solutions needed for mass collaboration.

Supply Chain Management: An International Journal, 7(3/4), 136-145.

House, J. (1977). *A model for translation quality assessment.* Tübingen, Germany: Gunter Narr.

Howard, G. S., Bodnovich, T., Janicki, T., & Liegle, J., et al. (1999). The efficacy of matching information systems development methodologies with application characteristics—An empirical study. *The Journal of Systems and Software, 45*(3), 177-195.

Hurd, J., & Isaak, J. (2005). IT standardization: The billion dollar strategy. *International Journal of IT Standards and Standardization Research, 3*(1), 68-74.

Iacono, S., & Kling, R. (2001). Computerization movements: The rise of the Internet and distant forms of work. In J. Yates, & J. Van Maanen (Eds.), *Information technology and organizational transformation: History, rhetoric, and practice* (pp. 93-136). Thousand Oaks, CA: Sage Publications.

IEEE. (1986). *IEEE 1003 trial use standard—Portable operating system for computer environments.* New York: IEEE.

IEEE. (2005). *IEEE procedures, project status and history.* Retrieved September 19, 2005, from http://standards.IEEE.org

IETF working group guidelines and procedures, RFC 2418. (1998). Retrieved from http://www.ietf.org/rfc/rfc2418. txt

IFAN strategies and policies for 2000-2005. (2000). Retrieved from http://www.ifan-online.org/

IFAN. (1997). *IFAN memento 1997.* Geneva, Switzerland: International Organization for Standardization.

Iivari, J., Hirschheim, R., & Klein, H. K. (2001). A dynamic framework for classifying information systems development methodologies and approaches. *Journal of Management Information Systems, 17*(3), 179-218.

International Electrotechnical Commission (IEC). (2003). *Cahier des charges et types d' editing.* Unpublished manuscript.

International Electrotechnical Commission (IEC). (2006). *The electric century.* Geneva, Switzerland: Author.

International Organization for Standardization (ISO). (2000). *ISO 9000:2000. Quality management systems: Fundamentals and vocabulary.* Geneva, Switzerland: Author.

International Organization for Standardization (ISO). (2000). *ISO 9001:2000. Quality management systems: Requirements.* Geneva, Switzerland: Author.

International Organization for Standardization/International Electrotechnical Commission (ISO/IEC). (2004). *Directives for the technical work of ISO/IEC JTC 1 on information technology.* Geneva, Switzerland: Author.

International Organization for Standardization/International Electrotechnical Commission (ISO/IEC). (2004). *Part 1: Procedures for the technical work.* Geneva, Switzerland: Author.

International Organization for Standardization/International Electrotechnical Commission (ISO/IEC). (2004). *Part 2: Rules for the structure and drafting of international standards.* Geneva, Switzerland: Author.

International Organization for Standardization/International Electrotechnical Commission ISO/IEC 17000. (2004). *Conformity assessment—Vocabulary and general principles.* Geneva: International Organization for Standardization (ISO).

Isaak, J. (1998). The role of government in IT standards. *IEEE Computer, 31*(12), 129,132.

Isaak, J. (2005). POSIX—Inside: A case study. *Proceedings of the 38th Annual Hawaii International Conference on System Sciences*, Waikoloa, Hawaii.

Isaak, J. (2006). The role of individuals and social capital in POSIX standardization. *International Journal of IT Standards & Standardization Research, 4*(1), 1-23.

Isinolaw Research Centre. (2001). Intellectual Property. Retrieved from www.isinolaw.com/jsp/ip/main.jsp?LangID=0

Iversen, E. J. (2000). Standardization and intellectual property rights: Conflicts between innovation and diffusion in new telecommunications systems. In K. Jakobs (Ed.), *Information technology standards and standardization: A global perspective* (pp. 80-101). Hershey, PA: Idea Group Publishing.

Jakobs, K. (2000). *Standardisation processes in IT: Impact, problems and benefits of user participation.* Braunschweig, Germany: Vieweg.

Jespersen, H. (1995). POSIX retrospective. *StandardView, 3*(1), 2-10.

Johnson, M. (2004). Business-to-business electronic commerce standards [Presentation]. Waltham, MA: Bentley College.

Jumpelt, R. (1961). *Die Uebersetzung Naturwissenschaftlicher Und Technischer Literatur.* Berlin, Germany: Langenscheidt.

Juran, J. M. (1988). *Juran on planning for quality.* London: The Free Press.

Kanellos, M. (06/05/2002,). Microsoft gets diplomatic in China. *News.com.* Retrieved from http://news.com.com/2102-1001-932927.html

Katz, M. L., & Shapiro, C. (1994). Systems competition and network effects. *Journal of Economic Perspectives, 8*(2), 93-115.

Katz, M.L., & Shapiro, C. (1985). Network externalities, competition & compatibility. *American Economic Review, 75*(3), 424-440.

Katz, M.L., & Shapiro, C. (1986). Technology adoption in the presence of network externalities. *The Journal of Political Economy, 94*(4), 822-841.

Kauffman, S. (1993). *The origins of order.* New York: Oxford University Press.

Khan, B.Z. (2002). Intellectual Property and Economic Development: Lessons from American and European History. Commission on Intellectual Property Rights. *Proceedings of the Conference on How IPRs could Work Better for Developing Countries and Poor People.*

Klemperer, P. (1987). The competitiveness of markets with switching costs. *Rand Journal of Economics, 18*(1), 138-150.

Klemperer, P., & Padilla, J. (1997). Do firms' products lines include too many varieties? *RAND Journal of Economics, 28*(3), 472-488.

Koch, C. (2002). It all began with drayer. *CIO Magazine, 15*(20), 1.

Kocks, C. (1997). Het échec van projectmanagement in de systeemontwikkeling. *Informatie, 39*(4), 6-11.

Kocourek, R. (1982). *La langue française de la technique et de la science.* Wiesbaden, Germany: Brandstetter.

Kraaijeveld, P. (2002). *Governance in the Dutch banking industry: A longitudinal study of standard setting in retail payments.* Rotterdam, The Netherlands: Rotterdam School of Management, Faculteit Bedrijfskunde.

Kranton, R. E., & Minehart, D. F. (2001). A theory of buyer-seller networks. *American Economic Review, 91*(3), 485-508.

Krechmer, K. (1998). The principles of open standards. *Standards Engineering, 50*(6).

Krechmer, K. (2000). The fundamental nature of standards: Technical perspective. *IEEE Communications Magazine, 38*(6), 70.

Krechmer, K. (2001). The need for openness in standards. *IEEE Computer, 34*(6), 100-101.

Krechmer, K. (2006). The meaning of open standards. *International Journal of IT Standards & Standardisation Research, 4*(1), 43-61.

Kriwet, C. (1997). *Inter- and intraorganizational knowledge transfer.* Bamberg, Germany: Difo Druck.

Krugman, P. (1991). History versus expectations. *Quarterly Journal of Economics, 106*, 651-667.

Kuilboer, J. P., & Ashrafi, N. (2000). Software process and product improvement: An empirical assessment. *Information and Software Technology, 42*, 27-34.

Kumar, N. (2002). Intellectual Property Rights, Technology and Economic Development: Experiences of Asian Countries. *RIS Discussion Paper*, 25-2002.

Langlois, R. N. (1992). Transaction-cost economics in real time. *Industrial and Corporate Change, 1*(1), 99-127.

Langlois, R. N. (2002). Modularity and technology and organization. *Journal of Economic Behavior and Organization, 49*(1), 19-37.

Lee, D.R., & McKenzie, R.B. (2001). Technology, Competition, and Antitrust Enforcement. *USA Today Magazine, 3*(129), 2670, 26.

Lee, S. C., Pak, B. Y., & Lee, H. G. (2003). Business value of B2B electronic commerce: The critical role of inter-firm collaboration. *Electronic Commerce Research and Applications, 2*, 350-361.

Lei, J. (2002). *ChinaByte Report: E-government project – an opportunity for Linux*. Retrieved from http://it.sohu.com/88/40/article17114088.shtml

Lewis, O. (2001). China: Protecting the Knowledge-based Economy and Intellectual Property Rights. Country Panel IV. *Proceedings of the Harvard Asia Business Conference.* Retrieved from www.fas.harvard.edu/~asiactr/haq/200102/0102a001cp5.htm.

Leydesdorff, L. (2001). Technology and culture: The dissemination and the potential "lock-in" of new technologies. *Journal of Artificial Societies and Social Simulation, 4*(3).

Lie, H. W., & Saarela, J. (1999). Multipurpose Web publishing using HTML, XML, and CSS. *Communications of the ACM, 42*, 95-101.

Liebowitz, S. J., & Margolis, S. E. (1990). The fable of the keys. *Journal of Law and Economics, 33*, 1-25.

Liebowitz, S. J., & Margolis, S. E. (1994). Network externality: An uncommon tragedy. *The Journal of Economic Perspectives, 8*(2), 133-150.

Liebowitz, S. J., & Margolis, S. E. (1995). Are network externalities a new source of market failure? *Research in Law and Economics, 17*(1), 1-22.

Liebowitz, S. J., & Margolis, S. E. (1995). Path dependence, lock-in, and history. *The Journal of Law, Economics and Organization, 11*(1), 205-226.

Liebowitz, S. J., & Margolis, S. E. (1996). Should technology choice be a concern of antitrust policy? *Harvard Journal of Law and Technology, 9*(2), 283-318.

Liebowitz, S. J., & Margolis, S. E. (1998). Network externalities (effects). *The New Palgrave's Dictionary of Economics and Law.* MacMillan.

Liebowitz, S.J., & Margolis, S.E. (1997). *Winners, losers & Microsoft: Competition and antitrust in high technology.* Oakland, CA: Independent Institute.

Linux International. (2005). *Linux history.* Retrieved September 19, 2005, from http://www.li.org/linuxhistory.php

Luitjens, S. (1997). Interorganisatorische informatioseringsprojecten bij de overheid. *Informatie, 39*(4), 18-22.

Lum, W., & Lau, F. (2002). A context-aware decision engine for content adaptation. *IEEE Mobile Computing, 1*, 41-49.

MacNeil, I. R. (1978). Contracts: Adjustment of long-term economic relations under classical, neoclassical, and relational contract law. *Northwestern University Law Review, 72*(6), 854-905.

Maillot, J. (1981). *La traduction scientifique & technique.* Paris: Technique & Documentation.

Marketing News. (2002). *The emergence of Linux in the fore of opportunities and challenges.* Retrieved from http://it.sohu.com/14/80/article204508014.shtml

Markus, M.L. (1983). Power, politics and MIS implementation. *Communications of the ACM, 26*(6), 430-444.

Martin, X., & Mitchell, W. (1998). The influence of local search and performance heuristics on new design introduction in a new product market. *Research Policy, 26*(7/8), 753-771.

Martinez-Costa, M., & Martinez-Lorente, A. R. (2003). Effects of ISO 9000 certification on firms' performance: A vision from the market. *TQM & Business Excellence, 14*(10), 1179-1191.

Mathews, R. (1994). CRP moves toward reality. *Progressive Grocer, 73,* 43-46.

Mathews, R. (1994). CRP spells survival. *Progressive Grocer, 73,* 28-30.

Matutes, C., & Regibeau, P. (1998). "Mix and match": Product compatibility without network externalities. *The RAND Journal of Economics, 19*(2), 221-234.

McIlroy, M. D. (1976). Mass produced software components. In P. Naur & B. Randell (Eds.), *Software engineering* (pp. 88-95). New York: Petrocelli Charter.

Milgrom, P., & Roberts, J. (1995). Complementarities and fit: Strategy, structure, and organizational change in manufacturing. *Journal of Accounting and Economics, 19*, 179-208.

Molina, A. (1994). Understanding the emergence of a large-scale European initiative in technology. *Science and Public Policy, 21*(1), 31-41.

Moore, G.A. (1991). *Crossing the chasm.* New York: HarperBusiness.

Morris, C.R., & Ferguson, C.H. (1993). How architecture wins technology wars. *Harvard Business Review, 71*(2), 86-96.

Naemura, K. (1995). User involvement in the life cycles of information technology and telecommunications standards. In R. Hawkins, R. Mansell, & J. Skea (Eds.), *Standards, innovation and competitiveness.* Hants, UK: Edward Elgar Publishing.

Nakamura, S. (1993). *The new standardization: Keystone of continuous improvement in manufacturing.* Portland, OR: Productivity Press.

Nandhakumar, J., & Avison, D. E. (1999). The fiction of methodological development: A field study of information systems development. *Information Technology & People, 12*(2), 176-191.

Nanfang Dushi Daily. (10/30/2002). The battle amongst Office software.

National technology transfer and advancement act of 1995. (1996). Pub. L. No. 104-113, 110 Stat. 775. Retrieved September 19, 2005, from http://standards.gov/standards_gov/index.cfm?do=documents.NTTAA

Nelson, M., Shaw, M., & Qualls, W. (2005), Interorganizational system standards: Development in vertical industries. *Electronic Markets, 15*(4), 378-392.

Nelson, P. (1970). Information and consumer behavior. *The Journal of Political Economy, 78*(2), 311-329.

Newmark, P. (1986). *Approaches to translation.* Oxford: Pergamon Press.

Nicolau, J. L., & Sellers, R. (2002). The stock market's reaction to quality certification: Empirical evidence from Spain. *European Journal of Operational Research, 142*(3), 632-641.

Nida, E. (2001). *Contexts in translating.* Amsterdam: Benjamins.

Nightingale, P. (2003). If Nelson and Winter are only half right about tacit knowledge, which half? A Searlean critique of "codification." *Industrial and Corporate Change, 12*(2), 149-183.

NIST. (1986). FIPS 151: Proposed federal information processing standard for UNIX operating system derived environments. *Federal Register, 51*(168), 90896-30897.

Nonaka, I. (1991). The knowledge creating company. *Harvard Business Review, 6*, 96-104.

Nonaka, I. (1994). A dynamic theory of organizational knowledge creation. *Organization Science, 5*(1), 14-37.

Nonaka, I., & Takeuchi, H. (1995). *The knowledge creating company.* New York: Oxford University Press.

Nord, C. (2005). *Text analysis in translation: Theory, methodology, and didactic application of a model for translation-oriented text analysis.* Amsterdam: Rodopi.

Noronha, F. (2002). Openness and sharing make Linux attractive. *Tribune,* 3 June 2003. Retrieved May 2, 2003, from www.school.net.th/linux/news/tribuneindia/

North, D. C. (1994). Economic performance through time. *The American Economic Review, 86*(3), 359-368.

NRENAISSANCE Committee. (1994). *Realizing the information future*. Washington, DC: National Academy Press.

Nuffield Languages Inquiry. (2000). *Languages: The next generation*. London: The Nuffield Foundation.

Office of Industry Liaison. (2000) Intellectual Property. The University of Western Ontario. Retrieved from www.uwo.ca/industry/intellectual/#IP

Ollner, J. (1974). *The company and standardization* (2nd ed.). Stockholm: Swedish Standards Institution.

Olson, M., Jr. (1965). *The logic of collective action: Public goods and the theory of groups*. Cambridge, MA: Harvard University Press.

Oly, M. P., & Slob, F. J. C. (1999). *Benchmarking bedrijfsnormalisatie: Een best practice voor de procesindustrie*. Rotterdam, The Netherlands: Erasmus Universiteit Rotterdam, Faculteit Bedrijfskunde.

OMB. (1998). *OMB circular A-119*. Federal participation in the development and use of voluntary consensus standards and in conformity assessment activities. Retrieved September 19, 2005, from http://www.eh.doe.gov/techstds.overview/omba119.pdf

Orlikowski, W. J. (2000). Using technology and constituting structures: A practice lens for studying technology in organizations. *Organization Science, 11*(4), 404-428.

Orlikowski, W. J., & Gash, D. C. (1994). Technological frames: Making sense of information technology in organizations. *ACM Transactions on Information Systems, 12*(2), 174-207.

Orlikowski, W. J., & Iacono, C. S. (2001). Research commentary: Desperately seeking "IT" in IT research—A call to theorizing the IT artifact. *Information Systems Research, 12*(2), 121-134.

Orlikowski, W.J. (1993). CASE tools as organizational change: Investigating incremental and radical changes in systems development. *MIS Quarterly, 17*(3), 309-340.

Palmer, J. W., & Markus, M. L. (2000). The performance impacts of quick response and strategic alignment in specialty retailing. *Information Systems Research, 11*(3), 241-259.

Parnas, D. L. (1972). On the criteria to be used in decomposing systems into modules. *Communications of the ACM, 15*(12), 1053-1058.

PASC. (2005). *PASC operating procedures and history*. Retrieved September 19, 2005, from http://www.pasc.org/plato

Pask, G. (1971). A comment, a case history, and a plan. In J. Reichardt (Ed.), *Cybernetics, art and ideas* (pp. 76-99). London: Studio Vista.

Paulk, M. C. W., Charles V., & Curtis, B. (1995). *The capability maturity model: Guidelines for improving the software process*. Reading: Addison-Wesley.

People's Daily. (12/18/1999). *Microsoft Loses Suit Against Beijing Company*. Retrieved from http://english.peopledaily.com.cn/19991218T105.html

People's Daily. (6/4/2001). *Kingsoft, Microsoft Contend for Supremacy in China*. Retrieved from http://english.peopledaily.com.cn/200107/04/eng20010704_74170.html

Perens, B. (1999). The open source definition. In C. DiBona, S. Ockman, & M. Stone (Eds.), *OpenSources voices from the open source revolution* (pp. 171-189). Sebastopol, CA: O'Reilly & Associates.

Perens, B. (n.d.). Open standards principles and practice. Retrieved from *http://xml.coverpages.org/openStandards.html*

Perera, C. (2006). *Standardization in product development and design*. In W. Hesser (Ed.), *Standardisation in companies and markets* (pp. 141-176). Hamburg, Germany: Helmut-Schmidt-University Hamburg.

Pinchuck, I. (1977). *Scientific and technical translation*. London: André Deutsch.

Pinsonneault, A., & Kraemer, K.L. (1993). The impact of information technology on middle managers. *MIS Quarterly, 17*(3), 271-292.

Pinsonneault, A., & Kraemer, K.L. (1997). Middle management downsizing: An empirical investigation of the impact of information technology. *Management Science, 43*(5), 659-679.

Polanyi, M. (1983). *The tacit dimension.* Gloucester, MA: Peter Smith. (Original work published 1966 by Anchor Books, Garden City, New York)

Portes, A. (1998). Social capital: Its origins and applications in modern sociology. *Annual Review of Sociology, 24*, 1-24.

Przasnyski, Z. H., & Tai, L. S. (1999). Stock market reaction to Malcolm Baldridge National Quality Award announcements: Does quality pay? *Total Quality Management, 10*(3), 391-400.

Public Law 104-113. (1995). *National Technology Transferer ad Advancement Act.*

Putnum, R. (2000). *Bowling alone.* New York: Simon & Schuster.

Pyzdrowski, R., & Donnelley, R. (2003). *E-enabling an efficient supply chain.* Retrieved from http://www.papinet.org

Quirk, R., & Greenbaum, S. (1973). *A university grammar of English.* London: Longman.

Rada, R., & Craparo, J. S. (2001). Standardizing management of software engineering projects. *Knowledge Technology and Policy, 14*(2), 67-77.

Raghunathan, S. (1999). Interorganizational collaborative forecasting and replenishment systems and supply chain innovation. *Decision Sciences, 30*(4), 1053-1071.

Ramasesh, R. V. (1998). Baldridge Award announcement and shareholder wealth. *International Journal of Quality Science, 3*(2), 114-125.

Ramiller, N. C., & Swanson, E. B. (2003). Organizing visions for information technology and the information systems executive response. *Journal of Management Information Systems, 20*(1), 13-50.

Rao, H., & Drazin, R. (2002). Overcoming resource constraints on product innovation by recruiting talent from rivals: A study of the mutual fund industry 1986-94. *Academy of Management Journal, 43*(3), 491-507.

Raymond, E. (1998). The Cathedral and the Bazaar. *Peer Review Journal on Internet*, 3(3). Retrieved from www.firstmonday.dk/issues/issue3_3/raymond/

Raymond, E.S. (2000, August 25). Homesteading the noosphere (Section 2). Retrieved from *http://www.csaszar.org/index.php/csaszar/interesting/the_open_source_reader*

Riordan, M. H., & Williamson, O. E. (1985). Asset specificity and economic organization. *International Journal of Industrial Organization, 3*(4), 365-378.

Röder, H. (2002). *Studie: Struktur und Marktanalyse der Holz verbrauchenden Industrie in Nordrhein-Westfalen* (Tech. Rep.). Retrieved August 25, 2004, from http://www.forst.nrw.de/nutzung/cluster/6_1.Absatzstufe.pdf

Rogers, E. M. (1983). *Diffusion of innovations* (3rd ed.). New York: The Free Press.

Romanow, K. (2004). *Collaboration roadmap: Success depends on process, not technology.* Frontline Solutions. Retrieved January 29, 2004, from http://www.frontlinetoday.com

Rosenkopf, L, Metiu. A., & George, V. P. (2001). From the bottom up? Technical committee activity and alliance formation. *Administrative Science Quarterly, 46*, 748-772.

Rosenkopf, L., & Tushman, M. L. (1998). The co-evolution of community networks and technology: Lessons from the flight simulation industry. *Industry & Corporate Change, 7*(2), 311-346.

Ross, J. (2003). Creating a strategic IT architecture competency: Learning in stages. *MIS Quarterly Executive, 2*(1), 31-43.

Rothstein, E. (2001). Wronging Microsoft. *Commentary, 9*(112), 2, 46.

Sager, J., Dungworth, D., & McDonald, F. (1980). *English special languages: Principles and practice in science and technology.* Wiesbaden, Germany: Brandstetter.

Saloner, G., & Farrell, J. (1985). Standardization, compatibility and innovation. *RAND Journal of Economics, 16*(1), 70-83.

Samuelson, P. (1995). Software Compatibility and the Law. *Communications of the ACM, 38*(8), 15-22.

Sandoe, K., Corbitt, G., & Boykin, R. (2001). *Enterprise integration*. New York: John Wiley & Sons.

Schacht, M. (1991). *Methodische Neugestaltung Von Normen Als Grundlage Für Eine Integration In Den Rechnerunterstützten Konstruktionsprozeß*. Berlin, Germany: Beuth Verlag GmbH.

Schaefer, S. (1999). Product design partitions with complementary components. *Journal of Economic Behavior and Organization, 38*, 311-330.

Schilling, M. A. (2000). Toward a general modular systems theory and its application to interfirm product modularity. *Academy of Management Review, 25*(2), 312-334.

Schilling, M. A. (2002). Technology success and failure in winner-take-all markets: The impact of learning orientation, timing, and network externalities. *Academy of Management Journal, 45*(2), 387-398.

Schmauch, C. H. (1994). *ISO 9000 for software developers*. Milwaukee, WI: ASQC Quality Press.

Schreinemakers, J. F. (Ed.). (1996). *Knowledge management: Organization competence and methodology*. Proceedings of the Fourth International ISMICK Symposium.

Schumpeter, J. A. (1939). *Business cycles: A theoretical, historical, and statistical analysis of the capitalist process*. New York: McGraw Hill.

Seifert, D. (2003). *Collaborative planning, forecasting, and replenishment: How to create a supply chain advantage*. New York: AMACOM.

Shapiro, C., & Varian, H.R. (1999). *Information rules: A strategic guide to the network economy*. Boston: Harvard Business School Press.

Shapiro, S. (2001). Setting compatibility standards cooperation or collusion. In R. C. Dreyfuss, D. L. Zimmerman, & H. First (Eds.), *Expanding the boundaries of intellectual property*. Oxford, UK: Oxford University Press.

Sheremata, W.A. (2004). Competing through innovation in network markets: Strategies for challengers. *Academy of Management Review, 29*(3), 359-377.

Shneiderman, B. (1992). *Designing the user interface: Strategies for effective human-computer interaction* (2nd ed.). Reading, MA: Addison-Wesley.

Shneiderman, B. (2000). Universal usability. *Communications of the ACM, 43, 84-91.*

Shurmer, M., & Swann, P. (1995). An analysis of the process generating de facto standards in the PC spreadsheet software market. *Journal of Evolutionary Economics, 5*(2), 119-132.

Simon, H. A. (1962). The architecture of complexity. *Proceedings of the American Philosophical Society, 106*, 467-482.

Simons, C. A. J., & De Vries, H. J. (2002). *Standaard of maatwerk: Bedrijfskeuzes tussen uniformiteit en verscheidenheid*. Schoonhoven, The Netherlands: Academic Service.

Slob, F. J. C. (1999). *Bedrijfsnormalisatie: De schakel tussen tacit en explicit knowledge*. Rotterdam, The Netherlands: Erasmus Universiteit Rotterdam, Faculteit Bedrijfskunde, Vakgroep Management van Technologie en Innovatie.

Smith, C.S. (2000). Fearing Control by Microsoft, China Backs the Linux System. *The New York Times*. Retrieved from www.hartford-hwp.com/archieves/55/424.html

Smith, I.W. (2003). The Status of Open Source and Free Software in Malaysia, especially Government/Policy. *Proceedings of the CICC/NECTEC Asian OpenSource Symposium*.Retrieved May 2, 2003, from www.asiaosc.org

Smith, I.W. (2003). Asia Opensource Symposium, Thailand Day 2. *Proceedings of the Asian OpenSource Symposium*. Retrieved May 2, 2003, from www.asiaosc.org

Soderstrom, E. (2002). Standardising the business vocabulary of standards. *Proceedings of the SAC 2002*, Madrid, Spain.

Söllner, F. (2001). *Die geschichte ökonomischen denkens*. Berlin, Germany: Springer.

Sommerville, I. (2004). *Software engineering* (7th ed.). Harlow, United Kingdom: Addison Wesley.

Soteriou, A. C., & Zenios, S. A. (2000). *Searching for the value of quality in financial services* (Working Paper). Philadelphia: University of Pennsylvania, The Wharton Financial Institutions Center.

Stallman, R. (1999). Saving Europe from Software Patents. 20 :11 UTC, http://features.linuxtoday.com/news_story.php3?ltsn=1999-05-16-003-05-NW-LF.

Stallman, R. (1999). 15 Years of Free Software, 17 :23 UTC, http://features.linuxtoday.com/news_story.php3?ltsn=1999-03-17-003-10-NW-LF.

Steinmueller, W.E (2001). ICT and the possibilities for leapfrogging by developing countries. *International Labour Review*, *140*(2), 193-210.

Stelzer, K., Mellis, W., & Herzwurm, G. (1997). A critical look at ISO 9000 for software quality management. *Software Quality Journal, 6*(2), 65-79.

Stockheim, T., Schwind, M., & König, W. (2003). A model for the emergence and diffusion of software standards. *Proceedings of the 36th Hawaii International Conference on System Sciences* (pp. 59-68).

Susanto, A. (1988). *Methodik Zur Entwicklung Von Normen*. Berlin, Germany: Beuth Verlag GmbH.

Swanson, E.B. (1994). Information systems innovation among organizations. *Management Science, 40*(9), 1069-1092.

Swanson, E.B. (2002). Talking the IS innovation walk. *Proceedings of the IFIP WG8.2 Working Conference on Global and Organizational Discourse about Information Technology,* Barcelona, Spain.

Swanson, E.B., & Ramiller, N.C. (1997). The organizing vision in information systems innovation. *Organization Science, 8*(5), 458-474.

Szyperski, C., Dominik, G., & Murer, S. (2002). *Component software: Beyond object-oriented programming* (2nd ed.). New York: ACM Press.

Takahashi, S., & Tojo, A. (1993). The SSI story: What it is and how it was stalled and eliminated in the international standards arena. *Computer Standards and Interfaces, 15*(3), 523-538.

Takahashi, T., & Namiki, F. (2003). Three attempts at "de-Wintelization": Japan's TRON project, the US government's suits against Wintel, and the entry of Java and Linux. *Research Policy, 32*(9), 1589-1606.

Tam, K., Hui, K.L. A choice model for the selection of computer vendors and its empirical estimation. *Journal of Management Information Systems, 17*(4), 97-124.

Taylor, P. (1999). *Hackers: Crime in the Digital Sublime*. London and NY: Routledge.

Teece, D. (1986). Profiting from technological innovation: Implications for integration, collaboration, licensing and public policy. *Research Policy, 15*(6), 285-305.

Teichmann, H. (2000). IEC work on terminology and fundamental concepts (IEC/TC 1 and 25). *Conférence pour une Infrastructure Terminologique en Europe* (pp. 43-46).

Teichmann, H. (2000). Situational dimensions of IEC standards. *TermNet News, 67*, 6-12.

Teichmann, H. (2003). Efficiency of IEC technical committee secretaries. *EURAS Yearbook of Standardization, 4*, 125-145.

Teichmann, H. (2003). Translating IEC texts. *Journal of the International Institute for Terminology Research, 11*(1), 1-135.

Teichmann, H. (2004). Market study on adequate language quality of IEC international standards. *Aachener Beiträge zur Informatik, 36*, 53-61.

Teichmann, H. (2005). Ethnolinguistic aspects of international standardization. *Aachener Beiträge zur Informatik, 37*, 71-78.

Teichmann, H. (2006). Linguistic shortcomings in international standards. *International Conference on Terminology, Standardization and Technology Transfer* (pp. 46-58).

Terziovski, M., Samson, D., & Dow, D. (1997). The business value of quality management systems certification: Evidence from Australia and New Zealand. *Journal of Operations Management, 15*(1), 1-18.

Tesfatsion, L. (2002). Agent-based computational economics: Growing economies from the bottom up. *Artificial Life, 8*, 55-82.

Thayer, L. (1968). *Communication and communication systems.* Homewood, IL: Richard D. Irwin Inc.

The RIS News Glossary. (2004). RIS News. Retrieved May 21, 2004, from http://www.risnews.com/Glossary/glossary.html

Tingling, P., & Parent, M. (2002). Mimetic isomorphism and technology evaluation: Does imitation transcend judgment? *Journal of the AIS, 3*(5), 113-143.

Tornatzky, L.G. & Klein, K.J. (1982). Innovation characteristics and innovation adoption-implementation: A meta-analysis of findings. *IEEE Transactions on Engineering Management, 29*(1), 18-45.

Torvalds, L. (1991). Linux History. *Linux International.* Retrived from www.li.org/linuxhistory.php

Toth, R. B. (Ed.). (1990). *Standards management: A handbook for profits.* New York: American National Standards Institute.

Truman, G. E. (2000). Integration in electronic exchange environments. *Journal of Management Information Systems, 17*(1), 209-244.

Turowski, K. (2000). Establishing standards for business components. In K. Jakobs (Ed.), *Information technology standards and standardisation: A global perspective* (pp. 131-151). Hershey, PA: Idea Group Publishing.

Tushman, M. (1977). Special boundary roles in the innovation process. *Administrative Science Quarterly, 22*(4), 587-605.

Tushman, M. L., & Anderson, P. (1986). Technological discontinuities and organizational environments. *Administrative Science Quarterly, 31*, 439-465.

Tushman, M.L., & Nadler, D. (1986). Organizing for innovation. *California Management Review, 28*(3), 74-92.

Tushmann, M. L., & Murmann, J. P. (1998). Dominant designs, technology cycles, and organizational outcomes. *Research in Organizational Behavior, 20*, 231-266.

UAOS. (1991, January 27). User alliance for open systems. *Overcoming Barrier to Open Systems Information Technology.* McLean, VA.

Ulrich, K. T. (1995). The role of product architecture in the manufacturing firm. *Research Policy, 24*(3), 419-440.

Ulrich, K. T., & Ellison, D. J. (1999). Holistic customer requirements and the design-select decision. *Management Science, 45*(5), 641-658.

UNCTAD. (2002). E-Commerce and Development Report 2002. *Proceedings of the United Nations Conference on Trade and Development (UNCTAD).*

United States District Court for the District of Columbia Civil Action No. 98-1232 (TPJ).

Unter, B. (1996). The Importance of Standards to HP's Competitive Business Strategy. *ASTM Standardization.*

Unter, B. (1996). The importance of standards to HP's competitive business strategy. *ASTM Standardization News, 24*(12), 13-17.

Updegrove, A. (1995). Consortia and the role of the government in standards setting. In B. Kahin & J. Abbate (Eds.), *Standards policy for the information infrastructure* (pp. 321-348). Cambridge, MA: MIT Press.

Updegrove, A. (2004). Best practices and standard setting (how the "pros" do it). In S. Bolin (Ed.), *The standards edge, dynamic tension* (pp. 209-214). Ann Arbor, MI: Bolin Communications.

Updegrove, G. (2005). China, the United States and standards. *Consortium Standards Bulletin, IV*(4). Retrieved from *http://www.consortiuminfo.org/bulletins/apr05.php#editorsnote*

Valente, T. W. (1995). *Network models of the diffusion of innovations*. Cresskill, NJ: Hampton Press.

Varian, H. R. (2001). *Economics of information technology*. Berkeley, CA: University of California.

Verband der Automobilindustrie e.V. (2003). *Auto jahresbericht*.

Verity Consulting. (1995). *Strategic standardization: Lessons from the world's foremost companies*. New York: ANSI.

Verity, J. (1997). Collaborative forecasting: Vision quest. *Computerworld, 31*, 12-14.

Verkasalo, M., & Lappalainen, P. M. (1998). A method of measuring the efficiency of the knowledge utilization process. *IEEE Transactions on Engineering Management, 45*(4), 414-423.

VICS. (2004). Collaborative planning, forecasting and replenishment (CPFR®). Retrieved June 15, 2004, from http://www.cpfr.org

VICS. (2004). CPFR technical specification. Retrieved June 15, 2004, from http://www.cpfr.org

VICS. (2004). CPFR® deployment scenarios. Retrieved June 15, 2004, from http://www.cpfr.org

VICS. (2004). Supply chain initiatives. Retrieved June 15, 2004, from *http://*www.cpfr.org

Vinay, J.-P., & Darbelnet, J. (1995). *Comparative stylistics of French and English*. Amsterdam: Benjamins.

von Burg, U., & Kenney, M. (2003). Sponsors, communities, and standards: Ethernet vs. token ring in the local area networking business. *Industry and Innovation, 10*(4), 351-375.

Von Hahn, W. (1983). *Fachkommunikation*. Berlin, Germany: de Gruyter.

von Weizsäcker, C.C. (1984). The costs of substitution. *Econometrica, 52*(5), 1085-1116.

Wade, J. (1995). Dynamics of organizational communities and technological bandwagons: An empirical investigation of community evolution in the microprocessor market. *Strategic Management Journal, 16*, 111-133.

Walli, S. (1995). The POSIX family of standards. *StandardView, 3*(1), 11-17.

Walras, L. (1954). *Elements of pure economics or the theory of social wealth*. London: Allen & Unwin.

Weiser, M. (1991). The computer for the 21st century. *Scientific American, 265*, 94-104.

Weiser, M. (1993). Some computer science issues in ubiquitous computing. *Communications of the ACM, 36*, 75-84.

Weitzel, T., Wendt, O., Westarp, F. von, & König, W. (2003). Network effects and diffusion theory: Extending economic network analysis. *The International Journal of IT Standards & Standardization Research, 2*, 1-21.

Wen, Z. (2004). *Reform and change: An introduction to China standardization*. Paper presented at the 11th International Conference of Standards Users IFAN 2004, Amsterdam.

Wendt, O., & Westarp, F. von. (2000). *Determinants of diffusion in network effect markets*. Proceedings of the IRMA International Conference, Anchorage, AK.

Wenström, H., Ollner, J., & Wenström, J. (2000). *Focus on industry standards: An integrated approach*. Stockholm, Sweden: SIS Förlag.

West, J. (1999). Organizational decisions for I.T. standards adoption: Antecedents and consequences. In K. Jakobs, & R. Williams (Eds.), *Proceedings of the 1st IEEE Conference on Standardisation and Innovation in Information Technology* (pp. 13-18). Aachen, Germany.

West, J. (2003). How open is open enough? Melding proprietary and open source platform strategies. *Research Policy, 32*(7), 1259-1285.

West, J. (2004). What are open standards? Implications for adoption, competition and policy. *Proceedings of the Standards and Public Policy Conference*, Chicago.

West, J. (2005). The fall of a Silicon Valley icon: Was Apple really Betamax redux? In R. Bettis (Ed.), *Strategy in transition* (pp. 274-301). Oxford: Blackwell.

West, J. (2006). The economic realities of open standards: Black, white and many shades of gray. In S. Greenstein, & V. Stango (Eds.), *Standards and public policy*. Cambridge: Cambridge University Press.

West, J., & Dedrick, J. (2000). Innovation and control in standards architectures: The rise and fall of Japan's PC-98. *Information Systems Research, 11*(2), 197-216.

West, J., & Dedrick, J. (2001). Proprietary vs. open standards in the network era: An examination of the linux phenomenon. In *Proceedings of the 34th Annual Hawai'i International Conference on System Sciences*, 5011.

Williams, R. (1999). ICT standards setting from an innovation studies perspective. Proceedings of the SIIT '99 1st IEEE Conference on Standardisation and Innovation in Information Technology. In Jakobs and Williams (Eds) *Standardisation and Innovation in Information Technology* (pp. 251-262). Piscataway: IEEE.

Williams, S. (2002). Profits from piracy. *Salon*. Retrieved from http://salon.com/tech/feature/2002/09/26/piracy_unlimited/print.html

Williamson, O. E. (1975). *Markets and hierarchies: Analysis and antitrust implications. A study in the economics of internal organization.* New York: Free Press.

Williamson, O. E. (1985). *The economic institutions of capitalism.* New York: The Free Press.

Williamson, O. E. (1996). Efficiency, power, authority and economic organization. In J. Groenewegen (Ed.), *Transaction cost economics and beyond* (pp. 11-42). Boston: Kluwer Academic Publishers.

Wilss, W. (1982). *The science of translation.* Tübingen, Germany: Gunther Narr.

Winckler, R. (1994). *Electrotechnical standardization in Europe.* Brussels, Belgium: CENELEC.

Winter, W. (1990). *Bedrijfsnormalisatie.* Delft, The Netherlands: Nederlands Normalisatie-instituut.

Witt, U. (1997). "Lock-in" vs. "critical masses": Industrial change under network externalities. *International Journal of Industrial Organization, 15*(6), 753-773.

Wright, S. (2004). *Standards for translation assessment and quality assurance.* Kent State University.

Wu, M.-W., & Lin, Y.-D. (2001). Open source software development: An overview. *IEEE Computer, 34*(6), 33-38.

Wynekoop, J. L., & Russo, N. L. (1993). System development methodologies: Unanswered questions and the research-practice gap. *Proceedings of the Fourteenth International Conference on Information Systems*, Orlando, FL.

Yang, Y. H. (2001). Software quality management and ISO 9000 implementation. *Industrial Management and Data Systems, 101*(7), 329-338.

Yin, R.K. (1994). *Case study research, design and methods* (2nd ed.). Newbury Park, CA: Sage Publications.

Zhao, C. (2003). OSS Movement of Redflag. *Proceedings of the Asian Open Source Symposium.* Retrieved May 2, 2003, from www.asiaosc.org

Zhu, K. (2004). The complementarity of information technology infrastructure and e-commerce capability: A resource-based assessment of their business value. *Journal of Management Information Systems, 21*(1), 167-202.

About the Contributors

Kai Jakobs joined Aachen University's (RWTH) computer science department as a member of the technical staff in 1985. Since 1987, he has been head of the technical staff for the chair of Informatik 4 (communication and distributed systems). He holds a PhD in computer science from the University of Edinburgh. Earlier research interests and activities concentrated broadly on the user friendliness of communication networks. This included, for instance, the support of group communication, user-friendly naming, directory services, and services to be provided by the underlying infrastructure. More recently, his interest shifted to various aspects of standards and the standardization process, focusing on the individual's role in standards setting, and on the pros and cons of user participation in this process. In addition, he has been working on a number of projects on various aspects of information networks. Jakobs is coauthor and editor of a textbook on data communication and, more recently, nine books on standardization processes in IT. More than 170 of his papers have been published in conference proceedings, books, and journals. He has been on the program committees of numerous international conferences, and has also served as an external expert on evaluation panels of various European research and development programs on both technical and socioeconomic issues.

***

Michel Avital is an assistant professor of information systems at Case Western Reserve University. His research focuses on the sociotechnical aspects of information technologies and emphasizes a positive stance toward our capacity to construct better organizations and technologies. Michel has published articles on topics such as information systems design, creativity, knowledge sharing collaborative action, and appreciative inquiry. Michel is guided by the premise that information technologies are agents of organizational and social innovation, and that their consideration is vital to our success. For more information, visit http://avital.case.edu.

Ulrich Blum, born 1953, is a professor of economics. Following his PhD (1982) and his habilitation (1986) at the University of Karlsruhe, he was visiting professor and Lynen Scholar of the Alexander von Humboldt Foundation at the University of Montreal in 1986 and 1987 and in 1987 and 1988. From 1987 to 1991, he was a professor of economics at the University of Bamberg. In 1991 he was appointed a professor of economics at the Dresden University of Technology and became its founding dean for the new Faculty of Economics and Management. There he held the chair of economics and political economy until 2004, when he became president of the Institute for Economic Research in Halle, one of the five national economic research centers of Germany. Since 2005, he has been the convener of a working group on the Future Landscape of European Standardization and heads the commission on the German Standardization Strategy. His academic interests are transportation science, regional economics, institutional economics, and industrial organization theory.

Bart Brusse studied sociology and mass communication at the University of Nijmegen, The Netherlands. He started his career in the cable industry where he worked in strategy, programming, and business development. In early 2000, he founded ConTeSt consultancy and worked with a variety of companies on several digital media projects. On behalf of CENELEC and ETSI, he produced several reports on standardization in interactive digital television for the European Commission. From early 2004 to early 2007, he managed the COPRAS project on behalf of CEN, which focused on improving interfacing between ICT research and standardization. Currently, Brusse is active in DVB where he is the secretary to the organization's working groups focusing on interactive TV and security. In addition, he focuses on new media development, as well as on the development and implementation of associated technology policies.

Albert Feilzer studied dentistry at the University of Amsterdam. In 1989 he obtained his PhD. From 1982 to 1998, he worked part time in a dental clinic in Amsterdam as a dentist general practitioner and part time as an academic researcher at the Department of Dental Materials Science of the Academic Center of Dentistry Amsterdam (ACTA). In 1998 he changed to a full-time occupation at ACTA. Since July 2001, he has worked, besides at ACTA, for one day a week as a professor of standardization in the Department of Management of Innovation and Technology at the Rotterdam School of Management (Erasmus University, Rotterdam). Since January 2002, he has been chairman of the Platform Kenbaarheid van Normen en Normalisatie, a project to improve the visibility and awareness of standards and standardization, initiated by the Dutch Ministry of Economic Affairs.

G. Keith Fuller received his PhD from the University of British Columbia and is currently an assistant professor of management science at the Odette School of Business at the University of Windsor, Ontario. He also has a BAS and MEng from the University of Toronto. Prior to returning to university for his PhD, he worked as a consultant in the IS field for nearly 20 years.

Matt Germonprez is an assistant professor at the University of Wisconsin, Eau Claire. His research is in computer-supported group practices. He relies on the examination of individual and group actions to determine how technology is tailored and used to resolve distorted communications. Secondary streams of study include theory use, mobile computing, and interface design. He teaches object-oriented programming, distributed computing, and network design. He has been published in *Communications of the AIS*, *Organization Studies*, and *IFIP Transactions*.

Heiko Hahn, born 1972, studied business administration and management at the Universities of Bayreuth and Sheffield before becoming a researcher at the University of the Federal Armed Forces in Munich. He holds an MBA from Bayreuth University and a PhD from the University of Augsburg. His main research interests include systemic- and market-related aspects of component-based software development.

John Hurd has served as manager for Digital/Compaq's Industry Standards and Consortia Group, as well as chair of the ITIC (Information Technology Industry Council) Standardization Policy Committee.

Mingzhi Li received his PhD from the University of Texas at Austin and is currently an associate professor of economics at the School of Economics and Management, Tsinghua University. From 1995 to 1999, he worked as a research associate at the Center for Research in Electronic Commerce of the University of Texas at Austin in the United States. Prior to joining the University of Texas at Austin, he worked for the State Information Center of China for 3 years. Professor Li's research has led to publications in international journals including *Computational Economics*, *Information Economics and Policy*, *Information Technology and International Development*, and *Electronic Commerce Research*. He focuses his formal academic work on microeconomic theory, industrial organization, and electronic commerce.

Kai Reimers is currently a professor of information systems at RWTH Aachen University with the Faculty of Business and Economics. From 1998 to 2003, he worked as a visiting professor at the School of Economics and Management, Tsinghua University, sponsored by the German Academic Exchange Service (DAAD). He earned a doctorate in economics from Wuppertal University and a venia legendi at Bremen University. He has published in international journals including *European Journal of Information Systems*, *Electronic Markets*, *Electronic Commerce Research*, *Communications of the AIS*, and *Journal of Information Technology*. He authored or coauthored five books in the field of information systems. His main fields of research are interorganizational systems, IT management, and IT standards.

Michael Schwind (schwind@wiwi.uni-kl.de) holds a PhD from the economics department of the Johann Wolfgang Goethe University, Frankfurt, Germany. His dissertation thesis, titled *Dynamic Pricing and Automated Resource Allocation for Complex Information Services: Reinforcement Learning and Combinatorial Auctions*, combines methods of artificial intelligence and multiagent technology for the development of advanced economically inspired allocation methods for resources in distributed information systems. He is currently assistant professor of business information systems and operations research at the Technical University Kaiserslautern, Germany. His research interests include diffusion effects in networks with different topologies and combinatorial allocation methods in supply chain management.

Nikhil Srinivasan is a doctoral candidate at the information systems department in the Weatherhead School of Management at Case Western Reserve University. His research focuses on the study of social networks in IT-mediated environments. His doctoral work examines the effects of social network structure and characteristics on decision outcomes in large collectives. In addition to that, his research interests extend to emerging technologies such as the semantic Web in the context of knowledge systems, communities, and social networks. For more information, visit http://www.home.case.edu/~nxs77.

Tim Stockheim (stockheim@is-frankfurt.de) holds a PhD from the Economic Department of the Johann Wolfgang Goethe University, Frankfurt, Germany. His dissertation thesis titled *Supply Network Optimization: Coordination Based on Economic Scheduling, Negotiation, and Trust* is based on 4 years of research on multiagent systems in the manufacturing logistics domain. He cofounded the start-up VARLOG and currently works as an IT consultant. His research interests include applied operations research, supply chain management, and logistics.

Hans Teichmann was employed as an engineer by several Swiss companies and by General Electric Co. in Schenectady, New York. Subsequently, from 1981 to 2000, he worked at the central office of the IEC (International Electrotechnical Commission) in Geneva, Switzerland, successively as an engineer, principal engineer, and technical group manager. He holds a degree in electrical engineering from the Technical University Darmstadt, Germany, and an MS (engineering) form Union College, Schenectady, New York. At present, he is a PhD candidate at the Rotterdam School of Management (Erasmus University, Rotterdam, The Netherlands).

Klaus Turowski, born 1966, holds the chair of business information systems and systems engineering at the University of Augsburg. Prior to assuming his current position, he was a visiting professor at the University of the Federal Armed Forces, Munich, and assistant professor at the University of Magdeburg. He earned a diploma in industrial engineering and management from the University of Karlsruhe and a doctorate in business informatics at the University of Münster. He was a visiting professor at the University of Tartu (Estonia) and had further teaching assignments at the University of Darmstadt and the University of Konstanz. Besides his theoretical background, he has been working in various consulting projects. His main research areas are component-based enterprise applications and interorganizational systems.

Ilan Vertinsky received his PhD from the University of California, Berkeley, and is the Vinod Sood Professor in international business at the Sauder School of Business at the University of British Columbia. He is also the director of the Center for International Business Studies and the W. Maurice Young Entrepreneurship and Venture Capital Research Centre.

Henk J. de Vries (1957) is an associate professor of standardization at the Erasmus University, Rotterdam School of Management, in the Department of Management of Technology and Innovation. From 1984 until 2003 he worked at the Dutch national standards institute NEN (Nederlands Normalisatie-Instituut). His responsibilities included performing as committee secretariat, training, consultancy, setting up a training department, developing informative publications, and research and development. He has worked at Erasmus University since 1994, and has been working full time there since 2004. Henk received a PhD with his thesis *Standardization: A Business Approach to the Role of National Standardization Organizations*. His research and education concern standardization from a business point of view. For more information, visit http://web.eur.nl/fbk/dep/dep6/members/devries for more details.

Kilian Weiss (k.weiss@mederi-research.com) is cofounder of the start-up Mederi Research, which develops optimization software in the medical domain. He is responsible for strategic development and marketing. His research interests include applied operations research, stochastic modeling, and standardization effects in networks.

Index

H

I

K

L

M

N

O

P

R

S

T

U

V

W